Praise for *Magisteria*

'This book, though, is surely [Spencer's] magnum opus. It is astonishingly wide-ranging . . . and richly informed . . . So much complex history, theology and science could be heavy. What lightens the book is its clarity and the effervescent writing.'

Sunday Times

'With patience, balance and deep learning, Spencer . . . dismantles the myths that have accumulated around Galileo Galilei, Charles Darwin and other scientific figures . . . Filled with wit and wisdom.'

Philip Ball, *TLS*

'Fascinating . . . prepare to read something genuinely fresh in what can be an extremely hackneyed debate.'

New Scientist

'Tremendous . . . [Spencer's] survey of more than two millennia to the present day is consistently well-informed, witty and merciless to those wanting easy headlines. Every journalist would benefit from reading this substantial but very useful text, but all its readers will emerge better informed – and perhaps even saner.'

Diarmaid MacCulloch, *Prospect*

'Highly readable . . . Spencer convincingly shows how, until the modern period, religion largely supported the sciences of the day.'

Financial Times

'Illuminating . . . Even (or especially) those readers inclined to disagree with him will find his narrative refreshing . . . [Spencer] is one of Britain's most astute observers of religious affairs . . . He offers an engaging tour of the intersection of religious and scientific history.'

Economist

'A fascinating tour through a history of a difficult relationship, the fate of which is still unclear.'

Wall Street Journal

'Magisterial and brilliant.'

Professor John Milbank

'Easily the best exploration of the complex relation between science and religion I have ever read. As exemplary in his even-handedness as in his patient research . . . I suspect it will become the classic work on its subject.'

Iain McGilchrist, author of *The Master and his Emissary*

'Sweeping and comprehensive . . . A compelling act of myth-busting.'

Nancy Marie Brown, author of *The Abacus and the Cross*

'Spencer's historical portrait is erudite and wide-ranging . . . necessary.'

Literary Review

MAGISTERIA

The Entangled Histories of Science and Religion

Nicholas Spencer

ONEWORLD

Typeset by Hewer Text UK Ltd, Edinburgh
Printed and bound in Great Britain by Clays Ltd, Elcograf S.p.A.

Oneworld Publications
10 Bloomsbury Street
London WC1B 3SR
England

Stay up to date with the latest books,
special offers, and exclusive content from
Oneworld with our newsletter

Sign up on our website
oneworld-publications.com

MIX
Paper from
responsible sources
FSC
www.fsc.org   **FSC® C018072**

The book is dedicated to John Hedley Brooke,
a model of scholarship, erudition, generosity and friendship.

'It is dangerous to show man too clearly how much he resembles the beast, without at the same time showing him his greatness. It is also dangerous to show him too clear a vision of his greatness without his baseness. It is even more dangerous to leave him in ignorance of both.'

Pascal, *Pensées*

Contents

Part 4: The Ongoing, Entangled Histories of Science and Religion

Introduction
The Natures of the Beast

The story of a story

Three tense exchanges. Three quick-witted responses.

Rome, 22 June 1633. A 69-year-old man is forced to kneel before an assembly of religious judges. He looks frail and broken by months of inter-rogation. He holds a lit candle in one hand and in the other a pre-prepared text. The statement denies his life's work and compels him to 'abjure, curse and detest' a scientific theory that he has spent thirty years promoting and that he knows to be true. Failure to comply will result in his torture. Reluctantly, he reads the statement in full and then, as he struggles to his feet and leaves the room, mutters under his breath *eppur si muove*. 'Still it moves.'

Oxford, 30 June 1860. Nearly eight hundred people are crammed into a cavernous but stuffy library. Having sat through a tedious lecture, they are fractious, anticipating a lively exchange between two eminent Victorians concerning the scientific theory of the moment. They are not disappointed. After his introductory remarks, one of them, a prince of the established church, makes a jibe about whether his opponent would prefer ape ancestry on his grandmother's or grandfather's side. The other, a pugnacious scientist not cowed by episcopal condescension, whispers to a friend 'The Lord hath delivered him into mine hands', before replying that he would rather be descended from an ape than a bishop. The crowd is outraged, or elated. Either way, there is uproar. One woman faints.

Dayton, Tennessee, 20 July 1925. Three thousand people stand in front of a courtroom. The nation is listening on the radio, and millions will read about the proceedings in the following day's newspapers. They are watch-ing two men play intellectual cat and mouse on a temporary stage erected on the court lawn. One, a 65-year-old politician, perhaps the most famous in the country, is answering questions about the Bible. Or rather, he is trying to. The other, three years older, and the country's most famous lawyer, is inching closer to his prey. 'Do you believe the story of Noah's flood to be literally true?' Yes, sir. 'When was that flood?' I would not attempt to fix a date. 'About 4004 BC?' That is the accepted estimate. 'Don't you know?' I never made a calculation. 'What do you think?' I do not think about things I do not think about. And then the killer line. 'Do you think about things you do think about?' The courtyard, and the nation, erupt into laughter, and the politician on the stand withers in the heat and humiliation.

The abjuration of Galileo, the Huxley–Wilberforce debate and the Scopes 'monkey' trial have long since passed from history into myth, and stand as iconic encounters in the endless war of religion against science. At first, religion wins, but only by compelling the greatest scientist of the age to deny the truth that the earth moves around the sun; hence Galileo's parting aside. Almost 250 years later, religion, no longer able to resort to the threat of torture, turns to mockery but meets its match in the form of biologist Thomas Huxley, who ably defends Darwin's new theory of evolution against the ignorant, browbeating Bishop Samuel Wilberforce. Finally, in the American South, religion, now firmly on the defensive, is publicly humiliated before a huge audience and retreats bruised and bloodied but vowing vengeance. From such material has a popular history of hostility and conflict, of comprehensive victory and humiliating defeat, been spun.

Unfortunately, as is frustratingly often the case with history, the truth is rather more complex and convoluted than the myths. Neither the specific encounters nor the master narrative hung from them turns out to be the morality play that has passed into the popular imagination. As we shall see when we return to these heated encounters in Parts 2, 3 and 4 of this book, the reasons for, events leading to, exchanges within and meanings of each of these affairs have little in common with their mythological interpretations.

Galileo never said *eppur si muove*. Huxley didn't (quite) tell Wilberforce he'd rather be descended from an ape than a bishop. And although Darrow did ask Bryan whether he thought about things he thought about, the Scopes trial was about a good deal more than evolution. In short, there are stories behind and within the story of each of these famed battles. The single, coherent narrative we have been sold fragments, on closer inspection, into a mess of variously connected tales. There is no such thing as a – still less *the* – history of science and religion.

Developing and destroying a myth

The reason we have a single history is, generally, because of the intellectual climate at the end of the nineteenth century and, specifically, because of the man to whose lecture Wilberforce and Huxley were, in theory, responding.

John William Draper was an eminent American chemist who also fancied himself as an intellectual historian. Son of a Methodist convert, he left Britain aged twenty, though he took his father's views of Rome with him. Having achieved scientific eminence, he turned to writing. His 1874 book *History of the Conflict Between Religion and Science* compressed the story of two allegedly clearly defined and self-evident entities into a simple narrative of relentless argument. In actual fact, Draper focused his aim on the Catholic Church – he more or less excused Protestantism, Eastern Orthodoxy and Islam from his critique – in a way that resonated powerfully with his largely Protestant audience, many of whom were worried about the Vatican's authoritarianism. Draper's *History*, combined with another large polemic written by Andrew Dickson White twenty years later, and the boundless energies of the publisher Edward Youmans, took the gospel of science and religious warfare to an unprecedentedly wide audience, and embedded the narrative in an anxious Protestant mind.

The academic study of the history of science was still in its infancy at the time and it wasn't until a generation or so later that scholars began to pick away at Draper and White's 'conflict narrative'. An important monograph, written by a young sociologist, Robert Merton, in 1938, argued that Puritanism had made a decisively important contribution to the birth of modern science, but the narrative only began to unravel fully half a century later. From the late 1980s, a growing band of historians (and sociologists)[1] began to undermine the idea that there is a single controlling metaphor for the long history of science and religion, let alone one of relentless warfare. Reality turned out to be much more entangled and much more interesting.

Some of this was down to the ordinary everyday process, and progress, of historical research. Some of it was a question of historians emerging from the shadows of what was once deemed obvious.[2] But some was also down to entirely new discoveries, a few genuinely ground-breaking. In 2018, Salvatore Ricciano, a postgraduate student from the University of Bergamo, was searching through the Royal Society archives when he stumbled across the original of a crucial but apparently lost letter that Galileo had written, but had then retracted and doctored, during his first clash with the papacy in 1615–16.

Around the same time, a keen-eyed American scholar noticed a previously unknown and remarkably full account of the Huxley–Wilberforce debate, published in the *Oxford Chronicle and Berks and Bucks Gazette*. Up until then, the famous debate had been known only through a handful of brief newspaper accounts and some gossipy letters, which meant that no one had ever really been sure what the two protagonists had said to one another, or how the audience had reacted. The new source finally put the uncertainty to rest.

Such new scholarship and fresh discoveries undermined many of the myths that have long disguised themselves as history in the field. For example, the science of Christendom was considerably more sophisticated than most people give it credit for; medieval science is not a contradiction in terms after all.[3] Nicolaus Copernicus never imagined that his theory was a threat to his religion. Senior Church figures were initially positive about heliocentrism. Almost nobody thought the Copernican decentring of the earth demoted or degraded humans, as Freud later claimed. Giordano Bruno was not made a martyr on account of his science. Galileo's trial was as much about Aristotle, the Protestant threat and his soured friendship with Pope Urban as it was about heliocentrism. Catholic science did not disappear after Galileo. Early scientific societies, such as London's Royal Society, were not anti-religious. Newton wrote considerably more about theology, which he judged far more important, than he did about science, and his science did not banish God from the universe (in that respect, Newton was not really a Newtonian). The Enlightenment – or at least the Enlightenment outside France – was the period of closest *harmony* between science and religion. Much of the early work of geology was conducted by clergymen, most of whom managed to accommodate the newly extended history of the earth into their faith without too many tears. Darwin did not lose his faith on account of evolution – or not exclusively on account of evolution – and to the end of his life he denied that evolution was incompatible with theism. The Huxley–Wilberforce debate was not about science vs. religion or even narrowly about evolution vs. Genesis. The Scopes trial was as much about eugenics as it was evolution. And so on and so forth. The facts that everyone knew about science and religion turned out not to be facts after all.

Myth-busting is helpful and can be fun but it can still leave a rather negative impression in the mind. Religion wasn't quite as destructive of science as we have been led to believe. Hallelujah! Rejoice! In actual fact, for much of history, religion wasn't just 'not at war' with science, but it actively supported it, serving to legitimise, preserve, encourage and develop scientific ideas and activities.

So, again, a few examples. There *was* an Islamic 'Golden Age' of science between the eighth and twelfth centuries and although some scholars have claimed that its originality came from the Greek thought that Muslims inherited, the fact is that Islamic scholars did make significant original contributions to their classical inheritance.

A few centuries later, the first great scientific flourishing in medieval Europe, before the eruption of Aristotelianism in the thirteenth century, witnessed a small group of Christian scholars – self-designated *physici* – develop a concept of nature that was rationally ordered, consistent, quantifiable, comprehensible and capable of analysis through scepticism and methodological naturalism.

The longstanding metaphor of God's two books – of scripture and nature; of his Word and his Works – created a powerful argument for the study of the latter. 'Science' was a theologically sanctioned – indeed theologically commanded – activity.

The development of the experimental method in the early seventeenth century, and in particular Francis Bacon's contribution to early science, was closely linked to the Protestant understanding of the Fall of mankind, in which human *cognitive* abilities were judged to be as damaged as our moral and spiritual ones. In effect, humans could not be expected to think their way to the truth, but had to feel, to test, to experience, to experiment their way there.

The early years of 'modern' science, as we have come to know it, during which time it promised much but achieved little, were lived under the protection of theology.

There are plenty of other examples peppered through this book but these five underline how, particularly in its formative centuries, religion acted as midwife to science. For much of the time, the relationship of science and religion has not only *not* been one of relentless conflict but it has also been characterised by profitable collaboration.

For *much of the time*; not, it should be stressed, at *all* times. There is no merit in demolishing one simplistic and unjustifiable narrative – of constant conflict – only to replace it with an equally simplistic and unjustifiable narrative of constant amity.

So, one final set of examples. Islamic science did decline after the thirteenth century, and it did so in part for theological reasons. The Church banned the teaching of Aristotle in Paris in 1277 because it was judged a threat to theology. Sixteenth-century Protestants pitted the book of Joshua against Copernicus's heliocentrism and found the latter wanting. The Catholic Church did threaten Galileo with torture, prohibit his books and ban the teaching of heliocentrism for nearly two hundred years. The Church in France, in particular, sought to suppress biological ideas in the eighteenth century for fear of what they did to the idea of God-created life. Geology was judged unbiblical by many in the nineteenth century. Darwin was roundly attacked by many Christian correspondents, including many clerics. The Huxley–Wilberforce debate and the Scopes trial were not *only* about the theory of evolution but they were still about it. And today, millions upon millions of Protestants reject Darwinism, as do an increasing number of Muslims. Whatever else this is, it is not a picture of unspoiled harmony.

What is a human, and who gets to say?

We are left with a bit of a mess. The histories of science and religion have not been dominated by straightforward conflict. But nor have they been tales of uncomplicated concord. Is there any order in all this? Are there any plots to be discerned amid the chaos of events?

Historians of science and religion have struggled to find or name any. In academic circles, 'complexity' rules.[4] In writing this book, however, it became clear that, for all their undeniable 'complexity', the histories of science and religion do converge, repeatedly, on two issues. Whether we were in the classical Mediterranean world, tenth-century Baghdad, thirteenth-century Paris, seventeenth-century Rome, eighteenth-century France, nineteenth-century Oxford, twentieth-century Russia or twenty-first century Silicon Valley, two particular themes kept on arising from the noise.

The first is the question of authority. Who has the right to pronounce on nature, the cosmos and reality? For much of the time there has been broad agreement on this, but when there has been conflict between science and religion – when the Church Fathers disagreed with the philosophers in the ancient world, rationalist theologians with religious scholars in the Abbasid caliphate, theologians and philosophers in medieval Paris, Baconian and Aristotelian scientists in the early modern period, clerical naturalists and scientists in the Victorian age, creationist Protestants and eugenicist Darwinians in the Deep South – it has usually been about this issue. Much of the Galileo affair was about the first great shift in such authority, when scholars like Galileo began to assert their right to judge the way of the natural world. The same can be said of the Huxley–Wilberforce debate, which was about newly professionalised scientists (like Huxley) knocking old-school natural philosophers (like Wilberforce) off their perch. And a similar point can be made of the Scopes trial, in which Bryan, 'the Great Commoner', focused the question on what level of authority resided in 'the people'. The question of authority served as a lightning rod for disagreement – and never more so than when combined with the second theme.

That is, the nature and status of the human. Time and time again, it is the concept of the human – our makeup, origins, purpose, dignity and uniqueness (or lack thereof) – that emerges from the debate. Time and again, when it seemed as if people were arguing about the power of the planets, the composition of the body, the order of the cosmos, the design of nature, the origin of life, the age of rocks or the development of species, they were really talking about the nature of the human beast.

On reflection, this makes sense. Only the most obtuse – whether religious or non-religious – believe that science can pronounce authoritatively on the question of whether God exists. Certainly, it may gesture in a particular direction – towards the God of Abraham, or the God of the Philosophers, or the God who hides himself, or the God who is simply the invention of the anxious human imagination. People do and always will disagree on this. But those who think science can judge definitively on the God question are getting their physics and metaphysics muddled up.

The nature of the human is different. Human existence is open to scrutiny in a way that God's is not, and what we think of the human profoundly influences what we think of the divine. If humans are – or more precisely, if they are *only* – puppets of the stars, or beasts of the field, or 'man machines', or accidental primates, or creatures of their desires, or marionettes of their genes, they are not really the kind of creatures envisaged by the world's religions. Most of those religions recognise that humans *are* material beings – physical, animal, evolved, genetic – but they claim that we are not *only* or *merely* such beings. We have some moral or spiritual or eternal or transcendent or divine dimension to us too. Throughout history, when science has claimed that humans are material beings, some religious thinkers have shrugged and some have shrieked. But when it has claimed that humans are *only* material beings, they have all shrieked. In this way, the messy histories of science and religion have repeatedly converged on the question of the nature of the human beast – or, more precisely, the natures of the beast, because what is at stake is not so much how we understand ourselves as whether there is more than one way of doing so.

What (or who) is the human, and who (or what) gets to say? These two questions run through the histories of science and religion like rivers through a landscape. Not every feature on the landscape we will pass through can be explained by the course of these rivers. Sometimes science and religion *have* been about the interpretation of empirical evidence or the reading of holy texts. Sometimes, a cigar is just a cigar. Moreover, even when these rivers – of the human and of authority – have been in clear view, they have not always been equally important. In the centuries before science and religion assumed anything like their modern form – say up until around 1600 – the question of authority was the more significant feature, if only because few people imagined that science, or natural philosophy, could cast any doubt on human spiritual identity. After that, authority would remain an issue but it would increasingly be the nature(s) of the human beast that would assume significance.

The two rivers finally met in the later nineteenth century – when science was professionalised and Darwin revealed our evolved history – and that was, not coincidentally, the moment at which the 'warfare' narrative was born. But thereafter, even as the professional status and authority of scientists was settled, science's capacity to redescribe humans and to remake their society and planet, meant that whether in Tennessee, Vienna, Moscow, Arkansas or Silicon Valley, science and religion would still engage in lively conversation about human beings, our nature and our future.

To repeat, then: the histories of science and religion are many and messy and it would be misleading to cram Hypatia of Alexandria, Muhammad ibn Musa al-Khwarizmi, Maimonides, Adelard of Bath, Robert Grosseteste, Nicolaus Copernicus, Galileo, Bacon, Margaret Cavendish, Boyle, Newton, La Mettrie, David Hartley, William Buckland, Darwin, Rifa'a al-Tahtawi, Dixon and White, Faraday and Maxwell, Jennings and Bryan, Einstein and Dirac, Freud, Frazer and Evans-Pritchard, Yuri Gagarin and John Glenn, George Price and Richard Dawkins, and Alan Turing and Ray Kurzweil into the same neat coherent narrative. We have to see the histories of science and religion in all their heterogeneity and complexity.

But those histories do repeatedly converge on the questions of who we are and who gets to say, and while it is quite possible to enjoy the stories of science and religion without paying attention to these two questions, it is impossible to understand the lie of the land, or the argument of this book, without them.

Entanglement

The first part of *Magisteria* takes the story from the classical world to 1600 (not, note, to the popular 'birth of science' in 1543). It covers the period when everyone worshipped god(s), and when educated minorities studied the world in considerable detail and with much sophistication, but when neither science nor religion were in anything like their modern forms. Beginning in the classical world, it moves to Islamic Baghdad and Spain, to the world of North African and medieval European Judaism, to Christendom, and finally to the exceedingly slow spread of Copernicanism through the sixteenth-century (but essentially still medieval) mind. There were moments of tension and aeons of harmony through these centuries, although given that neither 'science' nor 'religion' had reached anything close to their modern identities during this period, all talk of 'science and religion' here teeters on the edge of anachronism.

The book then narrows its focus to Western Europe in Parts 2 and 3, where science as we now know it emerged. Part 2, 'Genesis', looks at the period in which modern science developed in the seventeenth and eighteenth centuries and shows how European religion helped conceive, nurture and develop the new philosophy. Part 3, 'Exodus', explores the nineteenth century and recounts how science drifted away from its religious parent, sometimes with parental blessing, sometimes having to fight for its independence. This was the period in which the myth of conflict was born – and sometimes for good reason. Caught in a pincer movement between critical revisions of their holy scriptures and a palpable loss of intellectual authority, many believers lashed out at what the new profession of science claimed about the world. But it was also the period in which there was sometimes good reason to do so, as some scientists, confident in their ability to understand and improve human nature and society, and inhabiting imperial civilisations that sought to do just that, rather overstepped their authority.

Part 4 widens the geographical lens again, and takes the story from around 1900 to the present day, covering a period when, after most (though not all) of the authority disputes had been settled, science and religion found themselves in various inconclusive, sometimes beneficial, sometimes fractious, conversations. Whether it was anthropologists plumbing the depths of the past, Freudians the depths of the mind, or Soviet cosmonauts the depths of space, religion and science seemed inextricably entangled, and this sense of entanglement only seemed to grow as evolutionary biologists declared that humans were basically genes, neuroscientists that they were basically brain activity and Silicon Valley technoutopians that they were basically algorithms. As we steer our way through a trembling new century, eagerly awaiting the arrival of our robot overlords, the millennia-long story of science and religion seems more interconnected than ever before.

The late, great American palaeontologist Stephen Jay Gould claimed, towards the end of his life, that science and religion were 'non-overlapping magisteria': distinct fields of human activity that need not – should not – encroach on each other's territory. It was an intervention born of good intentions, hoping to bring a ceasefire to an exhausted local conflict between fundamentalists who disbelieved Darwin and fundamentalists who disbelieved anything that wasn't Darwin. Whatever its merits as a description of how science and religion *should* interact, Gould's model patently does not work when it comes to history. The 'magisteria' of science and religion are indistinct, sprawling, untidy and endlessly and fascinatingly entangled.

PART 1

SCIENCE AND RELIGION BEFORE SCIENCE OR RELIGION

The pagan mathematician Hypatia is dragged away to be butchered by a Christian mob. The Enlightenment turned Hypatia into an early martyr in the alleged conflict between science and religion. In reality, her death was the result of grubby power politics in the ever-rioting city of Alexandria.

ONE

THE NATURE OF NATURAL PHILOSOPHY: SCIENCE AND RELIGION IN THE ANCIENT WORLD

'Her quivering limbs were delivered to the flames': placing a murder

In March 415, an old woman was dragged from her chariot and hacked to death by a mob of enraged Christians. Hypatia was about sixty. She was considered one of the finest mathematicians, astronomers and philosophers of her age. She was returning home from a customary ride through Alexandria when the mob attacked. They dragged her to a church where she was stripped and sliced up with what the historian Edward Gibbon said were oyster shells. Her body was then dragged to the city outskirts where it was burned. The murder took place during Lent.

Hypatia's murder is, for the ancient world, what Galileo's inquisition or the Scopes trial are to the early modern and modern: testimony to the eternal clash between science and religion. When it began to filter into the European cultural bloodstream, around 1,300 years later, there was little doubt about what it signified. The Irish philosopher and freethinker John Toland wrote a long historical essay with a title that left little room for suspense: *Hypatia or, the History of a Most Beautiful, Most Virtuous, Most Learned, and in Every Way Accomplished Lady; Who Was Torn to Pieces by the Clergy of Alexandria, to Gratify the Pride, Emulation, and Cruelty of their Archbishop, Commonly but Undeservedly Titled, St. Cyril.* Fifteen years later, the French sceptic Voltaire wrote about her 'bestial murder' by 'Cyril's tonsured hounds', putting the crime down to her persistent belief in the pagan gods and her commitment to the rational laws of nature, adding with customary wit that 'when one strips beautiful women naked, it is not to massacre them.'[1] A generation later, Edward Gibbon luridly recounted how Hypatia was 'inhumanly butchered' by the 'troop of savage and merciless fanatics' before her flesh was 'scraped from her bones' and 'her quivering limbs were delivered to the flames.'[2] A century later, Anglican clergyman Charles Kingsley dedicated a whole novel to her, entitled *Hypatia or the New Foes with an Old Face*.

The life and death of Hypatia had something for everyone. For Toland, it was pungent with anti-clerical potential. For Voltaire and Gibbon, it underlined how superior was the cool, rationalist logic of the classical world to the frantic, faith-fed hysteria of Christendom. For Charles Kingsley the tale underlined the sinister wickedness of Catholicism. And for John Draper, who recounted her fate in graphic detail in his *History of the Intellectual Development of Europe*, it was yet another example of religion crushing science. 'Henceforth science must sink into obscurity and subordination. Its public existence will no longer be tolerated.'[3] Unfortunately for those with axes to grind, Hypatia's death had almost nothing to do with science, religion or philosophy.

Hypatia was one of the few celebrated female philosophers and mathematicians of antiquity, although hardly a feminist in any modern sense. She prided herself on her sexual abstinence, clothed herself in modest philosophic attire and once warned off an amorous student by brandishing her sanitary towel in his face and lambasting him for preferring the flesh over the intellect. Although a pagan she was in no way devout, preferring philosophical heights over cultish lowlands. Her students comprised a cross section of society, included numerous Christians, two of whom would become bishops, and one, Orestes, who was prefect of Alexandria and future governor of Egypt. Her teaching and her circle were highly elitist. 'What can there be in common between the ordinary man and philosophy?' asked one of her disciples.[4] Her pearls of wisdom were not to be cast before the city's swine.

Alexandria was one of the great cities of the empire, packed with churches, temples, mathematical, medical, catechetical and rabbinic schools, theologians, philosophers, plebs and mobs. It was notoriously violent, frequently succumbing to riots and high-profile lynchings. It was also going through a period of rapid, enforced and widely resented social change, as the emperor prohibited pagan practices and began to enforce Christian worship. Theophilus, the city's patriarch, had been attacking its pagan cults for years but the emperor Theodosius' edicts emboldened him to assault the Serapeum, the city's main pagan temple, an action that degenerated into all-out war. Theophilus had no interest in Hypatia and she had none in the defence of the Serapeum, but the affair showed how the city lived permanently on the edge of anarchy.

Theophilus' nephew and successor in 412, Cyril, made his uncle appear positively tolerant. Once in office, he turned first against opposing Christian factions, then against those he deemed insufficiently orthodox and then against the city's Jews.

The prefect Orestes resented Cyril's intrusion into the city's politics, and he arrested and tortured one of his men. Cyril looked to his monks for support, one of whom wounded the governor, who duly had him tortured and then put to death. Orestes then turned, fatally, to his philosophical friend for support. Hypatia was highly respected, but was among the city's elite rather than its populace. The rumour quickly spread among the people that she was a sorcerer, who was responsible for Orestes' hatred of the patriarch. The people responded as they knew best.

Hypatia's murder was condemned by chroniclers, pagan and Christian alike, but her memory soon faded, and was resurrected only centuries later by Enlightenment savants who had a point to prove about Catholics, or monks, or reason, or science. But Hypatia was no more a martyr for 'science' than she was a victim of 'religion'. Rather she was a casualty of the perennially ferocious power politics of Alexandrian society, although this is unlikely to have been much comfort to her as the Christian mob stabbed her to death with what the chronicler Socrates Scholasticus informs us were actually broken bits of pottery.

'All things are full of gods': *scientia* and *religio*

The true story of Hypatia's death should caution us against polemical readings from history. To be sure, Cyril, Orestes and the baptised pack of Alexandria emerge from the story as little more than manipulative, dogmatic, violent hooligans, but to elevate the sorry story into a

principled conflict between science and religion is to misread it badly –
and anachronistically.

Hypatia was, by all accounts, a fine astronomer and a first-rank mathematician, but mathematics and astronomy in the ancient world were not
what they are in the modern. Both were part of a wider, philosophical
search for truth, beauty and the good life. Geometry was a doorway to
divinity; understanding the heavens a way of grasping what was immutable and holy, elevating the mind and life towards perfection. Hypatia's
work may not have been as tangled up with mysticism and divination as
her father's, an equally formidable mathematician, but it was still, on one
level, deeply religious. Whatever else, it was not science as we use the word
today.

It is hard to overemphasise the significance of this point. Neither
'science' nor 'religion', as we commonly understand those terms, existed in
the classical world – or indeed in the Islamic, Indian, Chinese or medieval
European worlds. Each of these cultures had more or less organised,
systematic and rational schools for the investigation of the natural world,
but that investigation was not an independent discipline until well into the
nineteenth century.

Our modern word 'science' derives from *scientia*, meaning, simply,
knowledge. *Scientia* was a step on the way to *sapientia*, meaning wisdom
or discernment. The study of nature and the cosmos were entangled with
the wider objects of philosophy, such as identifying the true way to live
and worship. Particularly in the world of late antiquity in which Hypatia's
Neo-Platonic philosophy was dominant, *scientia* meant grasping the eternal and necessary truths of the cosmos so as to approach transcendent
perfection in this life.

'Science' for its own sake was not entirely unknown. Seneca remarked,
in his *Natural Questions*, that motivation for studying nature was 'not gain
but wonder', the sheer pleasure of encountering nature that 'captivates
people with its own magnificence'.[5] However, even Seneca, as a Stoic
philosopher, recognised that the ultimate purpose of studying nature lay
in the effect it had on calming the human mind. The purpose of natural
philosophy was to inform human life, ethics, religion and politics.
Understanding the natural order helped people understand the moral and
divine order.

Thus, the end of astronomy, at least according to its most famous practitioner, Ptolemy, was moral and spiritual formation. Hypatia studied geometry and the stars in order to realise unchanging perfection in her own life. Nature offered similar lessons. The philosophical sect of Stoics thought the good life was that lived in accordance with nature. Epicureans, by contrast, believed that understanding nature would free humans from the fear of death or a painful post-mortem fate. Rather than being a detached, neutral and objective discipline trying to comprehend nature in its own right, natural philosophy was primarily a means of forming human beings and the societies in which they lived.

Classical science (I'll drop the scare quotes now) was not, therefore, disinterested. Nor was it naturalistic. Methodological naturalism, the idea that only natural explanations can account for natural phenomena, is one of the characteristics of modern science. In contrast to this, the divine was everywhere in classical science. The Pre-Socratic philosophers, to whom science is often traced, were reacting against the mythic worlds of Homer and Hesiod, in which the gods had a direct, frequent, uncontrollable and capricious relationship with the world. But the world of these philosophers was hardly godless. The early philosopher Anaxagoras was banished from Athens for claiming that the sun was a fiery mass of molten metal and the moon was an earth-like body that reflected the sun's light, but he also thought that a divine, cosmic mind ordered and controlled all of creation. Pliny the Elder begins his encyclopedia *Natural History* with a paean to 'a deity, eternal, immeasurable'.[6] Scientific Hippocratic medicine rose in parallel with the rise of the religious cult of Asclepius. Physicians had no problem in explaining illness through divine and natural causes simultaneously. The classical world's best-known medical researcher, Galen, saw purposive agency behind the apparently flawless design of human and animal bodies. Even the greatest of ancient scientists, Aristotle, who posited a naturalistic chain of causation for everything in heaven and earth, drew on direct divine causation as a means of accounting for the motion of the stars, which he could otherwise not explain. God was called on to effect creation, to govern it, to intervene when 'naturalistic' explanations failed and as the presiding glory to which it all pointed. Natural philosophers would trivialise and minimise the power of gods but they rarely denied it altogether. All things were indeed 'full of gods', as the early philosopher Thales proclaimed.

In this way, natural philosophy merged into (what we might again anachronistically call) religion. Religion, or *religio*, in the classical world had more to do with piety, the correct forms of life and worship, than with doctrine or belief. Even among Christians, who, unusually, formulated their beliefs as creeds, religion was a matter of proper practice and worship. 'Pure religion', as the New Testament book of James says, is 'to visit the fatherless and widows in their affliction'.[7] 'What true religion reprehends', St Augustine wrote more critically, is the superstitious practice of 'sacrifice . . . offered to false gods'.[8]

As Augustine intimated, the practice of true religion depended in large measure on whom or what was being worshipped. Indeed, he went on to reason, 'it makes no difference that people worship with different ceremonies with the different requirements of times and places, if what they worshipped is holy'.[9] If, however, humans worshipped the created and the profane – false gods, earthly idols, imperial powers, the manifestations of nature – their *religio* was false.

'We have no need for curiosity beyond Christ Jesus': ambiguous attitudes

Seeing science as part of philosophy, and philosophy as a way of understanding human purpose, destiny and the right way to live in this god-soaked world helps us understand the complex and sometimes contradictory relationship that Christians (and Jews) of the classical world had with it.

There was, of course, no single relationship. Some Christian bishops sat at Hypatia's feet while others turned a blind eye to her murder. As a rule, most of the Church Fathers, as the leading scholars of the early Church have come to be called, had a positive if critical attitude to the philosophical world in which they lived. The earliest, Justin Martyr, called philosophers 'truly holy Men' and considered reason a divine gift.[10] Clement of Alexandria called philosophy 'the work of divine providence'.[11] His contemporary, Origen, was so steeped in pagan philosophy, as one of his students recalled, that he 'required us to study philosophy by reading all the existing writings of the ancients . . . with the utmost freedom we went into everything and examined it thoroughly, taking our fill and enjoying the pleasures of the soul'.[12]

Such respect was not ubiquitous. Tatian the Syrian launched an attack on pagan schools.[13] 'Shun all heathen books,' warned the author of the third century *Didascalia apostolorum*. Perhaps most famously, the Latin Church Father Tertullian asked pointedly what Athens – standing for pagan philosophy – had to do with Jerusalem, the religion of Jews and Christians. 'We have no need for curiosity beyond Christ Jesus, no investigation beyond the Gospel. When we believe [the Gospel], we need give credence to nothing else!'[14] It was the kind of myopic, self-righteous bombast that delighted critics then, as it does now. 'Let no one educated, no one wise, no one sensible draw near,' mocked the hostile pagan Celsus.[15] 'Ours are reasoned arguments,' claimed the last pagan emperor, Julian. 'All your reason can be summed up in the imperative "Believe".'[16]

The early Christian understanding of natural philosophy took its place within this fraught relationship with pagan reason. Studying the natural world was instructive: useful but dangerous. Augustine, in his misleadingly named *Literal Commentary on Genesis*, admitted that pagan philosophers knew a great deal about:

> the earth, the heavens, and the other elements of this world, about the motion and orbit of the stars and even their size and relative positions, about the predictable eclipses of the sun and moon, the cycles of the years and the seasons, about the kinds of animals, shrubs, stones, and so forth.

Such knowledge was not to be dismissed as, it seems, some Christians were wont to do. 'It is a disgraceful and dangerous thing,' Augustine thundered, 'for an infidel to hear a Christian, presumably giving the meaning of Holy Scripture, talking nonsense on these topics.' Christians swam in the same intellectual waters as pagans. Pronouncing ignorantly on scientific subjects merely gave their opponents ammunition.

In the same book, Augustine ruminated on the idea that 'the whole of creation has certain natural laws.' Drawing on Stoic and biblical ideas, he argued that nature has germs or 'seedlike principles' that governed its unfolding. This was not a world of direct divine causation, miracles aside.

Rather, God chose to apply secondary forces 'to bring forth . . . what has been created'.[17]

Such positive ideas and emollient tone were not the whole story, however. Augustine could accept natural philosophy in as far as it was a handmaiden to the deeper knowledge of God and his purposes. But beyond that, it could be pointless or even harmful. Curiosity could be a distraction and a kind of spiritual pride. This was especially so when it came to astrology. Astrology was a science, our distinction between it and astronomy being wholly absent in the classical world. Ptolemy had written both the *Almagest*, which dominated European and Middle Eastern astronomical thought for 1,300 years, and the *Tetrabiblos*, which described the philosophy and practice of astrology.

The Church Fathers excoriated astrology. Not only did it accord quasi-divine agency and power to the objects of creation, rather than to the creator, but it undermined human freedom and therefore the human role as creatures uniquely commissioned to do God's will. In their opposition, we see the seeds of the broader 'human' tension that would develop later, when science and religion emerged as distinct categories.

By astrological reckoning, human actions were governed by the stars rather than their own agency . It meant that humans could not be subject to praise, blame or judgement. Accordingly, astrology was commonly criticised and mocked by the Church Fathers, Augustine even deploying the logic of 'twin studies' to make his point. Astrologers, he argued, 'have never been able to explain why twins are so different . . . [indeed] twins are often less like each other than complete strangers . . . [despite being] conceived in precisely the same moment'.[18]

Many Fathers were happy to disentangle astrology from astronomy, but for others the taint could not be shifted. And even if it didn't necessarily lead to intellectual perdition, it could still prove a fatal distraction. Knowledge of the heavenly bodies, Augustine wrote in *On Christian Doctrine*, wasn't a 'superstition' but it nonetheless 'renders very little, indeed almost no assistance, in the interpretation of Holy Scripture'.[19] 'Because of this disease of curiosity', he wrote acerbically in his *Confessions*, 'men proceed to investigate the phenomena of nature' even though 'this knowledge is of no value to them . . . they wish to know simply for the sake of knowing'.[20]

This was, in fact, a longstanding critique, and not unique to Christians. A number of philosophers expressed the conviction, attributed to Socrates, that theoretical interest in heavenly phenomena was a waste of time, a distraction from the modes of philosophy that really helped attain the good life.[21] Still, the Church Fathers added a particular pungency to it. Ambrose was bishop of Milan, roughly contemporary with Hypatia. He was highly educated and much admired by Augustine, who remarked on his unusual habit of reading silently.[22] But he was also profoundly hostile to natural philosophy, which he saw as an arrogant and ultimately worthless attempt to plumb God's mysteries. Ambrose rejected the lawfulness of creation that Augustine – and indeed the scriptures – hinted at. The book of Genesis had revealed an order to creation and the Book of Wisdom taught that God 'had ordered all things in measure and number and weight', a text to which the physicist James Clerk Maxwell would allude centuries later.[23] Ambrose, however, rejected such an idea as inimical to God's absolute power and quoted instead Jesus' words from Mark 14:36 ('Nothing is impossible to you') as a corrective. Referring to the book of Job – somewhat bizarrely, given how that book shows a greater interest in the working of the natural world than any other in the Hebrew scriptures – Ambrose reasoned that 'God clearly shows that all things are established by His majesty, not by number, weight and measures.'[24] It was an approach that threatened to poison science at the root, and one that Islam would also wrestle with six hundred years later. By Ambrose's reckoning, it was not only foolish and dangerous to measure and study the created world by means of human reasoning. It was also impious and ultimately pointless. 'I believe that all things depend on His will, which is the foundation of the universe.'

'Do not let anyone ask if Moses is writing a work of astronomy': from the ruins

Ambrose's attitude to natural philosophy would seem to give succour to those who see in the triumph of Christianity the death sentence for classical science. Natural philosophy was murdered, and Ambrose and the other Church Fathers were in the mob.

This, like the popular accounts of Hypatia's death, is an oversimplification. There is some truth in the idea that the calculated demotion of natural philosophy among the more prominent Christian thinkers of late antiquity helped to weaken it. Science might make a fine intellectual handmaiden, but she should never get above her station. Nevertheless, we rather exaggerate the power of the Church Fathers if we imagine that they alone killed off ancient science.

In reality, natural philosophy was already in a parlous state by the fourth century. The subject had rarely if ever enjoyed imperial imprimatur, political patronage or institutional support. With the exception of medicine and to a lesser and diminishing extent astronomy, the activities of natural philosophy served no obvious personal, practical or social benefit. Such sciences had long existed solely on the funds and initiative of inquisitive individuals and their occasional patrons.

The greatest natural philosophers of the ancient world, supremely Aristotle, had been pioneering enquirers, investigating the natural world and open to new ideas about what they found there. However, long before the first Church Father had put pen to parchment, Aristotle had ossified into an authority, empirical science had become a rarity and natural philosophers were complaining about widespread indifference. As early as the first century, Pliny the Elder was grousing about the lack of serious scientific research in his time, and the indifference and ignorance of his fellow Romans. By the time of Ambrose and Hypatia, the greatest scientific developments lay centuries in the past, and scientific work comprised primarily in commenting on them. Ancient science had become a threadbare garment, quite independently of Augustine and Ambrose picking holes in it.

As it turned out, all this was immaterial, as the political and economic collapse of the Western empire in the fifth century would shred the garment altogether. Cities, libraries and schools were razed, imperial authority destroyed and centuries of learning lost to the elements. The idea, still repeated round atheist campfires today, that this loss was primarily the result of angry, ignorant Christians, like those who did away with Hypatia, has been firmly debunked, and in any case is undermined by the desperate attempts of monasteries to preserve what they could of that culture.[25] From Vivarium in southern Italy to Monkwearmouth-Jarrow in Northumbria, monks copied and exchanged the decaying manuscripts of antiquity, Christian and pagan alike.

What they salvaged from the ruins of antiquity was impressive. The eighth-century cleric and scholar Alcuin of York mentions works by Aristotle, Cicero, Lucan, Pliny, Statius, Trogus Pompeius and Virgil in his library in York, and cites Horace, Ovid and Terence in his correspondence. He was unusually educated but not anomalous. Benedict Biscop, the abbot of Monkwearmouth–Jarrow, made the perilous journey to Rome five times in the seventh century to collect books for his library.

Some of these books were scientific, at least in the philosophical sense of the word. Boethius, the philosopher executed in 524 by King Theodoric for alleged conspiracy, translated Aristotle, and wrote on mathematics, music and possibly geometry. Isidore, a seventh-century archbishop of Seville, wrote on mathematics, astronomy and natural history in his *On the Nature of Things*, the title deriving from the pagan Lucretius. The Venerable Bede, monk-scholar of Monkwearmouth and best known for his *Ecclesiastical History of the English People*, wrote on the reckoning of time, and also, reworking and critiquing Isidore, *On the Nature of Things*.

Modern readers are unlikely to be impressed by the scientific content of these books but the circumstances of their composition beg a more indulgent assessment. Isidore reflected the pagan scientific consensus about the structure, content and workings of the cosmos. He did not attempt to adjust the picture to fit a theological agenda, on the grounds, all too often ignored over coming centuries, that the scriptures did not offer an alternative, definitive or revealed template of nature. The one area in which he did disagree, along with all the Church Fathers and numerous Jewish thinkers, was on the question of the eternity of matter, proclaimed as a fact by Aristotle and an orthodoxy in the classical world, but one that was hard to square with the biblical account of creation *ex nihilo*. It was a disagreement that would run, albeit in some rather counterintuitive ways, all the way to the twentieth century.

Bede agreed with Isidore on this but was not above correcting him with passages from Pliny's 'delightful book', the *Natural History*. In his *Ecclesiastical History*, Bede offered readers a detailed description of Britain's geographical position and its natural resources, 'its plant and animal life, its hot and salt springs and its minerals'. Elsewhere, he linked tides to the phases of the moon, recording how the time of tides varies across the days, and from place to place, and even hinting 'that he was part of a network of scientific information exchange'.[26] Both Bede and Isidore sought to undermine belief in omens and the idea that natural forces operated with their own agency and intent, as opposed to under God's ordered direction and power. None of this constituted a radical departure from the natural theology of antiquity or a significant scientific step forward. But it does underline how, even in the darkest of centuries to afflict the former Western empire, an interest in natural philosophy was kept flickering faintly in monasteries across the continent.

The fall of Rome had much less of an impact on the Eastern empire than on the West. Alexandria itself continued as the intellectual capital of the empire, and in the process became Ground Zero for one of the most enduring myths about science and religion.

A few years before John Toland first popularised the myth of Hypatia's death, another tale of ancient Alexandrian madness came to light. *Christian Topography* was a remarkable book – part travelogue, part scientific geography – that came from the pen of a sixth-century Alexandrian merchant and monk, Cosmas Indicopleustes, whose surname literally means 'the man who sailed to India'. In it, Cosmas recorded various, more or less reliable, reports of the topography, trade, flora and fauna of the East, as far as Sri Lanka. He also forcefully argued that the world was flat.

This was too perfect for Enlightenment wits, and Gibbon twice skewered Cosmas for his attempt to 'confute the impious heresy of those who maintain the earth is a globe' as opposed to a 'flat oblong table'. By Gibbon's reckoning, Cosmas's view was simply that of the 'scriptures' – the actual 'study of nature [being] the surest symptom of an unbelieving mind' – and therefore also the orthodox view of all believers.[27]

Influential as Gibbon was, it was not until 1828, when the American short-story writer Washington Irving published a heavily romanticised biography of Christopher Columbus, that the myth of the flat earth – that it was a belief widely, indeed universally, held on the basis of firm ecclesiastical authority – gained a wider audience. It has proved almost as hard to shake off as the warfare metaphor itself.

As with Hypatia, there is enough in the flat-earth myth to make it credible. Cosmas didn't work alone. The fourth-century Church Father Lactantius was a supporter, and Cosmas also drew on the ideas of theologians from Antioch, traditionally a rival to Alexandria, who believed that a serious 'Christian cosmology' should replace pagan ideas *wholesale*. It was precisely because pagans supported sphericity that Cosmas advocated a flat earth, specifically a parallelogram surrounded by seas on four sides and shaped, overall, after the model of the tabernacle in the Old Testament.

In actual fact, much of what Cosmas wrote on the matter was directed against his fellow Christians, who considered his views embarrassing. Belief in the earth's sphericity was nearly universal in the ancient world, not least in Alexandria where Eratosthenes, director of the city's famous library, had once calculated the circumference of the earth to within about two percent. To Cosmas's frustration, Christians held the view every bit as much as pagans, and so it was upon them that his ire was trained. 'It is against such men my words are directed, for divine scripture denounces them . . . Were one to call such men double-faced he would not be wrong, for, look you, they wish both to be with us and with those that are against us.'[28] Cosmas wrote, in effect, like the sixth-century equivalent of a modern Young Earth Creationist, whose supreme contempt is reserved not for Darwinian atheists, who could hardly be expected to know better, but for fellow Christians who appear to have given up entirely on the word of God.

One such backsliding Christian was John Philoponus. John was an inhabitant of Alexandria and an exact contemporary of Cosmas, although there is no evidence that they knew one another. He was also one of the most important scientist-philosophers of late antiquity, commenting extensively and approvingly on Aristotle, while rejecting his ideas on the eternity of the world. At about the same time that Cosmas published his flat-earth theory, Philoponus was writing a commentary on the first chapter of the book of Genesis, entitled *On the Creation of the World*. This accepted the sphericity of the earth without question and, without naming him, dismissed Cosmas's bad science. More interestingly, however, John also rejected Cosmas's bad theology, specifically the idea that Moses (then believed to be the author of Genesis) was writing about astronomy at all. 'No one considering the systematic treatment of nature by later writers is going to ask Moses' scripture . . . what has been thoroughly researched on these subjects by specialists,' he wrote. 'That was not the excellent Moses' intent.' Excellent Moses, by John's reckoning, was 'chosen by God to lead people to knowledge of God', not of nature.[29]

'Do not let anyone ask if Moses is writing a work of astronomy or a technical treatise on natural causes. This is not the scope of theologians or for leading people to knowledge of God, but rather a job for specialized workers: for every field intends a useful purpose for human life.'[30]

Unfortunately, John was not quite as good as his word on this. Having dismissed Moses' astronomical credentials, he then spent an inordinate amount of time attempting to show how the sphericity of the earth, and the other scientific doctrines of his day, could in fact be found in the pages of scripture. 'As I have shown that Moses' cosmogony agrees with extant reality', he argued, going further to claim that the most highly reputed astronomers of the classical world, Hipparchus and Ptolemy, 'took their points of departure from Moses' writings'.

In effect, having disarmed Moses and the very idea of biblical science in order to prevent them from falling into the wrong hands, John was then happy to re-weaponise them for his own purposes. In so doing, he was to demonstrate one of the most serious issues that has repeatedly dogged religion in its millennia-long relationship with science. Somewhat counterintuitively, one of the biggest problems has been too much harmony rather than too much disagreement.

The problem of disharmony is visible, in glorious technicolour, in Cosmas's ideological, a priori dismissal of science in the name of eternal, sacred, textual truths. It's a problem with which we are all familiar, even if, as we shall repeatedly see, it has tended to be a minority position, from the Flat Earthers in Cosmas's day to Young Earthers of our own. Such angry denunciations of the science of the day gain much publicity, but little authority. In this regard, the science and religion debate has been much like a swimming pool, with most of the noise up at the shallow end.

In the long run, the cosmological arguments of *Christian Topography* disappeared. The book was never translated into Latin (one of the reasons it was slow to come into the consciousness of the West), was not widely read in the Eastern empire and was usually criticised when it was. When Photius, a ninth-century patriarch of Constantinople and senior cleric of the Byzantine empire, reviewed Cosmas's treatise he roundly mocked his ideas.

In reality, the greater danger lies at the opposite end of the spectrum: not in Cosmas's point-blank 'oppositionalism', but in automatic agreement, a position that is sometimes called 'concordism'. Instead of rejecting science, in the name of religion, this approach zealously baptises it, taking the scientific orthodoxy of the day and reading it back into the pages of holy scripture. But it is a mistake, as well as fundamentally dishonest. Reading the earthly sphericity into the Bible or the Qur'an is no more accurate or authentic than finding heliocentrism or evolution there. As John Philoponus argued, at least at first, the cosmology of such holy scriptures is almost always assumed rather than revealed, and their teachings demand few articles of scientific orthodoxy except, as we shall see, when it comes to human beings. Finding Aristotle, Copernicus, Darwin or Einstein in Genesis or in Revelation is essentially fraudulent.

And dangerous. Time and again, as James Clerk Maxwell would warn an overzealous bishop 1,500 years after John Philoponus, when the science changes, the religious structures that had been built on it teeter and topple. And science does change, its fluid and evolving nature making all attempts to locate religious doctrine within scientific 'facts' a perilous business.

Marrying the science of the age would repeatedly leave religion an embittered widow.

'Preparing Medicine from Honey', from a manuscript of an Arabic translation of *De Materia Medica* by Greek physician Dioscorides. Opinions are divided today, as they were in the nineteenth century, about how derivative or how original was the work done in the 'Golden Age' of Islamic science.

TWO

A FRAGILE BRILLIANCE: SCIENCE AND ISLAM

'There was no more science worth mentioning in Islamic countries': the case against Islamic science

In early 2007, the *Times Literary Supplement* saw a terse exchange about the history of Islam and science between theoretical physicist Steven Weinberg and historian of Islam and science Jamil Ragep.[1]

In a review of Richard Dawkins' *The God Delusion*, Weinberg explained that Islam had 'turned against' science in the twelfth century, in the wake of the philosopher Abû Hâmid Muhammad ibn Muhammad al-Ghazâlî, known to history as al-Ghazâlî (b. mid-eleventh century), from which point 'there was no more science worth mentioning in Islamic countries'. This drew an indignant response from Ragep who accused Weinberg of ignoring three or four generations of scholarship that had revealed 'scores of Islamic scientists' between the twelfth and eighteenth centuries who had, among other things, proposed the idea of pulmonary circulation, built large-scale astronomical observatories and 'developed the mathematical and conceptual tools that were essential for the Copernican revolution'. Weinberg was unbowed and retorted that Copernicanism owed nothing to later Islamic science and that al-Nafis's work on pulmonary circulation 'had no effect in the Islamic world'. The debate ended without concord or amity.

The history of science and religion, as it emerged in nineteenth-century Europe and America, was often a proxy for the history of Western progress. As a result, the tale of Islam and science played an important but ambiguous role. Some historians were eager to recognise Islamic scientific achievements and use them as a stick with which to beat Christian (or, better still, Catholic) obscurantism. Gibbon, never slow to puncture ecclesiastical pretensions, lauded the contribution of Muslims to astronomy, chemistry and medicine. The Prussian explorer and naturalist, Alexander von Humboldt, praised Arabic contributions to science, calling the Arabians 'the proper founders of the physical sciences'.[2] John William Draper extolled the twelfth-century polymath Ibn Rushd (known to Europeans as Averroes), and argued that Islamic theology enabled the development of the laws of nature far more readily than did Christian theology, with its unsustainable commitment to miracles.

Still positive, if somewhat more tepid, William Whewell wrote in his *History of the Inductive Sciences* that the 'Arabs discharged an important function in the history of human knowledge, by preserving and transmitting to more enlightened times, the intellectual treasures of antiquity'. Whewell's 'Arabs' were to be congratulated not for being original thinkers but as reliable messengers. Rather more critically, the French linguist and biblical scholar, Ernest Renan, proclaimed in a famous lecture at the Sorbonne in 1883 that although the Muslim world was intellectually superior to Christendom between about 775 and 1250, this was clearly an anomaly. The truly enlightened caliphs of this time were 'barely Muslims'. This science was 'called Arab' but what, Renan asked rhetorically, 'did it have that was Arab in reality?' Muslim countries 'plunge[d] into the saddest intellectual decay' after their Golden Age. Philosophy was 'abolished'. And the reason for this decay, unlike for the half-millennium of intellectual ascendancy, *was* intrinsically Islamic:

> From the beginning of his religious initiation, at the age of ten or twelve years, the Muslim child, until then still quite aware, suddenly becomes fanatical, full of a foolish pride in possessing what he believes is the absolute truth, happy with what determines his inferiority, as if it were a privilege. This senseless pride is the radical vice of the Muslim.[3]

For their part, those Muslim historians who began to write about science and Islam from the nineteenth century took a rather different angle. Some seized on the positive European narrative. When the Ottoman journalist Ahmed Midhat translated Draper's *History* into Turkish in 1895, he commented (at length) on how the book would prove that Islam was 'the most scientific of religions'.[4] Sixty years earlier, the Egyptian intellectual Rifa'a al-Tahtawi had argued that Europeans were simply carrying the scientific torch lit by Muslims. These sciences, his contemporary the Ottoman diplomat Mustafa Sami wrote, 'are our true heritage'.[5]

Either way, whether scholars were American, European, Arabic, Christian, secular or Islamic, the history of science and Islam could not but be read through the wide-angle lens of modernity, power, colonialism and civilisational clash. Something similar might be said, in a decade of Islamist terrorism and the Iraq war, of the exchange between Weinberg and Ragep. If, as Renan concluded his lecture, science 'will serve only progress . . . which is inseparable from respect of mankind and freedom', writing about the way in which Islam either abetted or impeded it involved more than writing simply about the history of science. There could be no bystanders in this conversation.

'O Philosopher . . . What is good?': golden years

That there is a conversation to be had about Islamic science has never been doubted, even by the most devout Islamophobe. From the ninth century onwards, Islamic territories, in particular but not exclusively the Abbasid caliphate based in Baghdad, boasted scientific thought and achievements that matched anything in the classical world. In astronomy, through accurate instruments, precise and repeated observation and a willingness to disentangle the subject from astrology, Islamic astronomers reached the point of critiquing and correcting the Ptolemaic system with models that are, in the eyes of many scholars, mathematically equivalent to those of the Copernican system, without actually being heliocentric. By adopting Hindu notation, the decimal place and the concept of zero, Islamic mathematicians made significant advances in algebra, geometry and trigonometry. Al-Khwarizmi, one of the earliest and perhaps the greatest Islamic mathematician, introduced ideas of positional notation in his early ninth-century *Book of Addition and Subtraction According to Hindu Calculation*, and contributed to the study of equations in his book on Al-jabr, thereby giving us 'algebra', and, via his own name, 'algorithm'. In medicine, a series of massive medical encyclopedias synthesised and systemised Greek knowledge. At the same time, Muslim physicians did original work, particularly in ophthalmology, and the thirteenth-century Syrian physician al-Nafis corrected Galen's authoritative teaching by describing how blood in the right ventricle of the heart reached the left by way of the lungs rather than by diffusing between ventricles. In optics, the seven-volume treatise by eleventh-century scholar Al-Haytham (Alhazen) circulated widely in Europe (more widely than it did in Islamic lands, in fact), whence it became the foundation of the work of Christendom's greatest optics scientists, such as Roger Bacon and Theodoric of Freiberg.[6] Even in chemistry, still centuries away from becoming a recognisable scientific discipline, Islamic thinkers displayed considerable curiosity and ingenuity, as evidenced by the sheer number of modern chemical terms in English – alkali, alchemy, alembic, alcohol, amalgam, benzoic, borax, camphor, elixir, etc. – that derive from the Arabic of this period. The legacy of this so-called 'Golden Age' of Islamic Science – a double-edged epithet for its insinuation that the achievement was in some way anomalous – is impressive by any measure.

The question, however, is less about whether Islamic thinkers made a contribution to science than how significant, how distinctive and, above all, how Islamic that contribution was. The implication lying behind Weinberg's, Whewell's and, in particular, Renan's criticism was that there was something about the authoritative natures of Islamic culture and theology that ultimately suffocated the scientific spirit.

The case for the prosecution sounds, at least at first, formidable. First, there is the argument from geography. During its first century, Islamic culture showed next to no interest in philosophy or science. Despite being based in Damascus, the Umayyad caliphate between 661 and 750 was indifferent to classical learning, something that changed only when the centre of Islamic gravity moved to Baghdad in 750. Baghdad was a new city in an old empire. Incorporating the ancient Persian empire, abutting the Byzantine empire in the West, open to Indian influences from the East and encompassing important intellectual outposts of Hellenism like Gundeshapur, the Abbasid caliphate stood at a significant intellectual crossroads.

The Abbasid elite settled among and intermarried with the old Persian nobility, in the process absorbing certain longstanding norms, such as the idea that the Persian 'king of kings' should serve as a patron to scholarship. Abbasid caliphs sought to consolidate their authority by adopting this role, which also enabled them to assert an intellectual superiority over Byzantium, an equally venerable civilisation whose emperors were increasingly seen as having allowed the tradition of Hellenic thought to wither on the vine. Accordingly, the natural sciences were known as the 'foreign sciences' among Muslims, as against the so-called 'Islamic sciences' which were devoted to the study of the Qur'an, the traditions of the Prophet Muhammad *(hadith)*, legal knowledge *(fiqh)*, theology *(kalam)*, poetry and the Arabic language. The philosophical and scientific turn in Islamic history was, in short, an alien appropriation as part of a geopolitical and cultural calculation.

Second, there is the argument from people. The Abbasid caliphate absorbed the work of a wide range of scholars from different cultures and religions, who became the foundation of the 'Golden Age'. A few were respected for their own work but most secured their reputation for translation. Of the sixty-one Greco-Arabic translators for whom we have names, fifty-nine were Christians, the majority from the Syrian Church.[7] Syriac translators did not work alone and Greek science was not the only source, but there is little doubt that, without their input, there would have been no Islamic scientific age, golden or otherwise.

Third, and most significantly, is the argument from content. The story goes that one night, Caliph al-Ma'mun, the seventh Abbasid caliph (r. 813–33), had a dream about Aristotle. According to one version of the story, the caliph found himself standing before a man 'of reddish-white complexion with a high forehead, bushy eyebrows, bald head, dark blue eyes and handsome features' who identified himself as the great philosopher. Awed and joyed, al-Ma'mun asked him, 'O Philosopher . . . What is good? . . . Then what? . . . Then what?' The order of the philosopher's responses was significant: 'Whatever is good according to the intellect . . . Whatever is good according to religious law . . . Whatever is good in the opinion of the masses.'[8]

This account of the dream inadvertently confirms the views of those who argue that these centuries of luminous science were in spite of rather than because of Islamic authority. Aristotle elevated rational knowledge over and above religious law, even to the point of being its arbiter. Al-Ma'mun apparently concurred. Reason rather than religion, so the argument goes, lay at the heart of Muslim science.

For those who have eyes to see it this way, the movement demonstrated quite how much Islamic science was in fact Greek. This view was further underlined by the fact that Arabic, which had been instantiated as the official language of administration by the Umayyads, had virtually no technical or mathematical vocabulary to accommodate many Greek texts. The result was that translators were often compelled simply to transliterate obscure Greek words in the attempt to capture an alien thought world.

The conclusion drawn from these arguments, then, is that for all its glory, the Islamic 'Golden Age' of science and philosophy was not especially Islamic and rested, instead, on borrowed culture, borrowed scholars and borrowed learning. When there was no more borrowing to be done and Greek science, mathematics and medicine had been fully absorbed, the gold began to tarnish and there was, to coin a phrase, no more science worth mentioning in Islamic countries.

'Not as an invading force but as an invited guest': the case for Islamic science

The argument seems persuasive at first but, on closer inspection, it unravels. For all that the Abbasid caliphs were influenced by their immediate context, geography was not as decisive as all that. The Islamic lands stretched from the Atlantic to India and had various centres of intellectual achievement. The region of Al-Andalus, for example, was three thousand miles west of Baghdad yet still boasted an impressive tradition of medicine, astronomy and philosophy.

Al-Andalus was under the Umayyad dynasty, which further underlines how Islamic science wasn't a uniquely Abbasid affair. The Umayyad caliphates in Damascus might not have invested in classical learning or translation but they did lay the foundations for it. Mu'awiya, the first Umayyad caliph, started a collection of books and through the detailed study of, and rational reflection on, language and texts, and a high respect for the book and for '*ilm*', or knowledge, the earliest Islamic thinkers provided the raw materials that were fashioned by their successors. Although the flowering of intellectual life under the Abbasids in the ninth to the twelfth centuries was the high point, it did not appear *ex nihilo*.

Nor did it disappear *in nihilum* either, as Weinberg and others claimed. This question is dogged by ignorance. In astronomy alone, Arabic manuscripts outnumber those in Greek and Latin put together for the medieval period, and most of them remain untranslated and/or unstudied.[9] There seems little prospect that new scholarship will unearth figures of the stature of al-Khwarizmi or Ibn Sina among dynasties like the Mamluks of Egypt, the Timurids of Iran and the Ottomans of Turkey, which flourished in the later medieval period, and to that extent, the narrative of scientific and intellectual *decline* appears to be correct. But decline is not death, and Ragep is right to point out that the more we learn about this period of Islamic history, the more it seems that the decline was not as precipitous as once thought.[10]

Such arguments – against geo-cultural determinism of Islamic science, and against the idea of its precipitous decline after 1200 – go some way to undermining the view that the Islamic 'Golden Age' was little more than Greek science in Islamic clothes. However, they also somewhat miss the point and ignore the catch-22 in which this whole debate places Islam.

When Aristotle replied to al-Ma'mun by putting 'whatever is good according to the intellect' before 'whatever is good according to religious law', the case *against* Islamic science says that this shows how science demands the relegation of religion in order to succeed. By this logic, science was only possible because Greek reason took the starring role, and Islamic religious authorities – the theologians and legal scholars – were pushed to the wings.

But if Aristotle were to have reversed the order, the prosecution would simply have pointed out that an Islamic story that showed how religious law must be elevated above the intellect simply proved how science and rational thought are incompatible with, because subordinate to, religion. Either way, by treating religion and science or, in this dreamy scenario, religious law and the intellect, as alternative and *competing* entities, the conclusion is preordained. It's either scientific freedom or religious authority: you can't do both.

In reality, the open, receptive and pluralistic attitude of Abbasid Islam to the learning of the Greek world is one of the strongest arguments *for* Islamic science and the (potential) concord between science and religion. There is, it's important to recognise, no such thing as *mere* translation. The task requires a determination to find, purchase, transport, house and understand texts, and a willingness to find and fund translators to do work with no obvious or immediate benefit. Above all, it requires an openness to foreign cultures and traditions of learning, and a genuine readiness to listen and learn from them.

Nor was this limited to the Abbasid court in Baghdad. Al-Andalus, the region of Islamic Spain that lasted for seven hundred years, witnessed a rich dialogue between Islamic and Jewish scholars, as we shall note in the following chapter. It was, or at least was often, the same story in the Fatimid caliphate in North Africa. There is a story from mid-tenth century Cairo of a debate between Ibn Ridwan, 'a self-taught physician, with an enormous ego and a quick temper', and Ibn Butlan, a Nestorian Christian physician lately arrived from Baghdad. The debate was about a topic in Aristotelian biology, but the issues of the social status and authority of the state chief physician Ibn Ridwan and the parvenu Ibn Butlan were never far away. The debate rapidly turned rancorous. Each party shifted from criticising his opponent's arguments to targeting his opponent by condemning his position and his learning, or lack thereof. But in spite of this and the ample opportunity afforded, it is telling that neither party at any point attacked the other's religion.[11] If successful intellectual activity was characterised by openness to 'foreign' traditions, Islam was very successful.

'Foreign sciences' required, and for the most part received, religious and political approval and support. Abbasid caliph Harun al-Rashid founded a library in Baghdad at the end of the eighth century, to house the growing number of texts and, a generation later, his son and successor-but-one, Caliph al-Ma'mun, expanded the institution into what would become the famous 'House of Wisdom'. Destroyed in the Mongolian siege of Baghdad in 1258 and leaving no physical traces today, there is significant dispute over its size and function, with claims that it housed hundreds of thousands of volumes and functioned not only as a library but also as a fully-fledged research centre. Exaggerated as these claims may be, the

House of Wisdom, exceptional but not unique as a centre of learning, was undoubtedly an important intellectual centre, as much for what it symbolised – official religious support for the 'foreign sciences' – as for what went on there.

Less common, though no less spectacular, were observatories. From the 820s, when al-Ma'mun ordered the construction of the first in Baghdad, astronomical observatories spread across the Islamic world. Nine state-sponsored ones are known. Some were large, expensive and impressive. The Maragha Observatory in northern Iran accommodated numerous astronomers, mathematicians and instrument makers, various maps and spheres and a large library. The Ulugh Beg Observatory in Samarqand, which flourished in the early fifteenth century, was built on Maragha's plans and produced some of the most accurate tables and measurements of the medieval period.

Ironically, one of the problems faced by Islamic science was too much enthusiasm. Al-Ma'mun was a keen supporter of Mu'tazila, a 'rationalist' tradition of theology that championed reasoned enquiry, receptivity to foreign science and a highly flexible reading of the Qur'an. His enthusiasm tipped over to outright persecution by the end of his reign, the following fifteen years becoming known as the *mihna* or inquisition, during which Mu'tazilites instituted a violent campaign against more conservative religious scholars. Religious leaders who advocated a literal reading of the (uncreated) Qur'an, and who judged it, along with the *hadith*, a sufficient source for human wisdom, were imprisoned or killed. The campaign would leave scars and, within decades, generate and be used to justify a reaction against it, in the name of theological orthodoxy. The coincidence of science, religion and authority would be a perilous business, right into the twentieth century, though not necessarily in the ways expected.

In addition to these various state-sponsored translations, libraries, observatories and intellectual crusades for science, there were important and specific theological injunctions that aided the cause of science. On occasion, the idea was voiced that studying nature was a way of glorifying God by recognising his handiwork in creation, although this was rhetorically rarer than it would be in Christian Europe, in part because Islam had a more uncompromising view of God's transcendence, and in part because such 'natural theology' carried with it the implication that there could be something lacking in God's revealed word.[12]

More importantly and pervasively, there were needs relating to the practice of Islamic faith that spurred scientific work. From the earliest days, Muslims were required to face Mecca when saying their daily prayers. The direction in which they should face – the *qibla* – naturally varied according to location, and calculating it was far from straightforward. Ongoing efforts to do so with ever-greater accuracy, in a religious empire that stretched five thousand miles, acted as a catalyst to Islamic astronomy, geometry and trigonometry.

In a similar way, Muslims were obligated to pray at five scheduled times throughout the day, and determining these involved understanding the varying appearance, disappearance, path and altitude of the sun. The resulting science of timekeeping boasted several eminent scientists, al-Khwarizmi among them. By the thirteenth century, the role of *muwaqqit*, or official timekeeper, had emerged in parts of the Islamic empire, a position that perfectly straddled the science–religion 'divide'.

The idea, then, that Greek science was an alien, invasive or simply parasitical presence within Islamic intellectual life is untenable. There was a powerful (if not, as we shall see, constant or reliable) religious and political imprimatur for the new learning. There were elements of Islamic religious practice and ritual that positively catalysed certain disciplines. And there were many exemplary thinkers who not only appropriated 'foreign sciences' but integrated them within Islamic thought. As the eminent historian of science in Islam, A. I. Sabra, once put it, Greek science entered the world of Islam 'not as an invading force . . . but rather as an invited guest'.

'The problems of physics are of no importance for us': the points of tension

Rebalancing the scale in this way still leaves an outstanding question. If Islamic authorities had been so hospitable to natural philosophy, and if there were so many brilliant proto-scientists accommodating and developing Greek learning in Islamic territories, why was the promise of the ninth to the twelfth centuries not fulfilled in some kind of Islamic scientific revolution? Why did Baghdad not produce Isaac Newton and Robert Boyle?

When one starts picking at the details of Islamic science, even during the early centuries, it becomes clear that while there was genuine intellectual, religious and political support for Greek science, it could be localised and fragile. The authorities did welcome Greek science as an invited guest, but not all of them, and not consistently.

Caliph al-Ma'mun was soon succeeded by Caliph al-Mutawakkil (r. 847–61), who not only rejected his predecessor's more rationalist approach to theology but replaced his rational inquisition with a theological one of his own. Al-Kindi, one of the greatest philosophers of the mid-ninth century, was beaten up and had his library confiscated in the process. The remarkable polymath Ibn Sina remarked that, when growing up in the last years of the tenth century, he learned Hindu arithmetic not in any formal educational setting but from an Indian greengrocer. The tenth-century mathematician al-Sijzi mentioned in passing that in his region people held it lawful to kill geometers. The following century, the authoritative theologian and philosopher al-Ghazâlî wondered, acerbically, quite how many mathematicians a community actually needed. The brilliant twelfth-century mathematician al-Maghribi wrote about how he was unable to find anyone to teach him Euclidian geometry when he moved out of Baghdad. Even observatories, one of the crowning glories of the Islamic support of science, were often founded and patronised for their astrological potential over and above their astronomical.

Such tales suggest that in spite of the speed and extent of the Islamic reception of 'foreign sciences', that welcome could be distinctly ambivalent in places. Moreover, it could be ambivalent not simply in the way that all cultures through history have been ambivalent towards new and alien ways of thinking, but for reasons that were coherent and religiously principled.

In the first instance, there could be problems with the *conclusions* at which rational science apparently arrived. The Aristotelian commitment to the eternity of the world, which had vexed earlier, Christian, philosophers, and the seemingly rational conviction that the stars had power over human affairs were both incompatible with Islamic belief. In different ways, both ideas undermined God's non-negotiable omnipotence. If, as the Qur'an proclaimed, 'God has power over all things' (Q48:21), how could anyone reason that the cosmos or the stars had their own autonomy?

There could also be problems with the *premises* on which such conclusions were built. The natural philosophy inherited from the ancient world presupposed a broadly naturalistic realm. Causality ran from one object or event to another, rather than each being actively and purposively directed by God or another divine agent. As we saw in the previous chapter, this was rarely the completely naturalistic realm imagined by modern science, but it was nonetheless a realm that could be studied on its own terms. This was an assumption rather than a conclusion, but it could be a troubling one, subtly subversive of the idea of divine control. If nature ran itself, what was God's role?

This problem led to a third, which was really at the heart of such tensions. Where did *intellectual authority* reside? The Islamic world of the ninth century witnessed fierce disagreements about the un/createdness of the Qur'an but there was rather less controversy about its unique status as the revelation of God. To be a Muslim was to recognise and honour this belief. But that still left questions about its sufficiency for human knowledge. If the Qur'an was indeed 'a detailed explanation of all things' (Q12:111), what need was there for other modes of intellectual enquiry? The foreign sciences were not, of course, based on the Qur'an. What then was their status? Did they simply offer tools of logic and reason by means of which God's revelation might be better understood? Or did they contribute anything new, anything that could not be known through revelation? Was there more than one road to truth and, if so, might the main path of revelation be, in theory, superfluous?

Questions of intellectual authority were never far from those of wider authority, and such concerns about the epistemological status of foreign science naturally cashed out in a fourth area of tension, that of social and political authority. Islamic society, like every other, had various centres of authority, all of which were necessarily connected, albeit in different ways, to the ultimate authority of God and the Qur'an. *Ashab al-hadith* – adherents of the *hadith* – held the view that only the Qur'an and the *hadith* were acceptable authority for Islamic thought and practice. *Mutakallimun* – practitioners of *kalam* or philosophical theology – honoured the revelation of the Qur'an and the *hadith* and sought to clarify and systematise understanding of both through the traditions of reason inherited from the Hellenic world. *Falâsifa* – philosophers – were heirs to the recently translated Hellenic (Aristotelian) rationality, which they pursued on its own merits. The *Mu'tazilites* went further still and drew reason not only as a legitimate and necessary source of knowledge but as an authoritative partner in the dialogue with revelation, arguing that what was revealed should conform to that which was established by reason. *Fuqaha* – jurists – were experts in sharia, under which Islamic society operated, whose reasoning was based on the Qur'an, *hadith* and principles of analogy and of consensus. Sufis were mystics who championed a personal, experiential approach to God but crystallised their beliefs and practices from the twelfth century, in the process forming influential social orders. Individuals within all these different groups were often self-appointed. Their authority depended on reputation and persuasion, rather than any institutional basis. The consequence was that, in an already crowded and tense field of different, sometimes competing, authorities, there was often an incentive to undermine the foundations of rival groups. The foreign sciences could be framed as undermining the authority of the Qur'an and, therefore, the entire social and political order built on it.

These four factors – tensions with the conclusions of science, its premises, its intellectual foundations and its social manifestations – amounted to powerful counter-currents that ran through Islamic intellectual life, always critiquing, sometimes condemning and occasionally forsaking the foreign sciences that had been appropriated so willingly. Thus, in opposition to Mu'tazilism, the early tenth-century theologian al-Ashari developed a doctrine of what came to be known as occasionalism. The idea here was that all events and processes within creation, whether natural or human, were wholly caused and directed, moment by moment, at an atomic level, by God. Indeed, the world was effectively recreated by God at every moment, no subsequent moment of time having any causal link to the one that preceded it. By this reckoning, as Weinberg said in the article that sparked his dispute with Ragep, 'a piece of cotton placed in a flame does not darken and smoulder because of the heat, but because God wants it to darken and smoulder.'

Such teaching was in contradiction with the naturalistic and rationalistic presuppositions of Greek science. It didn't *necessarily* sound the death knell for rational and scientific enquiry as it could be argued that because God's control of creation was faithful and reliable, human rational investigation was therefore possible. However, the way in which such theology undermined the internal coherence, rationality and indeed legitimacy of such enquiry made the rational sciences harder to sustain in its wake. It was this point that Pope Benedict XVI was gesturing towards in his famous and controversial Regensburg lecture in September 2006. In Muslim teaching, he observed, God is 'absolutely transcendent', and 'his will is not bound up with any of our categories, even that of rationality.'[13]

In his occasionalism, al-Ashari was developing a latent strand of thought within Islamic theology. By far the most influential advocate of this view, whose views on burning cotton Weinberg quoted, was al-Ghazâlî. Director of the Nizâmiyya madrasa in Baghdad, one of the first institutions of higher education in the Islamic world, a confidant of sultans and well placed at the caliphal court in Baghdad, al-Ghazâlî renounced his positions in middle age, dogged by the growing conviction that proximity to power compromised his spiritual and moral health. Prostrate before the tomb of Abraham in Hebron, he vowed never again to serve the political authorities.

Al-Ghazâlî was no insular, religious 'fundamentalist'. He argued against necessary causality, in terms that would not have been alien to David Hume seven hundred years later – 'the connection between what is habitually believed to be a cause and what is habitually believed to be an effect is not necessary' – and he did so on sophisticated grounds. Despite the title of his most famous book, *The Incoherence of the Philosophers*, he was not entirely opposed to Aristotelianism or the deployment of philosophical thought. When wrestling with the ever-thorny question of the eternity of the cosmos, he drew on John Philoponus's reasoning to counter the mighty Aristotle. He did, however, object to the sense of superiority among the *falâsifa*, in particular their conviction that reasoning was sufficient because it was sound – a belief he considered to be a risk to revelation, religion and law.

The *Incoherence* took twenty claims of the *falâsifa*, four pertaining to natural science and the rest to metaphysics, and demonstrated that they did not reach their own self-proclaimed standards of proof and were in fact constructed on unsustainable premises, rhetoric or simply the authority of the ancients. Although he didn't necessarily disagree with all the conclusions that the *falâsifa* reached, he did object to the authority on which they reached them. Reason alone was simply not enough. The conclusions that were wrong were mostly inconsequential. Some, however, such as the eternity of the world or the non-resurrection of the body, were serious and dangerous. Those who publicly taught them, he remarked at the end of the book, are apostate and merit death.

The *Incoherence* was refuted at length by Ibn Rushd in the following century in his *Incoherence of the Incoherence*, but the influence of the original book was irresistible. Al-Ghazâlî had articulated, powerfully and at length, a strong intellectual counter-current within Islamic thought, which grew in significance from the early twelfth century. Certain forms of natural philosophy and scientific enquiry remained acceptable, indeed commendable – medicine, astronomy, mathematics – but their imprimatur was instrumental, justified primarily for their religious utility. 'In the early centuries', according to A. I. Sabra, the scientist may have served as a medic or an astronomer, but 'he was also recognised and revered as a hakim . . . one who investigates any aspect of the world for the sake of knowing truth about the world.' That was not so later on, where the scientific space granted by religious authority *had* shrunk. As the historian Ibn Khaldun wrote in the fourteenth century, 'the problems of physics are of no importance for us in our religious affairs or our livelihoods. Therefore, we must leave them alone.'[14]

Al-Ghazâlî did not, and should not, have the final word on Islamic science. Not only was he not the sole, typical or representative figure – any more than al-Khwarizmi, al-Kindi or Ibn Sina were – but he also did not stop Islamic science in its tracks. Individuals never do, except according to the History of Great Men. What he did do was articulate a forceful religious critique of natural philosophy, of a kind that has always accompanied the study of the natural world.

This was not in itself a problem. Science is an inherently questioning, inherently conflictual practice; being critiqued should not be a problem, even when that critique is at the level of philosophical presupposition, as it was with al-Ghazâlî. The problem comes when there is no secure arena for that practice, and critique, of science; when such questions risk tipping over from being merely demanding to being potentially deadly. The authority of natural philosophy was ultimately too contingent, too dependent on that of other intellectual disciplines and institutions to survive such a threat intact. It was this, rather than al-Ghazâlî, that undermined Islamic science in the later medieval period.

The medieval Muslim world boasted many institutions of higher learning – al-Ghazâlî taught at one – but they tended to be *waqf* institutions, charitable endeavours endowed by private individuals to whose personal concerns they were beholden. Unlike in Christendom, there was no concept of the legal corporation, by means of which a group or guild of professionals, such as scholars, could come together and determine their activities in comparative freedom. As we shall note in chapter 4, the decentralisation of intellectual and institutional authority in medieval Europe helped protect all intellectual activity. For science, an inherently questioning and therefore inherently threatening activity, this was essential.

Madrasas existed across the Islamic world from the eleventh century. Also *waqfs*, these could and did have varying educational functions, which could include algebra, geometry, trigonometry, logic and astronomy. But their centre of gravity was firmly in the study of the Qur'an, *hadith* and sharia, rather than *falsafa* (philosophy), *kalam* or the natural sciences. Books on these subjects were copied and often held in libraries, and as we have seen, scholars remained familiar with them, studying and teaching the foreign sciences privately. But after al-Ghazâlî's powerful argument for their inferiority, they were at best secondary, and at worst suspect, authorities for knowledge.

Even observatories, which were very substantial investments in the scientific study of the cosmos, were vulnerable institutions. Indeed, it is striking how short-lived most were. The Maragha Observatory in Iran lasted for only fifty years. The Ulugh Beg Observatory survived for an even shorter period. It was founded in 1420 and destroyed thirty years later, being entirely lost to history until its foundations were rediscovered in 1908. The Malikshah Observatory of Isfahan survived from 1074 until the death of its founder, Sultan Malik-Shah I, less than twenty years later. And the Istanbul Observatory, in spite of making some highly accurate observations, was destroyed three years after it started, on the order of the sultan, having been the subject of antagonistic preaching.

None of this precluded good scientific work. Indeed, it was precisely those disciplines that retained some formal religious approval, like astronomy and medicine, that made the most impressive advances in the late medieval period. But its brilliance was fragile. Ultimately, the authority of natural philosophy in the Islamic world was vulnerable, and the sciences failed to acquire a sufficiently *secure* institutional setting to withstand powerful intellectual counter-currents and social pressures.

A page from Moshe ben Maimon's (Maimonides)
Guide for the Perplexed, which brought theology
into constructive dialogue with Greek philosophy.
Maimonides was incalculably important for mediaeval
Jewish science, revered but also criticised.

THREE

AMBIGUOUS AND ARGUMENTATIVE: SCIENCE AND JUDAISM

'Nature is so obedient to him': science in early Judaism

In about AD 39, a Jewish philosopher from Alexandria found himself confronting the most powerful man on earth. The emperor was Caligula, renowned for his paranoia, volatility and cruelty; the philosopher was Philo, a respected leader of the Jewish community in his home city, by then in his fifties. Philo had been chosen to represent his people against the charge of sedition. The target of recent mob violence in that endlessly rioting city, the Jews had been accused of dishonouring Caesar by refusing his statue in their synagogues. 'You cannot possibly have been ignorant of what was likely to result from your attempt to introduce these innovations respecting our temple,' Philo berated the emperor, at least according to his own account.[1]

The Jews refused to look upon anything that was created as being divine or worthy of worship. The emperor was just a man. The world was created. Nature was simply nature. It was an attitude that, many centuries later, the German philosopher Hegel would claim fed the roots of scientific enquiry. The Old Testament, he claimed, desacralised nature. Rather than feared or worshipped, it was to be controlled. God, he wrote, is represented 'as using the wind and the lightning as servants and messengers, Nature is so obedient to him'.[2] As with God, so with those made in his image. By dethroning the gods on earth – whether they were in temples, in humans or in nature – the Jews took the first and most important step on the road to science.

There is a certain appeal to this idea but it is, at best, a partial truth if only because it flattens the Jewish approach to science, which defies easy categorisation. When he was not defending his people against accusations of disloyalty, Philo was attempting to reconcile the Hebrew scriptures with Greek philosophy; the first Jew, of whom we know, to embark on that endeavour. He saw himself as a great mediator between the two traditions. He wrote a treatise on the creation stories of Genesis, which mixed literal and symbolic interpretations generously, and drew on Greek science in the process. Time, he argued, was a property of space, provably coeval with it.[3] Creator and creation could be understood through the concept of *logos* or reason. Philosophy, he contended, in an argument that the Church Fathers would make popular, was a 'handmaiden' for true religion. Here was a tradition of Judaism that found a serious dialogue partner in the science of the time.

Yet Philo was contemporary with the Qumran community in Palestine, which chose to live a life of complete isolation rather than endure any contact with Greek and Roman culture. And he was contemporary with a number of other sects that, while not going to the extreme lengths of the Qumran community, rejected the Hellenic culture, natural philosophy and all, in which they had lived for centuries. Athens, as far as they were concerned, had nothing to do with Jerusalem.

There was, in short, no single Jewish approach to Greek science at the time of Philo – or indeed after. For what marks off the early Jewish engagement with science from that of other religions was not simply that it was plural in itself – the same can be said for Christianity and Islam – but that it always took place in a plural context. At the time of Philo, the Jews of Alexandria rubbed shoulders, angrily, with the Greeks of the city, all of them living under Roman rule. Five hundred years later, the dominant religious culture was Christian although, in Alexandria at least, one still saturated with Greek philosophical ideas. By the ninth century the majority of Jews in the world were living under Islamic rule, speaking Arabic and enjoying direct access, and making direct contribution, to natural philosophy. In Spain especially, between the ninth and mid-twelfth centuries, Jews enjoyed particular intellectual freedom, translating a great deal of natural philosophy from Arabic. Thereafter, they continued to live as a minority in Christendom, more peripheral and vulnerable to the majority culture, although still making contributions when they could. In short, the histories of science and religion are never more complex– encompassing embrace and rejection, contribution and denunciation – than in the story of Judaism.

'Our Gods are as numerous as our towns': varied believing

For all Hegel claimed for them, the Hebrew scriptures are not quite as univocal as he imagined. In them, the natural world, including the heavenly bodies, is indeed desacralised – certainly in comparison with other religious cultures in the Ancient Near East. Nature is under God's sovereign control and the prophet Isaiah, among others, warned his people against consulting astrologers.[4] But at the same time, the book of Judges pictures the stars as alive and fighting for the Israelites, and the Jewish historian Josephus describes his contemporaries following astrologers into a revolt against the Romans.

Creation had an order to it. The opening chapter of Genesis was carefully structured, drawing out the pattern and order in the heavens and the earth. The book of Ecclesiastes poetically recited how everything had its due order and due time. The book of Proverbs recorded how God made the world through Wisdom. All this lent itself to a structured understanding of the world. God was a lawgiver. Creation followed his laws.

And yet, things were not obviously orderly. Even a cursory reading of the scriptures revealed a world that was crammed full of spirits, dreams, ghosts and divine intervention. The Hebrew Bible was largely uninterested in laying out, let alone defining, any particular model of the cosmos, and was content instead to adopt and adapt the models prevalent in surrounding cultures. The scriptures offered no scientific prescription to embrace or reject. In the words of scholar Noah Efron, 'It is not just that the Bible packages together conflicting views under the same covers. It also privileges views that conflict with one another . . . it remained an open book.'[5]

This helped generate a particularly flexible Jewish tradition when it came to natural philosophy. The Talmud, the collection of rabbinic

writings, commentaries and disputes that were set down by the early sixth
century, lent itself to such flexibility in its layered and discursive format:

> Rav Aḥa bar Yaʾakov strongly objects to this proof . . . perhaps the stars are
> stationary within the sphere like the steel socket of a mill . . . Alternatively,
> perhaps they are stationary like the pivot of a door . . . similarly, perhaps
> the constellations are stationary within a sphere . . . Therefore, Rabbi
> Yehuda HaNasi's statement is not necessarily true.[6]

The result was a range of different, sometimes apparently contradictory,
views of nature and the cosmos within both the Babylonian and the
Palestinian Talmuds. The cosmological discussion above, for example,
continues by comparing what 'the Jewish Sages' say about the sun and the
firmament to what 'the sages of the nations of the world' say, and then
concludes in a somewhat nonplussed way, 'the statement of the sages of
the nations of the world appears to be more accurate than our
statement.'[7]

This tension was reflected in the wider Talmudic attitude to what was
usually called 'Greek' or 'foreign' wisdom, or 'the wisdom of the nations'. Is
such wisdom acceptable? the Babylonian Talmud asks at one point. The
answer was yes: 'the fact that Rabbi Gamliel allowed half of his household
to study Greek wisdom indicates that it is permitted.' Except that the
answer might also be no. 'The members of the house of Rabbi Gamliel are
different, as they were close to the monarchy, and therefore had to learn
Greek wisdom in order to converse with people of authority . . . cursed is
the person who teaches his son Greek wisdom.'[8]

The importance of the Talmud in forming Judaism effectively canon-
ised this complex, shifting and plural attitude to science, an attitude that
was discernible among Jewish philosophers as they came to wrestle with
Greek science in Islamic and then Christian intellectual contexts. Saahdiah
Ben Joseph Gaon was born in Egypt in 882. Author of biblical, grammati-
cal, liturgical as well as mathematical and philosophical works, he has
been called the founder of scientific activity in Judaism for his works,
written in Arabic, which sought to demonstrate the compatibility between
the law of Moses and philosophical reason. He was, in effect, the first
Jewish thinker since Philo to bring science and religion into sustained

discourse. Saahdiah's commentary on Genesis entertained long digressions on astronomy and biology, and his major work, *The Book of Beliefs and Opinions*, draws on Stoic 'physics' and argues against the Aristotelian belief in a fifth element, or quintessence, in the heavens.[9] More significantly, the book argued that there were four independent sources of human knowledge – sensation, reason, logical inference and reliable tradition – and it (tentatively) offered *reasons* for the laws revealed by God.

This was far from uncontroversial, and not simply because of the ambivalent status of Greek science in the Talmud. By the time Saahdiah wrote, most Jews were living under Islamic rule. Their encounter with the Qur'an and, more precisely, the Islamic attitude to the Qur'an, had catalysed a Jewish reaction, which came to be known as Karaism. Impressed and influenced by the certainty Muslims found in their holy text, Karaites sought to bypass the discursive and circuitous approach of the rabbis, in favour of a more straightforward didactic and authoritative reading of the Torah. Treating the oral tradition of the Talmud, let alone the science and reason of the Greeks, as a source of authority was a betrayal of God's revelation, they argued. Saahdiah objected strongly to Karaism but it remained a significant movement for centuries.

Saahdiah moved to Baghdad in his forties, where he wrote his major texts. Baghdad was, as we have seen, the heart of the translation movement of Islamic science at the time, to which a number of Jews contributed. It was at the other end of the Islamic empire, however, in Al-Andalus, that Jews were most visible in scientific discussions.[10] From the mid-tenth century, under the reign of Caliph Abd al-Rahman III, through to the 1140s when the Almohad dynasty invaded, Al-Andalus reached levels of artistic, literary, philosophical and scientific sophistication that had not been seen in Europe for five hundred years. The Umayyad caliphate in Spain was receptive to different intellectual traditions and a number of Jews became prominent scientific figures. Records suggest that between a third and a half of physicians in Spain and Provence during these centuries were Jewish, despite the fact that Jews made up no more than ten percent of the population.[11]

The Jewish contribution was not limited to medicine. Familiar with Hebrew, Arabic, Latin and often Greek, Jewish thinkers were in a unique position to translate Greek-Arabic science not only for their own community but also for Christian scholars who were slowly becoming aware of it. They also ventured original contributions. The philosopher, mathematician and astronomer Abraham Bar Hiyya (d. c.1136) wrote a treatise on geometry which contained the earliest European algebra, while Abraham ibn Ezra (d. 1167) was a prolific poet who also wrote on arithmetic, astronomical instruments and astrology. Pretty much every single prominent Jewish thinker in Al-Andalus and beyond matched their scientific work with commentary on the Torah and the Talmud. Indeed, as we shall note, it was the ultimate and necessary justification for their scientific and philosophical work. Abraham ibn Daud, who died around 1180 in Toledo, spoke for his peers when he spoke, in a gloriously mixed metaphor, about his desire to 'grasp with both his hands two lights, in his right hand the light of his religious law, and in his left hand the light of his science'.[12]

Study of Jewish thought and especially science in medieval Europe is patchy, so generalisations are difficult, but it appears that the scholars of Al-Andalus were a stimulant to wider Jewish thought beyond Islamic borders, particularly in Provence. From the mid-twelfth century, Jewish scholars, many fleeing north from persecution and forced conversion under the Almohad dynasty, introduced scientific work to traditional Talmudic studies of Provence. Jewish translators, a disproportionate number from one family, the Tibbons, began to render into Hebrew the works of al-Farabi, Ibn Sina, Euclid, Archimedes and Ptolemy, as well as numerous commentaries on Aristotle's books on natural science and physics.

Neither the size nor the scope of this particular translation movement should disguise the fact that there was disapproval of the new science in both Al-Andalus and Provence. Controversy abounded. Some attacks were specific and scornful. The same Abraham ibn Daud who wished to grasp the light of his science in his left hand mocked those who wanted to learn algebra as being like a man who 'wanted to boil fifteen quarters of new wine so that it be reduced to a third, [and who] boiled it until a quarter thereof departed, whereupon two quarters of the remaining wine were spilled . . .' (the sarcastic story continues for quite a while).[13]

Behind such taunting, there was real concern. Abraham ibn Ezra records a dispute with the poet and physician Judah ha-Levy in which ha-Levy put it to him that if the best way to know God is through the study of his creation, why did God reveal himself to Moses as the God who liberated his people from Egypt rather than as the God who made heaven and earth?[14] Behind the seemingly abstruse biblical question lay a tension central to Jewish ambivalence towards Greco-Arabic science, which indeed would dog the history of science and religion, particularly in Enlightenment Europe. Was God better known – perhaps only known – through the study of nature or through personal or historical experience? Was God the God of Exodus or the God of nature? Was he, to quote the slip of paper that the French mathematician Blaise Pascal would sew into his coat centuries later, the God of Abraham, of Isaac, of Jacob, or was he the God of the Philosophers and the scholars?

This was a particularly acute question for the Jews of Al-Andalus and Christendom, who were an always vulnerable minority and for whom a degree of religious and intellectual coherence was not just desirable but essential. Natural philosophy exerted a certain centrifugal force on a community. Its ideas and discoveries were not guaranteed to harmonise with the community's foundational beliefs or to serve its common good. Science, by its nature, is an unpredictable and challenging activity, a potential threat to any political regime or culture. For medieval Judaism, the centrifugal force of natural philosophy had always to be balanced with the centripetal force of law and its study, on which common beliefs, life and security were founded.

This left a delicate balance, traceable back to the Talmud, of agreement and dispute. On the one hand, the community drew together round the authority of the Law and any Jewish thinker who wanted to study mathematics, or medicine, or natural philosophy needed also to be a scholar of the Torah. On the other, there was perpetual disagreement between and within different communities. Qalonimos ben Qalonimos, a Provençal translator and philosopher, lamented, in the introduction to one of many books, that 'each [Jewish] district upholds its own persuasion . . . each condemning the other saying, "I am afraid there is some heresy . . . God is not its God" . . . our Gods are as numerous as our towns.'[15] In short, there was throughout medieval Judaism an inherent suspicion of foreign science, but primarily for social reasons. Natural philosophy, at least in the wrong hands, might undermine the authority of the Torah, and without the Torah there would be no Jewish community.

'Apothecaries, cooks and bakers': Maimonides and the handmaiden

For all that there were disagreements between Jewish scholars and communities, there were relatively few schisms, sects or charges of heresy. The inherently dialogical and disputative nature of the Talmud allowed greater scope for disagreement. Judaism lacked the established hierarchy of a church or the coercive authority of a state, by means of which heretics could be policed and expelled. And, in any case, the religion had not really developed the kind of propositional or creedal dimension against which heresy could be defined. Ironically, the man who did most to introduce that dimension was also the one who did most to legitimise, and then limit, science among Jewish thinkers: the figure around whom, it is no exaggeration to say, the entire history of Jewish intellectual life in the later Middle Ages pivots.

Moses ben Maimon, usually known as Maimonides, packed more into his sixty-six years than is right or proper for any philosopher. Born in Cordoba, in Al-Andalus, in 1138, he was from a long line of Jewish sages, and an inhabitant of the most intellectually vibrant city in the West. It was not to last. The Almohad dynasty of North Africa invaded Spain in the 1140s and, in their determination to restore a pristine Islam of the Qur'an and *hadiths* free from alien encrustations, whether Greek or Jewish, compelled Jews to convert, depart or die.

By 1148, the Almohads had conquered Cordoba and for the next twelve years the Maimon family was peripatetic, before settling in Fez in Morocco. Maimonides lived there for five years, again under the Almohads, and there is some evidence that the young man feigned conversion in order to avoid further trouble. Either way, the family then left for the East in 1165, nearly drowning in a storm on the way. Maimonides would mark the anniversary of that day in solitude and prayer for the rest of his life. Disembarking in the then Christian city of Acre, whence he made a pilgrimage to Jerusalem, Maimonides finally settled in the city of Fustat, next to Cairo, in Egypt. Political instability followed him. The Fatimid caliphate which had ruled Egypt for two centuries was crumbling. Fustat was burned to the ground by crusaders three years after Maimonides arrived, and the dynasty was then replaced by Saladin and the Ayyubid dynasty three years later.

The Fatimid dynasty had been as amenable to the foreign sciences as the Umayyads in Spain. Cairo boasted one of the greatest libraries in the world, together with an institution called the Academy of Science. Alongside Baghdad in the East and Cordoba in the West, it was one of the three great seats of Islamic learning. This suited Maimonides well, and he taught mathematics, logic and astronomy there, and studied the Torah, while also serving the last Fatimid caliph, Al-Adid. He then managed to negotiate the regime change skilfully, acquiring a powerful patron. Drawing on his training in Al-Andalus, he worked as a court physician after the loss of his beloved brother and the family fortune on a trading trip to India.

Marrying into a prestigious family, and with his court connections, he rapidly became a centrepiece of the mixed Jewish community there, which included two rabbinic synagogues and a Karaite one. His son Abraham later recorded that although Moses did not attend either rabbinic synagogue on a regular basis but rather held prayer services in his own home, he was still appointed Head of the Jews, the final religious and judicial authority for the community. In that capacity – over five hundred of his judgments survive – he pronounced on issues of marriage, divorce, inheritance, property, tax, business partnerships, synagogue decorum, debts, conversion, apostasy, circumcision and menstruation. His reputation for law and wisdom spread across North Africa and to Palestine, Baghdad and Yemen.

It was not as a judge or a community leader that he was to be remembered, however, but as a scholar and prolific author, writing on logic, astronomy, astrology (which he dismissed) and medicine. His medical output sought to simplify and structure Galen's voluminous corpus but also included original works, such as *On Cohabitation*, written for Saladin's nephew, which detailed various aphrodisiac recipes for the young man, who was apparently being exhausted by the young women he felt obliged to satisfy.

His fame and authority rest on two texts above others. The first was his fourteen-volume collection and systemisation of Jewish law, the *Mishneh Torah*, which would become a touchstone for Jewish jurisprudence and secure his reputation as a uniquely important authority on the Jewish law. The second was his *Guide for the Perplexed*, written in his late forties, in which, much as Thomas Aquinas would do for Christians the following century, he sought to bring theology into harmonious dialogue with Greek philosophy and science. Maimonides had an open attitude to foreign sciences, arguing that the scholar should 'hear the truth from whoever says it'. He outlined three reliable sources of knowledge – from reasoning such as arithmetic or geometry; from the five senses; and from the prophets or the righteous. He held Aristotle in extremely high regard, contending that he had reached 'the upper limit of knowledge attainable by man', and thought that everything he said about the sublunary world was 'indubitably correct'.[16]

He was not beyond challenging Aristotle, however, and took issue with his conception of the celestial realm, in particular the Greek's understanding of the movement of the spheres, which was irreconcilable with the planets' observed paths. This more empirical approach also marks his writing on medicine, in which he criticised Galen and Hippocrates and argued that the effects of drugs can only be known by experience, and not a priori by reason. 'Experience', it should be noted, was still centuries away from becoming 'experiment'. Maimonides, like his medical peers, believed that experience had proved that those bitten by a crocodile would heal instantly if they put crocodile fat on the wound, or that hanging a peony round the neck would cure an epileptic, or, most worryingly, that dogs' excrement helped in cases of swelling of the throat.

As important as his endorsement of reason was his critique of it. Human reason was fundamentally limited. 'Regarding all that is in the heavens, man grasps nothing but a small measure of what is mathematical.'[17] Reason was intrinsically incapable of adjudicating on many issues, such as whether the world is indeed eternal, as Aristotle argued. Science could clarify, structure and explain what humans already knew (or thought they knew) but it was not adept at discovery. 'Maimonides' philosophy made no allowances for the idea of progress of human knowledge of nature . . . for investigations seeking new information of the physical world.'[18] Ultimately, its role was (again) as a handmaiden, or as Maimonides put it in a letter to Rabbi Jonathan ha-Cohen of Lunel, the foreign sciences are mere 'apothecaries, cooks and bakers', maids in the service of the Torah.[19] 'The deity alone,' as he wrote elsewhere, 'knows the full reality.'[20]

Maimonides' critical commitment to reason and science was not inherently problematic among his peers but it led him to an attitude to scripture that was. The scriptures, he argued, could not contradict reason or that which had been demonstrated by science. If they appeared to do so, it was because they were being read wrongly. This meant that literal readings should be taken figuratively or symbolically. 'Seeing', in the scriptures, thus meant 'understanding' rather than literally looking with your eyes. God 'being near' meant spiritual apprehension rather than physical proximity. Miracles were 'the coming into existence of something which is outside its normal and permanent nature' rather than a clear supernatural intervention in the world.[21] The creation story as written in Genesis and interpreted in various rabbinical and mystical texts was nothing other than Aristotelian physics written in story form. 'Parables and secrets' were the key to its understanding.[22]

The Bible, Maimonides repeatedly said, was written in the language of ordinary people. This was the 'doctrine of accommodation', as later Christian theologians would describe it. That was why God was described as corporeally as he was, complete with hands, eyes and sometimes wings. A literal interpretation of this was not only nonsensical but positively dangerous, as it led more or less directly to idolatry: people worshipping the stuff of creation rather than the creator. As a result, the purpose of such language was to get beyond itself, to a higher, more intellectual, more spiritual, more elite and more tentative understanding of God. Maimonides tended strongly towards what is sometimes known as the apophatic approach to the divine, describing God by slowly paring away false positives to get closer to the ultimately undefinable divine. It is an approach that commonly, and not just within Judaism, gravitated towards silence – as Maimonides recognised: 'silence with regard to You is praise.'[23] It was a subtle, philosophically sophisticated and scientifically receptive approach to the scriptures. But it was not to everyone's taste and irked many believers in 1200, as it does today.

'It was inevitable that the philosophers deny the Torah': the reaction

Maimonides' influence on later medieval Jewish thought was incalculable. His enormous personal authority and undoubted legal expertise meant that all subsequent Jewish scholarship was indebted to him. Disagree with him as they might, people could not ignore or dismiss him. As a result, his support for and integration of Greek science into Jewish religion was foundational, making it almost obligatory among Jewish scholars. But it was also ambiguous and contentious.

With regard to its ambiguity, Maimonides clearly limited science as much as he legitimised it. 'Along with the explicit and insisting affirmations concerning the importance of science, there is the view that it is only ancillary to metaphysics and that some questions cannot be subjected to scientific inquiry.'[24] Such arguments could be (and were) used as much by anti-scientific Jewish thinkers as by pro-scientific ones, the former deploying ideas about the weakness and inadequacy of human reason to argue for its redundancy or to defend literal readings of the creation story. Even when not deployed polemically, his view of science as merely the apothecary, cook or baker to scripture and theology placed strict restraint on scientific authority and autonomy. Science was vital but only for bigger purposes.

Maimonides' theological approach and its allegedly corrosive effect on the scripture and law – and therefore also on the community that was founded on that scripture – angered and upset many, even those schooled in religious disputation. 'Philosophy being a natural science, it was inevitable that the philosophers deny the Torah,' wrote Rabbi Asher ben Yehiel.[25] Soon translated into Hebrew (it was originally written in Arabic) and circulating around Jewish communities from France to Baghdad, the *Guide* provoked as much anxiety as it did respect.

It was banned in some places and burned in others. Rabbinic academies in the East condemned Maimonides, resenting his encroachment on their authority and accusing him of undermining the institutional foundations of Judaism. In Provence, a school of anti-Maimonides scholars denounced his work and, drawing on the help of the Inquisition, had his books publicly incinerated in the 1270s. A generation later, opponents sought to effect a wider ban on the science of the ancients, except for practical ones like medicine and astronomy, for fear of the effect it had on the authority of the scriptures. In reality, this ban was no more effective than the various ecclesiastical bans on teaching Aristotelian science we will encounter in the next chapter, and no one is known to have been persecuted on account of it.

In reality, the attacks on Maimonides and his *Guide* no more symbolised medieval Judaism's attitude to science than the similar attack on Aristotle by Bishop Tempier of Paris symbolised medieval Christendom's. On the contrary, both bans suggest – as bans so often do – that people *were* reading and absorbing Greek science and philosophy in numbers. Moreover, again as they often do, both bans ended up stimulating rather than suppressing interest, and catalysed scientific thought.

Ultimately, the Jews' attitude to Greek science was inextricably tied up with their increasing vulnerability in Europe. Driven from Al-Andalus in the twelfth century, and from England and southern Italy in the thirteenth, then expelled from France, Bavaria and Hungary in the fourteenth, and from Spain and Portugal altogether in the fifteenth, this was a community that was losing the political and intellectual freedom and security it had enjoyed under the Umayyad caliphate.

This did not altogether preclude further Jewish scientific engagement. Levi ben Gerson, known as Gersonides, the greatest Jewish natural philosopher after Maimonides, lived through the controversies over the *Guide* and the gradual erosion of Jewish freedoms. An accomplished mathematician, he wrote on arithmetic and geometry and even dedicated a work on trigonometry to Pope Clement VI, with whom he had good relations. He had a powerful commitment to accurate measurement and instruments (he has a crater of the moon named after him) and a confrontational attitude to scientific authority. When his own observations failed to concur with Ptolemy's, he said he preferred his own. And he had a robust appreciation of empiricism. 'No argument can nullify the reality that is perceived by the senses,' he wrote in his major work *The War of the Lord*, 'for true opinion must follow reality but reality need not conform to opinion.'[26]

Perhaps most significantly, he insisted, against the Maimonidean approach, that scientific investigation was its own justification. Rather than simply being a function of religious life, the study of nature permitted access to and knowledge of God. It had a dignity and purpose of its own. None of this was intended to undermine the scriptures, about which Gersonides wrote at length, but that was naturally how critics, of whom there were many, read it. His major philosophical work was renamed *The War against the Lord*.

And yet, in spite of his tough-minded scientific approach, Gersonides was, unlike Maimonides, a firm believer in astrology, which he credited with a determinism over human affairs. He was, in effect, a living embodiment of the complex ambiguities that ran through early Jewish engagement with natural philosophy, in which science, scripture, superstition, sectarianism and security all played a vibrant role. One of his last texts, the only one on astrology to survive, was about the conjunction of Saturn and Jupiter, which was projected to occur in March 1345. Gersonides predicted that there would be 'extraordinary evil' following the conjunction, with war, disease and death that would endure for a long time. Two years later, with Gersonides now having passed away, the Black Death arrived in Europe.

The frontispiece of Adelard of Bath's Latin translation
of Euclid's *Elements* from Arabic. Adelard was a
provocative genius, one of the twelfth-century *physici*
who put forward 'the earliest vision of the discipline
of science to appear in the West'. 'The visible
universe is subject to quantification,' he insisted.

FOUR

SCIENCE IN CHRISTENDOM

'A poignant lost opportunity': the so-called 'Middle' Ages

A few years before he died in 1070, the Spanish-Arab astronomer Said al-Andalusi wrote an early history of science. The *Book of the Categories of Nations* covers the Indians, Persians, Chaldeans, Greeks, Romans, Egyptians and Arabs. Conspicuous primarily for their insignificance are the Europeans, whom al-Andalusi places in a short chapter entitled 'Nations having no interest in science', and dismisses as being little better than animals.[1] True, these ignorant Europeans had been led by a rather mathematically minded pope, Sylvester II, who was instrumental in introducing the abacus, the astrolabe and Hindu-Arabic notation to Western Europe. But Sylvester had taken all his mathematics from Arabic Spain and, even then, his mathematical reasoning was primitive in comparison to al-Khwarizmi's, 150 years earlier. Europe's mathematical pope merely served to underline al-Andalusi's point. Christendom was hardly even a backwater when it came to science. Few disagreed at the time.

Or indeed since. When the cosmologist Carl Sagan's book *Cosmos*, which accompanied his wildly popular TV series, drew up a scientific timeline for the world, the centuries between 500 and 1500 were left blank, labelled only with the phrase 'poignant lost opportunity for mankind'.[2] The narrative persists today, with talk of the closing of the Western mind and the Christian destruction of the classical world. The Middle Ages is, by definition, the 'in-between' time, the period between the glories of antiquity and their resurrection at the Renaissance: a period when serious thought, let alone science, was simply non-existent.

The view dates back to the period itself, when the Italian poet Petrarch compared the ineloquent darkness of his age with the sophistication of Rome. The Reformation sharpened the polemical edge in this comparison. Francis Bacon spoke for good Protestants everywhere when, in *The Advancement of Learning*, he lamented the 'degenerate learning' of medieval scholastic philosophers who had an 'abundance of leisure . . . but small variety of reading, their minds being shut up in a few authors, as their bodies were in the cells of their monasteries'. If the human mind were to study the actual works of God, as Bacon argued, then it could be profitable indeed. If, conversely, it merely worked upon itself and its own ideas, all the time crushed by the overbearing authority of the Catholic Church, it would only produce 'cobwebs of learning, admirable for the fineness of thread, but of no substance or profit'.[3] The Middle Ages would necessarily become part of – the very foil to – modernity's self-image. The narrative would, in the later nineteenth century, have an incalculable impact on popular ideas of the history of science and religion, and remains alive and kicking today.

What, though, are we to make of Robert Grosseteste? In the early 2000s, Tom McLeish, at the time professor of Polymer Physics at the University of Leeds, attended a History and Philosophy of Science seminar on Grosseteste, led by James Ginther, from the Department of Theology. Grosseteste was a scholastic philosopher and bishop of Lincoln, born in the last quarter of the twelfth century. Not obvious fare for such a seminar, he had in his fifties written a series of scientific treatises on astronomy, the rainbow, tidal movements, the use of mathematics in the natural sciences and, finally, light, the topic of this particular seminar.

In line with the popular narrative of 'medieval science' – i.e. that the phrase is an oxymoron – McLeish's expectations were not high. 'The experience of that first reading was unforgettable,' he later wrote.[4] Through the fog of 'technical vocabulary of Aristotelian natural philosophy', which at the time escaped him, it became clear to McLeish that the treatise, *De Luce*, began with 'a critical assessment of the failure of classical atomism to explain the solidity and extension of condensed matter' before proceeding to invoke 'the mathematical concept of infinite series, in the description of the way that atomistic matter and light might interact'. Grosseteste was dealing with the ideas – if not the language or symbols – with which a twenty-first century professor of physics could identify.

The result of the seminar and the growing interest in Grosseteste was the 'Ordered Universe' project, an international interdisciplinary research project integrating history, physics, theology, psychology, English, engineering and Arab studies, and dedicated to translating, analysing and promoting Grosseteste's impressive scientific corpus. The project will publish all six of Grosseteste's treatises and has already generated a number of academic papers, such as 'A Medieval Multiverse?: Mathematical Modelling of the 13th Century Universe of Robert Grosseteste', published in *Proceedings of the Royal Society*, and 'Color-coordinate system from a thirteenth-century account of rainbows', published in the *Journal of the Optical Society of America*. By anyone's reckoning, this is science.

Grosseteste was a particularly impressive figure, not least because he rose from what appears to have been poverty and obscurity and managed to combine scientific work with a lively ecclesiastical career, which included clashing swords with both the archbishop of Canterbury and Pope Innocent IV. But he was far from the only medieval scientist to conduct meaningful research.

Roughly contemporary with Grosseteste were John of Sacrobosco, who popularised Hindu-Arabic numerals and the mathematics of Islamic thinkers in his *Algorismus* and whose treatise *On the Sphere* included a detailed outline of the earth's sphericity, and Albertus Magnus, the most original European biologist since Aristotle, who conducted original empirical work into animals, minerals and vegetables. A little later in the thirteenth century, Roger Bacon, a disciple of Grosseteste, made tentative steps towards experimental science in his study of optics. At around the same time, Peter Peregrinus of Maricourt in France performed experiments to determine the magnetic properties of the lodestone, and wrote an early treatise on magnetism, while Theodoric of Freiberg in Germany experimented with water-filled glass globes (as a model for water droplets in a cloud), as a way of demonstrating that the rainbow was caused by two refractions and an internal reflection through each droplet. The following century, Jean Buridan borrowed impetus theory from John Philoponus to explain projectile motion and the acceleration of free fall, while the mathematician and cosmologist Nicole Oresme anticipated Cartesian co-ordinates, discussed the possible rotation of the earth on its axis, dealt with the dynamics of motion and denounced alchemy as fraud.

Such figures are even less well known today than al-Khwarizmi and Ibn Sina, but they were hardly insignificant thinkers. In reality, the timeline of medieval science was not as poignantly blank as Carl Sagan had believed. Indeed, as Michael Shank and David Lindberg, the editors of the 700-page *Cambridge History of Medieval Science* wrote in their Introduction, Sagan's empty timeline reflects neither the historical truth nor even the state of historical knowledge in 1980, 'but Sagan's own "poignant lost opportunity" to consult the library of Cornell University, where he taught'.[5]

'Between you and me, reason only shall be judge': medieval physicists

The early scientific indifference identified by al-Andalusi notwithstanding, Latin Christendom would go on to formulate a formidable set of theological justifications and tools for the systematic study of nature and the cosmos. The scriptures presented reasons to study what was termed God's 'book of nature'. 'The heavens declare the glory of God; and the firmament sheweth his handywork', proclaimed Psalm 19. Studying creation was a legitimate, if limited, way of studying God, leading to both 'the exaltation of the creator and the perfection of our souls', according to William of Auvergne.[6] Beyond this basic reason, however, a small group of scholars in the twelfth century – Adelard of Bath, William of Conches and Thierry of Chartres foremost among them – began to bring principles of dialectic and of disputation to bear on their study of nature. Theirs was, in the words of one scholar, 'the earliest vision of the discipline of science to appear in the West'.[7]

These self-designated *physici*, a term previously reserved for physicians but now used to describe those engaged in natural philosophy, worked on the premise that creation reflected the ordered, rational mind of its creator. 'Nature is like its Creator,' Adelard of Bath reasoned, 'purposeful, logical and free from chaotic confusion.'

That being so, Adelard continued, 'there is nothing in nature either dirty or unsightly [except] whatever is contrary to the *ratio* of nature.' A rationally ordered creation meant searching for explanations *within* nature, rather than constant and unpredictable divine intervention.

Moreover, it invited a quantifiable natural explanation, with the *physici* even hinting at the possible mathematisation of nature. 'The visible universe is subject to quantification, and is so by necessity,' Adelard explained. 'The world would seem to have causes for its existence, and also to have come into existence in a predictable sequence of time [which] can be shown to be rational,' agreed Thierry of Chartres.

None of this should be construed as undermining God. 'I do not detract from the power of God,' Adelard said, 'for all that exists does so from him and by means of his power.' On the contrary, acknowledging the *ratio* of creation was to his glory, as to imagine that 'nature in itself is chaotic, irrational or made up of discrete elements' would be to imply something similarly irrational and arbitrary about God himself. William of Conches said the same thing, responding to a similar accusation robustly: 'on the contrary, it enhances [the power of God], for we have attributed to Him both the establishment of such a nature in the universe and the creation of the human body through the medium of that nature.' Mathematicians and physicists William Thomson and George Gabriel Stokes would be making precisely the same point eight hundred years later.

The *ratio* of the creator and his creation was mirrored in the *ratio* of the human mind. Berengar of Tours, the eleventh-century master of the cathedral school at Chartres, who popularised the dialectic reasoning that *physici* applied, had said that 'it is by his reason that man resembles God.' It was an increasingly widespread view. 'Man is not armed by nature nor is [he] naturally swiftest in flight,' Adelard explained. What he does have, though, 'is better by far . . . reason'. No tougher or faster than other creatures, and no better fitted with 'mere physical equipment', it was solely through his exercise of reason that 'he exceeds the beasts to such a degree that he subdues them.' The argument could have a distinctly Enlightenment confidence to it. 'Hope and you will discover a solution to the problem . . . since we must assume that all nature is based on a sure and logical foundation.' Reason was sufficient, both for discovery and for disagreement. 'Between you and me, reason only shall be judge.'

Confidence in reason did not necessarily preclude a degree of scepticism. Indeed, it arguably demanded it. According to the eminent eleventh-century French philosopher Peter Abelard, it was only through assiduous and frequent questioning that humans found true knowledge.

'By doubting we come to inquiry, by enquiring we learn.' When studying natural philosophy, this demanded an unnerving willingness to accept a degree of uncertainty rather than demanding the kind of proofs preferred by metaphysicians. Living with uncertainty and even ignorance, an attitude that would become essential to the entire scientific method, was vital. 'We beg our readers not to reproach us if we speak of things that are classified as visible [as if they were], probable and not necessary,' William of Conches entreated his audience.

Lack of certainty did not mean, and should not be, an excuse for lack of clarity, however. William of Conches was a highly respected teacher of rhetoric but he nonetheless insisted that the works of the *physici* should be clear, without literary embellishment or circumlocution. 'Anyone who finds himself put off by the dryness of our discourse would be much less apt to miss rhetorical embellishments if they fully understood our intention,' he explained. Many demanded rhetorical finesse from philosophers but, William said, sounding like Francis Bacon four centuries later, it comes at the cost of truth. 'Of those who boast of their honesty, we alone sweat out the truth,' he said, 'and we prefer to present that truth in a naked state rather than clothed in lies.' It was a rhetorical device itself, of course, but a disarmingly powerful one.

The call to search for natural explanations *within* nature had challenging implications. First, it meant that miracles could not be treated as an adequate mode of explanation. Effectively, the *physici* argued against the God of the gaps six centuries before the phrase was dreamed up, and they did so witheringly. Challenged to explain a locally believed miracle, William of Conches was scornful.

> I know what they will say: 'We do not know how this may be; we know that God can do it!' Wretches! . . . God does not do this just because He can. A peasant's saying is: 'God can make a calf from a tree-trunk.' [But] has He ever done so? Let us either show the reason [for a miracle] or its usefulness; or let them desist from declaring that it is a miracle.

This was strong meat for a culture that not only believed in miracles, as indeed William did, but *needed* them, understanding the miracle of the Mass as the keystone of the entire social edifice.

Second, it meant that the scriptures were not necessarily adequate to explain nature either. This was a potentially explosive point and the *physici* were careful in what they claimed. Again, none denied the revelation of scripture any more than they did the possibility of miracles. The question was *what* scripture was revealing. William of Conches argued that while Genesis asserted the existence of various beings, it did not explain *how* they came about. It was fundamentally uninterested in the mechanisms of nature, revealing instead its meaning. The investigation of causes was the proper task of the natural philosopher.

In this vein, Thierry of Chartres introduced his commentary on Genesis 1, saying that it was a study 'from the point of view of an investigator of natural processes and of the literal meaning of the text'. God in His wisdom had constructed the world to exhibit a rational and beautiful order, and Thierry planned to use the sciences to illustrate that and 'to show how it can be rationally accounted for'. William of Conches adopted the same line and was prepared to go even further, praising Plato's *Timaeus*, at the time the still pre-eminent classical work of science in Latin Europe, as superior in its causal explanations, and even going so far as to correct the scriptures in places for making physical assertions that were 'contrary to reason'.[8]

To his critics who complained that 'they do not find it written like this in Scripture', he explained that they failed to understand that 'the Christian writers say nothing about the philosophy of the world'. That was 'not because it is incompatible with faith', but simply 'because it has not much relevance to the building up of faith which is what they are concerned with'. It was certainly not right to contradict Bede or any other of the Church Fathers when it came to matters of faith. But, in matters of natural philosophy, 'if they err in any way it is permissible to assert something different'. The Fathers were indeed 'greater men than we are, yet they were men'.

The arguments of the *physici* were astonishing for their time and amounted to what we would today call a commitment to 'methodological naturalism', a conviction that nature proceeded along secondary or 'natural' causal lines and should be studied accordingly. 'Physics searches out and considers the causes of things as found in their effects, and the effects as derived from certain causes', said Hugh of St. Victor. 'The visible world is this machine,' he said in a metaphor that others would employ, 'that we see with our bodily eyes.' All this reached way beyond the somewhat grudging 'handmaiden' role for science of the Church Fathers. Thierry of Chartres claimed that the 'four kinds of rational disciplines that lead men to an understanding of God as Creator – namely, the demonstrations of arithmetic, music, geometry and astronomy' are 'as tools for theology . . . slight use'. Natural philosophy, the *physici* implied, had its own justification, its own authority, and didn't need to borrow one from the theology faculty.

'What else should authority be called but a bridle?': the coming of Aristotle

For all their brilliance, the *physici* made little lasting impact on medieval thought. Their ideas were threatening precisely because they circumvented the normal channels of intellectual authority. 'This generation has an innate vice, namely that it can accept nothing which has been discovered by contemporaries,' Adelard lamented. The intellectual life of the time ran on iron rails of *auctores*, written authorities, of enormous prestige. Merely citing a respected name, usually that of a Church Father, or text would be enough to convince readers that a statement was incontestably true. This was not the method of the *physici*, whose originality implied – or at least could be taken to imply – a contempt for all tradition and authority. 'An authority has a nose of wax,' Alan of Lille said. 'It can be twisted in any direction.' 'What else should authority be called but a bridle?' Adelard asked rhetorically. Scholars who placed themselves under such authority in this way were 'as brute beasts . . . not knowing whither or why they are led'. It was not a view calculated to win allies.

They were attacked. The ceaseless investigation into the natural world was fruitless and distracting, proclaimed Absalom of Saint-Victor.[9] It led Christians to suspect conclusions. 'Let no one impiously think – as certain ungodly men do – that things contrary to nature [i.e. miracles] cannot occur,' opined one anonymous critic. That, of course, was precisely *not* what the *physici* were saying but their naturalistic stance invited such misunderstanding and condemnation. Thierry of Chartres was forced out of his teaching post at Chartres. William of Conches was attacked for his views, and although Adelard of Bath seems to have been relatively unscathed, he did mention those who 'point their finger at me and impute to me madness'.

Ultimately, however, the *physici* were not so much crushed as washed away by the flood of Greek science and philosophy into Europe from the mid-twelfth century onwards. Latin Christendom had inherited only a few texts of natural philosophy from antiquity: works from Pliny and Boethius and, most importantly, Plato's *Timaeus*. Byzantium's general indifference to Greek science and Rome's growing friction with Constantinople meant that the West inherited no more during the first millennium. By the closing years of the tenth century, however, Arabic translations of Greek thought works were being rendered into Latin at the Monastery of Santa Maria de Ripoll in northern Spain.

The capture of Toledo in 1085 and of Sicily in 1091 opened this channel, as of course did the First Crusade, and over the next century intrepid scholars from across Europe began to realise what they had been missing. Adelard of Bath was among them, travelling to southern Italy and then to the Near East to acquire texts from those whom he called 'my Arab masters'. Few travelled as far but some dedicated their lives to the task of translation; Gerard of Cremona, for example, moving to Toledo, learning Arabic and rendering more than seventy texts into Latin over the course of his long life.

The natural philosophy of the classical and Islamic worlds, which had heretofore been known only by reputation if at all, dripped slowly into Christendom's bloodstream. Euclid's *Elements* and *Optics*, Ptolemy's *Almagest*, Ibn al-Haytham's *Optics*, Al-Khwarizmi's *Algebra* and the medical writings of Galen, Hippocrates, al-Razi and Ibn Sina were among the hundreds of scientific texts that appeared in European libraries for the first time in the twelfth century.

More important than any of these, however, was the natural philosophy of Aristotle, and those who commented on him. Known until then only by a few minor works, Aristotle was now read via his *Physics*, *On the Heavens*, *On Generation and Corruption*, *Meteorology* and *Posterior Analytics* – all translated by Gerard of Cremona – as well as his *On the Soul* and books on biology analysing the history, generation, parts, movement and progression of animals. Behind him there arrived an army of commentators and interpreters, including John Philoponus, al-Kindi, al-Farabi, Ibn Sina, al-Ghazâlî and Ibn Rushd, who, as Averroes, became better known in Europe than he was in Islamic lands.

As with the Islamic translation movement four hundred years earlier, there were multiple translations and huge numbers of copies made. Medievalists have identified 288 extant copies of Aristotle's *Meteorology*, 363 of *On Generation and Corruption* and 505 of his *Physics*. Altogether, there were, at a bare minimum, two thousand medieval Latin manuscripts of Aristotle's books in circulation.[10] The sheer volume and range of the scientific corpus translated from Aristotle and his late antique and Arabic acolytes says much about its reception, which was as enthusiastic as it had been in the Abbasid caliphate four hundred years earlier.

Although Aristotle's biological works were translated, they had less influence on Christendom than they did on the Islamic world, and it was his physics and cosmology that changed the medieval mind. The cosmos became Aristotle's, or rather his and Ptolemy's. Creation was divided into celestial and terrestrial realms. The celestial or superlunary realm was a place of fixed, eternal, incorruptible perfection. Made up of imperishable aether or quintessence (literally 'fifth essence'), it held the seven 'planets' (moon, Mercury, Venus, sun, Mars, Jupiter and Saturn) and the stars fixed beyond them, after which was the outermost, empyrean sphere where the blessed were believed to reside. Each of the planets was embedded in and carried by its own crystalline sphere, the stars were fixed in their own single sphere beyond the planetary orbits and each planetary sphere contained sub-spheres that were needed to explain the planets' 'wandering' paths. Harmony was not quite perfect, and the two classical astronomers disagreed on the nature of planetary orbits. Aristotle's spheres were concentric with respect to the earth, whereas Ptolemy's were eccentric, not centred on the earth, and epicyclical, having a mini-cycle within the larger orbit as a means of explaining why planets somehow appeared to move backwards.

The other – terrestrial or sublunary – realm was the arena of change, corruption and decay. It comprised the four elements of fire, air, water and earth. Each 'gravitated' to its own proper place in the terrestrial sphere. Together, the four combined to make the properties of heat, dry, wet and cold. All bodies in the terrestrial realm were made up of matter and form, and explicable in four ways: material (the matter from which it is made), formal (the shape or form it takes), efficient (the immediate source of its being) and final or teleological (the ends for which it exists).

For all his influence over medieval science, however, there were tensions with Aristotle, beyond those associated with Ptolemy and the movement of the planets. Some pertained to specific elements of his teaching, elements which had vexed the Church Fathers and Islamic philosophers. Others were rooted in the more profound question of intellectual authority, not dissimilar to that which had dogged the Abbasid caliphate in the tenth and eleventh centuries.

The very basis of Aristotelian natural philosophy was its apparent self-sufficiency, its ability to make authoritative statements about the cosmos on the basis of reason alone, independent of any form of revelation. More often than not, those conclusions were perfectly acceptable and wholly compatible with the scriptures. However, the very principle of intellectual independence concerned some theologians, who began to try to bridle this new scientific authority in the early thirteenth century. By the time they did, however, Aristotle had arrived back on the scene and was beginning to change everything.

'Nothing is better known because of knowing theology': fighting in Paris

Even in its darkest, poorest and most vulnerable years, Latin Christendom had had schools and libraries, albeit significantly less well stocked than those of the Islamic world. These were almost exclusively based in monasteries, and focused on scriptural and theological teaching. They were not entirely closed to classical learning, however. Benedictine scriptoria copied hundreds of pagan manuscripts; Bede, as we saw, knew Pliny and Plato's *Timaeus* was widely read. Nevertheless, classical philosophical resources were thin, and scientific ones thinner.

In time, cathedral schools were founded alongside monastic ones and when natural philosophy began to develop in an atmosphere of greater peace and prosperity from the twelfth century, it was there that scholars resided. Judging from the arguments of the *physici*, the cathedral schools were quite capable of producing original minds, but they were nonetheless primarily ecclesiastical establishments, inspired by ecclesiastical concerns, governed by ecclesiastical authorities and in no position to protect those who might take a different approach to learning. In this regard, they were not so very different from the madrasas that were spreading across the Islamic world at the time.

From the mid-twelfth century however, on the back of ecclesiastical reform and the codification of Church and Roman civil law, there developed the idea of a legally recognised and autonomous corporation, or *universitas*. Rather than being founded on the order of an emperor, monarch, pope or bishop, such corporations developed as grass-roots associations. Some were professional organisations, such as were founded by doctors and lawyers. Others were economic, such as the guilds that developed across Europe. Certain of them became religious orders or were charitable, while a few were civic or were academic and educational. They became Europe's first universities.

Universities were recognised and treated by law as a single person or agent, and granted rights accordingly. They had the right to own property and dispense of it as they chose, to owe debts, to sue and be sued, to be represented in court and to make their own rules and regulations. They did not disappear with the passing of any particular founder, benefactor or inspired teacher, but remained the same legal entity even as members changed. They were effectively self-governing institutions and although they had to answer to the same political and religious authorities as everyone else, they were not directly run by them. Perhaps most importantly, universities were not, unlike madrasas, set up, owned, run by or subject to particular benefactors, who could then dictate their operations and objectives. Rather, they were free, and legally protected, to develop their own curricula, to teach Averroes and Aristotle, to stock their libraries with natural philosophy and to argue on the basis of reason and logic, which is more or less what they did.

The University of Bologna became the first such foundation in the twelfth century, followed by Paris, Oxford, Salamanca, Cambridge and Padua in the early years of the thirteenth. By the end of the century there were about a dozen, and by 1500 around sixty across Europe. Perhaps most surprisingly, and contrary to the popular view, medieval universities were scientific institutions before they were theological ones. By best estimates, about thirty percent of the medieval university curriculum was focused on the natural world. In the words of historian Edward Grant, 'the medieval university provided to all an education that was essentially based on science . . . [indeed it] laid far greater emphasis on science than does its modern counterpart.'[11]

Universities inherited a curriculum comprising the seven 'liberal arts' – grammar, rhetoric and logic along with arithmetic, music, geometry and astronomy – from the cathedral schools, but this was transformed with the advent of Aristotle, particularly with regard to geometry and astronomy. Corporate freedom being what it was, different universities had different emphases but, as a rule, most universities familiarised their students with Euclid, Ptolemy, Aristotle and Averroes, as well as more contemporary works like John of Sacrobosco's treatise *On the Sphere*, alongside law, rhetoric, moral philosophy and metaphysics. Only then were students able to progress on to theology and, in reality, only a minority did so. Most universities had no theology faculty until the later fourteenth century.

Medieval universities were thoroughly Christian institutions, the foundation, purpose and corporate life of each coloured through by belief and scripture. *Dominus illuminatio mea*, 'the Lord is my light', the opening words of Psalm 27, may have become the motto of Oxford University only centuries later, but it could have served as such for every university in Latin Christendom. That said, even from their earliest days, there were clerics who resented the new universities' fiercely guarded corporate freedom – the new-found authority of academic masters who needed to be neither ordained nor tonsured, and their apparent sanctification of the pagan Aristotle.

In 1210, not long after his books had been rendered into Latin, a provincial synod of bishops at Sens banned the teaching of Aristotle's natural philosophy in the Paris area, on pain of excommunication. The ban was renewed, for the University of Paris, by the papal legate five years later, and then in 1231 Pope Gregory IX joined the fray, issuing a bull that declared that Aristotle's natural philosophy was not to be taught (again at Paris) until a papal commission had reported on it and 'purged [his books] of all suspected error'. For reasons unknown, the commission never submitted a report and the ban went into abeyance. By mid-century,

Aristotle was being taught publicly in Paris as he was in Oxford, Cambridge, Padua and elsewhere.

The tensions remained, however, and escalated in the 1260s, as John of Fidanza (Saint Bonaventure) and other conservative-minded theologians sought to place limits on pagan and Arabic learning. At first they met with little success but when they appealed to Bishop Tempier of Paris in 1270, he issued a condemnation of thirteen articles from Aristotle and Averroes. The conflict rumbled on. Giles of Rome published a volume entitled *Errors of the Philosophers*, which condemned a list of errors from non-Christian thinkers, including Aristotle, Averroes, Moses Maimonides and al-Ghazâlî. Finally, the new pope, John XXI, vexed by the ongoing disruption at Paris, ordered the bishop to investigate thoroughly. He needed no encouragement to clip these academic wings and within an improbably short period of time issued a condemnation of 219 propositions, prompting historians to believe he had been in the process of acting, papal mandate or not.

The Condemnation of 1277 was a ragbag list of denunciations without obvious coherence or order. It was targeted at the errors made by 'certain scholars at the faculty of arts . . . [who were] transgressing the limits of their own faculty', but who remained unnamed throughout. Modern attempts to pin particular heresies on particular scholars have proved largely fruitless. A number of the denunciations picked up on specific ideas that had been a problem for Abrahamic thinkers since the days of antiquity, most commonly Aristotle's assertion that 'the world as a whole was not generated and cannot be destroyed . . . but is unique and eternal', an idea (or rather a cluster of ideas) that was condemned in 27 of the 219 propositions.[12] Other articles condemned determinism, astrological beliefs and any limitations on human free will (the subject of about fifteen propositions), as well as the idea that the soul did not survive death and that humans were not subject to any form of post-mortem judgement (the final seven articles). The fear, it seemed, was not so much the general thrust of natural philosophy but the implications it might have on human nature, freedom and destiny.

These were all genuine points of tension for Christian thinkers in the thirteenth century but they grew from a soil in which a bigger problem lurked. Some of the condemnations have an almost idiosyncratic feel to them, such as article 49 which condemned the idea that God could not move the heavens or the earth in a straight line ('with a rectilinear motion'). Why, modern readers wonder, was this an issue that should vex the bishop of Paris and his theological friends? The reason is that Aristotle had apparently demonstrated that a vacuum, such as would be created by moving a planet in a straight line, was impossible. (Pagan) physics had spoken and God was obligated to conform. Rectilinear motion and the existence, or not, of vacuums was about who ran the universe. Such assertions, the theologians argued, constituted an unjustifiable limitation of God's power. In a similar way, they denounced the idea that God could not 'be the cause of a new act' or create more than one world. No one, as far as we know, actively believed in a wholly deterministic cosmos or in the existence of other worlds, but this was the kind of restraint on the divine against which theologians strained.

Beneath these specific and general objections, there was also something else going on. Aristotle had been taken to the heart of Christendom's new universities with remarkable rapidity. 'Why did a Christian society at the height of the Catholic Church's power readily adopt a pagan natural philosophy as the basis of an extensive educational programme?' the historian Edward Grant asked.[13] The answer, discomfiting as it is to us, is that medieval Christian thinkers were possessed of an intellectual breadth, generosity and inquisitiveness that we often find hard to acknowledge.

That recognised, the speed and depth of that adoption inevitably upset many people, and underneath the specific Aristotelian ideas condemned in 1277 was the underlying question about Aristotle's authority. Philosophers, most famously Thomas Aquinas, argued that there was no inherent tension in the relationship between (natural) philosophy and theology. Critics disagreed. Several of the condemnations focused on the error of what was known as 'double truth', the idea that 'things are true according to philosophy, but not according to the Catholic faith, as if there could be two contrary truths.'[14] Reason, such critics argued in line with the Church Fathers, could only be a handmaiden to theology. Yet, the intellectual autonomy and authority granted to reason by philosophers risked rendering revelation redundant, or worse, simply contradicted it.

Inevitably, swirling round this question of intellectual authority were issues of institutional, social and personal authority. Pope Alexander IV had ordered the reluctant secular masters of the University of Paris to accept Bonaventure as Master in 1256/57; sometimes academic independence isn't all it's cracked up to be. Bonaventure had written an entire treatise, *Retracing the Arts to Theology*, insisting that the secular subjects taught at Paris be made subordinate to theology (in which he had held the chair). It was a power struggle all too evident in the condemnations of 1277, which reek of the bad blood between the Parisian faculties. Thus, article 152 condemns the error that 'theological discussions are based on fables'; article 153 that 'nothing is better known because of knowing theology'; and article 154 that 'the only wise men of the world are philosophers'.[15] You do not need especially acute hearing to recognise the sound of scores being settled.

Reading these particular condemnations of Aristotelianism in 1277, one is reminded of the aphorism that the reason why academic disputes are so bitter is because the stakes are so low. That is unkind and no doubt unfair, at least in this case. Debates about the very nature of the world, the basis of human knowledge, personal (im)mortality, determinism, naturalism, divine providence and free will could hardly be more important. Be that as it may, the most significant dispute between religion and science in the entire medieval period was indeed bitter, but it was bitter primarily on account of the vexed question of where, and with whom, intellectual authority truly resided.

'According to the imagination': the birth of the thought experiment

The ecclesiastical condemnations of Aristotle in the thirteenth century did have an effect on medieval science, although ultimately a rather counterintuitive one. Initially, the impact was limited. Universities were independent and self-governing corporations, after all. When Aristotle was first banned in Paris, the University of Toulouse issued an invitation to those masters and students 'who wished to scrutinize the bosom of nature to the innermost parts' to come and study the pagan master there, in 'the second land of promise, flowing with milk and honey . . . [where] Bacchus reigns in vineyards'. It was an invitation worthy of any university prospectus.[16]

The Condemnation of 1277 was a more serious affair, and remained in effect throughout the fourteenth century, except for those articles that targeted Aquinas, which were soon nullified. Some scholars changed their tune. The Parisian scholar Peter of Auvergne, for example, wrote two commentaries on Aristotle's *On the heavens*, one before the Condemnation and one after, the latter positing very different views on the eternity of the world. More pervasively, 1277 placed a strong emphasis on God's (inscrutable) will and his absolute power, weakened confidence in the demonstrative certainty of reason and science and put the truths of natural philosophy at a lower level to those revealed truths of faith.

All of this sounds like a death knell for science (and for philosophy) although its impact should not be exaggerated. Aristotle, Averroes and others still formed part of the university curriculum in the fourteenth century and beyond; they were simply treated with greater caution and less authority than had been the case. In effect, the unstoppable force of Aristotelian natural philosophy in the thirteenth century met the immovable object of divine omnipotence in 1277. The object won, but rather than stopping the force, it redirected it. Ironically, the Condemnation of 1277 ended up having a(n unintended) liberating effect on medieval science.

Aristotle had declared – or at least was read as saying – that the world was eternal, that other worlds were not possible, that nature abhorred a vacuum, that the planets could only make circular not rectilinear motions, and so forth. The Church did not necessarily disagree with these points. After all, both Aristotle and the Bible agreed that there was only one world. But it did consider them to constitute an unwarranted limitation of divine power. If he wished, God could do any of these things, whether Aristotle liked it or not. So, scholars began to ponder what it would look like if he did. Such hypothetical possibilities became an important part of fourteenth-century scientific reasoning. In effect, the Condemnation of 1277 gave birth to the thought experiment.

Thus, for example, in the closing years of the thirteenth century, the Franciscan philosopher Richard of Middleton argued that if God had created worlds other than earth, each would naturally behave as earth did, each its own centre of a closed, circular system. That being so, it necessarily followed that no unique or privileged centre to the cosmos could exist.[17]

Richard no more believed in a non-geocentric universe than he did in the actual existence of other worlds, but the forceful emphasis on God's power in the 1277 Condemnation gave him cause to contemplate their possibility.

Such hypothetical constructs multiplied, under the rubric of *secundum imaginationem*, 'according to the imagination'. A vacuum being a possibility now that Aristotle's ban on it had been lifted, the natural philosopher and bishop of Lisieux, Nicole Oresme, pondered the existence of an 'intercosmic void' between our world and other possible worlds, while another fourteenth-century thinker, Thomas Bradwardine, actually proclaimed its existence. Others posed provocative questions. Would a stone placed in this void be capable of rectilinear motion? Would people in a vacuum be able to see or hear one another? Why wouldn't surrounding celestial spheres not collapse in through the void? Some hypothetical questions acquired a peculiarly modern feel to them. Were there different kinds of infinities such as God could create? Could one angel be in two places at the same time? Could two occupy the same space simultaneously? Did angels move between different spaces with finite or instantaneous speed?

There were risks in all of this. In the light of the Condemnation of 1277, which emphasised God's power, the contingency of creation and the ultimate inadequacy of human reason (and experience), a number of thinkers began to adopt a 'voluntarist' approach to God's relationship to his creation, one that could have a similar effect on natural philosophy as that risked by the 'occasionalism' of al-Ghazâlî. By this reckoning, there were no limits on divine freedom, which need not therefore be understandable through reason. In which case, the scientific study of nature was more or less pointless at the outset.

In general, however, these scientific thought experiments took place within the context of a well-established distinction between the *absolute* and the *ordained* power of God. The former emphasised that God, being omnipotent, could do whatever he wished; the latter contended that God operated only from within his preordained plan for creation. In effect, God had realised one of innumerable, initial possibilities for creation, and could now be relied on to stick with it. Christian natural theologians could thus have their cake and eat it. God's power was not to be limited by Aristotle but, at the same time, the (cautious and chastened) use of reason for natural philosophy was an acceptable, indeed profitable, activity, whatever Bishop Tempier might have thought.

In the early twentieth century, the French theoretical physicist and historian of science Pierre Duhem wrote a ten-volume history of classical and medieval science. It more or less single-handedly rescued medieval science from the self-serving condescension of Reformed thinkers and Enlightenment philosophers, and turned it into a respectable discipline. Duhem knew better. Thinking of the Condemnation and the strange new thought worlds it catalysed, he wrote that, 'if we must assign a date for the birth of modern science, we would, without doubt, choose the year 1277.'[18]

A page of Nicolaus Copernicus' *De revolutionibus orbium coelestium* with the sun at the centre of the solar system and the planets in orbits around it. Almost nobody had their faith challenged by heliocentrism in the sixteenth century. Indeed, almost no one actually read his book.

FIVE

1543 AND ALL THAT

'A memorable history of science and religion': not a revolution

The prolific science writer John Gribbin began his history of science with the year 1543 and the literary critic John Carey started his anthology of science writing with a book published in 1543. The year 1543 was when the Flemish anatomist Andreas Vesalius first published his *De humani corporis fabrica*, *On the Fabric of the Human Body*, which mapped out human bones, cartilage, ligament, muscles, veins and arteries with unprecedented accuracy. And 1543 was when the Polish canon, Nicolaus Copernicus published his *De revolutionibus orbium coelestium*, *On the Revolutions of the Celestial Spheres*, which described the mathematics of a heliocentric cosmos. 'The year 1543 in science and technology marks the beginning of the European Scientific revolution', Wikipedia informs us. So the year 1543 is a turning point.

All such 'turning points' are arbitrary but there is something particularly problematic about treating 1543 as Year Zero. This is partly because Copernicus first wrote about heliocentrism thirty years earlier. It is partly because he retained as much of the Ptolemaic cosmos – nested planetary spheres; finite space; uniform, eternal circular motion – as he jettisoned. It is partly because Vesalius's book drew from over a decade of anatomical study at Paris, Louvain and Padua. And it is partly because neither the scientific method nor the experiment played any role in the publications of 1543, whereas a principled commitment to methodological naturalism, which *is* essential to science, can be traced back to the twelfth century. In short, there doesn't seem to be that much special about 1543.

The date is especially misleading, however, when it comes to the history of science and religion, thanks largely to Sigmund Freud. In his *General Introduction to Psychoanalysis*, first delivered as lectures between 1915 and 1917, Freud wrote how 'humanity, in the course of time, has had to endure from the hands of science two great outrages upon its naive self-love.' The first of these was when we 'discovered that our earth was not the centre of the universe, but only a tiny speck in a world-system hardly conceivable in its magnitude', a revolution ever to be 'associated in our minds with the name "Copernicus" '.[1]

The observation was not original. The physician Emil du Bois-Reymond, under whom Freud had once planned to study, had made precisely the same point in a eulogy on Darwin forty years earlier. Nor was it disinterested. Freud claimed that in addition to Copernicanism and the second great 'outrage', Darwinism, there was now a third 'and most irritating insult . . . flung at the human mania of greatness', namely Freudian psychoanalysis.

Not original, disinterested or, as we shall see, accurate, Freud's observation was at least influential. The allegedly traumatic effect that Copernicus had on humanity – and in particular on humanity's religiously grounded self-importance – has passed into folklore. Given how the Copernican revolution is so often told in the light of its apparent culmination at the trial of Galileo ninety years later, the result becomes a self-fulfilling prophecy. The moment science was born, spitting hard truths about human insignificance, religion was there, opposing it. The result is a kind of *1066 and all that* history, a Memorable History of Science and Religion, comprising all the parts you can remember, including hundreds of Good Discoveries (among them Medicine and the Universe), three Bad Incidents (Galileo, Oxford, the Scopes trial), and two Genuine Dates 1543 (when Copernicus showed that the religious view of the universe was wrong) and 1859 (when Darwin showed that the religious view of humanity was wrong). If science begins with Copernicus (and Vesalius) in 1543, religious conflict is built into its foundations.

This is not to claim that nothing of significance happened in 1543. Copernicus's book did indeed mark a 'paradigm shift' – indeed, it became the archetypal paradigm shift – despite its being, at the time, unprovable and wrong in many details. Similarly, Vesalius's work marked an important juncture, in so far as it showed him parting company with the authoritative Galen and from written medical authorities in general, in favour of careful human anatomy. Both Copernicus and Vesalius thus represented important shifts, but neither effected a(n instant) revolution and neither shook religious belief to its core. Indeed, for quite a long time no one really noticed. Things only began to shift in the last quarter of the century, when the heavens literally did change and in a way that everyone could see.

'The theologians will easily calm down': the revolution will be postponed

Copernicus *was* like Darwin in one important respect. Both conceived their theories decades before they went public. Darwin developed evolution by natural selection more than twenty years before *The Origin of Species* was published, Copernicus landed on heliocentrism at least thirty years before *De revolutionibus*. Through Freud's lens, his delay looks like it was due to the fear of (ecclesiastical) consequences but, in reality, this was no more the case for Copernicus than it was for Darwin.

Born in Torun, in Poland, in February 1473, Copernicus studied at Jagellonian University in Krakow, reputed for its astronomy and mathematics, before moving to Bologna and Padua, where he read law and medicine, and then to the University of Ferrara, where he received a doctorate in canon law in 1503. He then returned to Poland to work as secretary and physician to his uncle, the bishop of Warmia, and to take up a position as canon at Frombork Cathedral, having been elected in his absence.

Mathematical astronomy was one intellectual interest among many, albeit a particularly passionate one. 'When a man is occupied with things which he sees established in the finest order and directed by divine management,' he said, in the carefully rhetorical preface to *De revolutionibus*, 'will not the unremitting contemplation of them . . . stimulate him to the best and to admiration for the Maker of everything?'[2] Like any good medieval natural philosopher, Copernicus knew that true science was an aid to a virtuous life.

According to his pupil, Georg Rheticus, a professor of mathematics at the Protestant University of Wittenberg who had travelled to Copernicus in 1539 and become his pupil, disciple and publicist, Copernicus was already lecturing on astronomy in Rome by 1500, and within a decade of his return to Poland he had generated his alternative to the received Ptolemaic system. There was, at the time, no significant anti-Ptolemaic tradition in medieval Christendom – certainly compared with Islam, where astronomers had been critiquing Ptolemy since the eleventh century and proposing (geocentric) alternatives since the thirteenth. This has led scholars to wonder what Copernicus may have owed to Islamic astronomy.

Copernicus cited five Islamic astronomers in *De revolutionibus* – Thabit ibn Qurra, Al-Battani, al-Zarqallu, Ibn Rushd and al-Bitruji – but as the last of them had died over three hundred years earlier, most readers have seen them as an important but essentially historical part of the story. Since the early twentieth century, however, scholars have pointed out that Copernicus also made use of mathematical models devised by later Islamic astronomers, in particular Nasir al-Din al-Tusi and Ibn al-Shatir, and a few have claimed that the lettering in the relevant diagram in *De revolutionibus* 'follows the standard Arabic lettering rather than what one might expect in Latin'. This has led some to conclude that Copernicus drew directly on Islamic mathematics and astronomy for his revision of Ptolemy, although no one has yet claimed that he took his heliocentrism from that source.[3]

The problem with this is that there is no direct, and precious little indirect, evidence of Copernicus encountering or adopting these models. Scholars have located a manuscript in the Vatican dating from about 1300, a Greek reworking of an Arabic treatise, which contains the important so-called 'Tusi couple'. However, no one has yet shown when, where or even whether Copernicus read this, or indeed any other relevant text of recent Islamic astronomy. By comparison, a number have posited that Copernicus's thought could equally have been catalysed by astronomers working in Padua, where he is known to have spent years or, indeed, could simply have been self-generated. Were historians of science to demonstrate that Copernicus had drawn directly on late medieval Islamic mathematical astronomy, it would – given the iconic nature of this moment – help bolster the reputation of late medieval Islamic science. But as things stand, the evidence is not yet strong enough.

His intellectual debts aside, by 1514 Copernicus had written a manuscript, known as the *Commentariolus*, or *Little Commentary*, in which he set out his ideas. 'A quire of six leaves of a theory asserting that the Earth moves whereas the Sun is at rest', reads the sarcastic-sounding library entry of a Krakovian physician, the first evidence we have of the new theory. The text circulated among a few close friends and it was years before the idea had become more widely known. When it had, the ecclesiastical response was broadly positive. In 1533, the German orientalist Johann Albrecht Widmanstetter presented the theory to Pope Clement VII and members of the curia, who were impressed enough to present him with a valuable Greek codex in which Widmanstetter noted the occasion. Three years later, following Clement's death, Widmanstetter had become secretary to Cardinal Nikolaus von Schönberg, who was equally impressed with the theory. He wrote to Copernicus asking for a copy of his astronomical manuscripts and urged him to publish, guaranteeing his patronage of the new planetary theory. Copernicus still delayed. Given how the European intellectual landscape had changed between 1514 and 1536, and would change still further between 1536 and 1543, this might be considered something of a 'poignant lost opportunity'.

The late medieval world in which Copernicus lived was, as we have seen, familiar with provocative thought experiments, including those about the rotation of the earth, but only a few thinkers, such as Jean Buridan and Nicole Oresme, took this idea seriously. Copernicus was clear that his idea was not simply a thought experiment. He wanted his theory to be taken as a model of reality.

The problem was that the idea faced significant intuitive and more subtle, scientific objections, which risked his humiliation if not his life. If the earth did move, particularly at the speed it would need to, why did it feel so still? Why was there not a constant gale blowing in people's faces? Why didn't the oceans slop about violently on the shore? Why did a falling object land below where it was dropped? Why did cannons fired to the east and west not automatically reach different distances?

Careful thought could defuse such problems but it did little to solve more technical ones. If the earth moved in a sizeable orbit, why was there no sign of any movement among the background stars, the so-called stellar parallax, as there should have been? Why did Venus not exhibit phases in the way the moon did? And why did Mars not appear commensurately brighter and dimmer? More sensitive observational equipment would eventually answer these questions – the differences were there, they were just too small to detect – but in the 1540s they constituted serious scientific objections.

No less problematic were more abstract or fundamental objections. Copernicus's theory was based on geometry, a relatively humble discipline in the medieval hierarchy. Overturning everything on the basis of geometry made little sense. His argument that the earth both revolved and orbited contravened a basic principle of Aristotelian physics, that a simple body can have only one motion proper to it. Most substantially, heliocentrism disrupted the fundamental Aristotelian division of the cosmos into sub- and superlunary spheres, on which pretty much everything rested. It was these objections that deterred Copernicus from going public.

When he finally did, it was under the influence of Georg Rheticus. Rheticus ventured into public discussion first. In 1540, he published a short pamphlet outlining Copernicus's ideas, *Narratio prima*, which was considerably clearer than his master's magnum opus would be. It drew no hostile fire and the two astronomers were encouraged. *De revolutionibus* was finally published three years later, supposedly as Copernicus was on his deathbed. By this time, however, the intellectual climate was significantly worse.

The Council of Regensburg, the final significant attempt to retain Church unity through theological dialogue between Lutherans and Catholics, had broken down in failure and acrimony in 1541. Worse, at least as far as Pope Paul III was concerned, Emperor Charles V had shown an alarming willingness to compromise with the Protestants. An already insecure papacy issued the papal bull *Licet ab initio*, setting up the Roman Inquisition in 1542 and, a year later, banned the printing and sale of all books unless authorised by the Church. This was not a good time to be publishing bold new ideas through some kind of Lutheran–Catholic partnership, even in a book dense with mathematical calculations, whose frontispiece warned off casual readers with the words of Plato's Academy: 'Let none ignorant of geometry enter.'

Rheticus and Copernicus tried to negotiate the minefield. *De revolutionibus* appended von Schönberg's letter of 1536, to suggest a patronage relationship that did not in fact formally exist (and never would, given that the cardinal had died in 1537). More significantly, it incorporated a fulsome dedication to Pope Paul III. This referenced von Schönberg's encouragement to publish, along with that of Tiedemann Giese, bishop of Chełmno, another of Copernicus's ecclesiastical supporters. It outlined the mathematical superiority of his system. And it made a powerful case against unwarranted judgement by 'babblers', who were ignorant of astronomy or who distorted the scriptures in a self-serving literalist way that invariably made them look as ridiculous as flat-earth Church Fathers. 'I disregard them even to the extent of despising their criticism as reckless.'[4] Copernicus came out fighting.

It was not only the Catholic party that needed assuaging. *De revolutionibus* may have been printed in the Protestant stronghold of Nuremberg and catalysed by a Protestant professor, who had been appointed by and was good friends with leading reformer Philip Melanchthon, but there were still dragons in Protestant waters too. Or at least one big dragon.

From the 1520s, Martin Luther had entertained guests and disciples at home in order to discuss matters of importance. This 'Table Talk' was subsequently published in 1566, and records how, on 4 June 1539, Luther dismissed the new astronomer as a 'fool' for upending the cosmos and proposing an idea that was incompatible with the plain meaning of scripture. Some have questioned the reliability of this entry, published a quarter of a century after the event, but Copernicus's ideas had been rumoured for some time by then and the fact that the remark was made around the time Rheticus travelled to meet Copernicus is surely significant.

Luther's objection was to be rehashed by many subsequent critics. The Old Testament book of Joshua records how, during a battle between Israel and the Amorites, 'the sun stopped in the middle of the sky and delayed going down about a full day', so as to enable Israel's victory.[5] This, of course, was technically compatible with either geo- or heliocentric systems but, Luther reasoned, 'Joshua ordered the Sun to stop and not the Earth', meaning that Copernicus had to be wrong. To this scriptural objection, others would be added. Did not the prophet Isaiah call upon God to make a shadow miraculously go back ten steps?[6] Did not Psalm 96:10 say 'The world is firmly established, it cannot be moved'?

These were not solely Protestant objections. The events of the previous twenty-five years had made Rome a good deal more inclined towards biblical literalism, nudging Catholics away from the medieval multi-level reading of scripture towards the plain (i.e. literal) meaning of the text. However, the objections found particular purchase among Protestants, who believed that *sola scriptura*, scripture alone, was sufficient in matters of faith and order. The Protestant Rheticus was alert to this and, probably in 1541, wrote a tract arguing 'Against the Irreconcilability of Sacred Scriptures and Terrestrial Motion'.

'The obscurities of nature, which we sense as the work of God, the almighty Architect, should be dealt with, not by making assertions, but by research,' Rheticus argued. 'Since God desires to be glorified in nature, there is no doubt that our study will be pleasing to Him.' The Bible, he insisted, was 'intended only to teach that which is necessary for salvation'. We do not study it 'as if Scripture were a philosophical textbook'. On scientific matters 'it does not speak apodictically [i.e. with necessary or absolute certainty]'. Reading it properly demanded an 'accommodationist' approach, meaning that God had accommodated the language of the scriptures so that they might be understood by ordinary people. The Bible was written not in the language of the lecture hall but of the field, farm and family.

Thus, the reformer John Calvin argued that astronomers investigated 'whatever the sagacity of the human mind can comprehend', whereas Moses, a teacher of 'the unlearned and the rude as of the learned' wrote in a 'popular style' which 'ordinary persons' endued with 'common sense' were able to understand. They were, in effect, writing in two different languages, which should not be played against each other. 'Astronomy is not only pleasant, but also very useful to be known', Calvin said, and its study should not be judged or condemned 'because some frantic persons are wont boldly to reject whatever is unknown to them'.[7] In the same vein, Rheticus wrote, the Bible 'adapted itself to the common way of speech or vulgar opinion'. Thus, references to the earth's 'stability' should be understood in terms of its integrity or reliability rather than its immobility. References to static suns or back-tracking shadows should similarly be understood as descriptions of appearance rather than statements about the mechanism of the cosmos.

Rheticus's arguments would be repeated as often as Luther's objections over the coming century, although his treatise never in fact saw the light of day.[8] Copernicus clearly shared his concerns and had written, in July 1540, to Andreas Osiander, another Protestant reformer and one of Rheticus's contacts, asking his advice on how to present heliocentrism vis-à-vis potential scriptural objections. Osiander proposed Copernicus side-step the whole issue by treating the theory as simply a hypothesis with no claim to reality. The 'theologians will easily calm down', he wrote in the same vein to Rheticus, 'if you tell them that various hypotheses could account for the same apparent motion and that you do not affirm the reality of them'.[9]

Copernicus was not impressed, recognising that such a strategy would undermine his fundamental idea, which stood right at the head of Book I. He ignored Osiander's advice. In the end, it made no difference. Rheticus arranged the publication of *De revolutionibus* with Osiander, who took it upon himself to include an unauthorised and unsigned preface which explained that the book's model was not intended as a description of reality but rather as a mathematical device to simplify planetary calculations. Unable to distinguish preface from book, which in any case Copernicus, in his dedicatory letter to Paul III, had directed his audience to read as 'mathematical things for mathematicians', early readers of *De revolutionibus* took it as an impressive exercise in mathematical modelling rather than a theory of heliocentrism.

In effect, Osiander's intervention allayed Copernicus's and Rheticus's concerns about a hostile literalist reaction, but did so by neutering their publication. His cure had effectively killed the patient. *De revolutionibus* was not, as the novelist Arthur Koestler once claimed, the book that nobody read. There are, today, at least six hundred copies in existence, many of which are heavily annotated. But it was the book that (almost) nobody read as its author had intended. The revolution was postponed.

'The indestructible sun [is] subject to destruction': the Wittenberg interpretation

Osiander's interpretation of *De revolutionibus* shaped the book's reception over the next generation. In a remarkably determined example of scholarship, the historian of science Owen Gingerich spent thirty years tracking down every single surviving copy of the first (1543) and second (1566) editions of *De revolutionibus*, among them those owned and annotated by Johannes Kepler and Galileo. Studying the marginal comments made by the book's first readers, he found that, almost without exception, students valued the book for its mathematical sophistication over and above any claims to scientific truth.[10]

This was certainly the dominant approach in Protestant lands. For all their stress on *sola scriptura*, Protestants tended to hold astronomy in high regard. It showed, like the Psalmist, how the heavens declared the glory of God. Melanchthon, in particular, prized natural philosophy, wrote an introductory textbook on the subject published in 1549, and instigated a strong astronomical tradition among Protestant universities. His textbook was critical of Copernicus but, in the same year, he also wrote how he had begun 'to love and admire Copernicus more', and, in later editions of the book he removed or revised his Copernican criticisms. Melanchthon's Copernicus, however, was fully Osiander's version of him. Protestant universities, with their newly founded chairs in astronomy, often filled with Melanchthon's own disciples, popularised *De revolutionibus* but only in what became known as the 'Wittenberg interpretation'.

The early Catholic response was more muted but also more critical. The two decades after *De revolutionibus* was published were dominated for Rome by the Council of Trent, a sustained attempt to respond to the Protestant challenge biblically, theologically, spiritually, ecclesiastically and educationally. Copernicus's theory was at no point discussed in any of the twenty-five sessions of the Council, underlining quite how peripheral it was to sixteenth-century Catholic concerns.

It was not entirely ignored. Early on, in 1544, the Dominican Bartolomeo Spina, Master of the Sacred Palace in Rome, had intended to write against the book but died before he had the opportunity to do so. After him, the challenge fell to a fellow Dominican, Giovanni Tolosani. Tolosani had written a large apologetic work in *On the truth of sacred scripture*, to which he added a series of appendices, the fourth of which criticised Copernicus (though it was not published until the twentieth century, again because of the author's death). Unlike the Protestants, Tolosani treated the book as its author had intended, realising that Osiander's introductory letter was a red herring. Peppering his critique with ad hominem attacks on Copernicus, Tolosani took up cudgels on a number of fronts, some scriptural, some scientific, but most rooted in his underlying commitment to Aristotle.

Copernicus's theory violated the hierarchy of disciplines, treating physics and theology as if they were subordinate to geometry and mathematics. It also violated the Aristotelian idea of the cosmos. The point here was not, as Freud thought, that decentring the earth served to knock human 'naïve self-love'. If anything, heliocentrism elevated the earth. In the words of the Renaissance humanist Giovanni Pico, in the geocentric cosmos the earth occupied 'the excrementary and filthy parts of the lower world'.[11] The planets and stars, by contrast, existed in eternal, unchangeable perfection. Placing the earth in the heavens and the sun at the centre, elevated the former at the expense of the latter. Tolosani was upset because, in his words, 'Copernicus puts the indestructible sun in a place subject to destruction.'[12]

His main problem was Copernicus's denial of Aristotle's physics. The tensions over Aristotle had dissipated since the late thirteenth century. Scholastic philosophers, supremely Thomas Aquinas, had successfully integrated the Greek's thought, which was now the lens through which Catholics read and understood their doctrine and theology. It was always going to be hard to overturn such a comprehensive system. Copernicus's ideas appeared to demand too much and offer too little. In Tolosani's view, there was simply not enough evidence for such a huge change. 'It is stupid to contradict an opinion accepted by everyone over a very long time for the strongest of reasons, unless the impugner uses more powerful and incontrovertible demonstrations.'[13] Scientifically speaking, it wasn't bad advice. As far as Tolosani was concerned, Copernicus had failed to prove his case.

And not just Tolosani: thirty years after *De revolutionibus* was published, it still had only one single informed advocate: Rheticus himself. By historian David Wootton's calculations, a decade later, by the early 1580s, this had risen to three. Rheticus had died, ceding his tiny Copernican territory to a German, Christoph Rothmann (who did not publish and eventually abandoned the theory), an Italian, Giovanni Benedetti, and an Englishman, Thomas Digges. By then, however, something had happened that had begun slowly to shift the ground beneath the astronomers' feet.

On 11 November 1572, a Danish nobleman by the name of Tycho Brahe noticed an object in the night sky. At first it looked like a star, albeit an unusually bright one. Over the following week, however, it grew brighter still so that at one point it was even visible during the day. It was found in the constellation of Cassiopeia and showed no signs of movement against the background of stars, implying it was located in Aristotle's superlunary sphere. But it was completely new, appearing out of nowhere that November, and fading away within a few weeks. That, however, was not possible, because nothing changed in the superlunary spheres. Aristotle said so.

Brahe was not alone. Through that November, the supernova – as we now know it was – was studied by many other stargazers, including Thomas Digges in England, who arrived at the same conclusion regarding its location. Brahe published a book about it, *De nova stella*, the following year. Three years after that, Digges published a new edition of the almanac written by his father, Leonard Digges, *A Prognostication Everlasting*, which mentioned the star and included the first defence of Copernicanism in English. 'New Stars' had been observed before by Chinese and Arabic astronomers but this was the first such event in Christian Europe and it suggested that the Aristotelian physics to which Christendom had been happily married was flawed.

Many, including Brahe, interpreted the star as a miracle or a sign from God, but however true this might have been it was not a sufficient explanation, particularly among Protestant stargazers, who were now self-consciously wary about modern miracles. In any case, there were more wonders to behold. Five years after Tycho's Nova, a comet appeared across Europe. These were familiar sights and easily accommodated within the Aristotelian universe, which understood them as sublunary (because changeable) events. Better measurements in 1577, however, showed that this new comet had, in fact, to be superlunary. Aristotle was wrong again.

Such a challenge to the great Greek's scientific authority was not entirely novel. The discovery of the New World had proved beyond doubt that the Greek concept of the world was fundamentally wrong, or at least partly ignorant. The ancients were not as omniscient as Europeans had come to think. Fifteen years before Tycho's Nova, Jean Péna, then Royal Lecturer in Mathematics in Paris, wrote in the preface to his Latin edition of Euclid's *Optics* that Aristotle's theory that the heavens were incorruptible was wrong. Such challenges were momentous but they paled into insignificance against the reformers' broader attack on Aristotle, whom they saw as the sinister spider sitting in authority at the centre of a whole corrupt Catholic web of lies. To undermine Aristotle, even only his physics, risked adding your voice to an ever louder Protestant chorus.

Not that a new approach to the heavens was solely a Protestant affair. Tycho's Nova and the comet of 1577 ran in parallel with the Catholic Church's revision of the ancient and increasingly erroneous Julian Calendar. Reform had been instituted at the Council of Trent but it was not concluded until 1582. The whole process had required careful observation and precise calculation of the heavens, led by first Aloysius Lilius and then Christopher Clavius, who was to become one of the most respected astronomers in Europe. The experience did not cause Catholic astronomers to reject geocentrism any more than the events of 1572 and 1577 caused them to reject Aristotle, but it did intensify debate and widen interest in new, Copernican ideas, dragging them from the shade of 'mere mathematics' in which Osiander's preface had left them.

In light of his new star, Brahe received from King Frederick of Denmark an estate on the island of Hven, where he built a research institute, Uraniborg, which incorporated an astronomical observatory so sensitive that it had to be housed in a basement to avoid being shaken by the wind. Here, a team of researchers charted the skies and collected data in unprecedented detail. Brahe was sympathetic to Copernicus, hailing him as a second Ptolemy, but could not go all the way with his heliocentrism. He too felt the power of the scriptural arguments against the earth's motion but these were not his major concern. Brahe could not persuade himself that the heavy, dense, material earth was capable of the perfect, smooth, eternal, circular motion that Copernicus posited. Uraniborg's new measurements demonstrated that the Polish canon's observations were inaccurate. More broadly, Brahe felt unable to jettison the entire Aristotelian model entirely on the basis of Copernicus's, admittedly impressive, mathematics. This was a question of science, rather than religion, critiquing science.

In its stead, Brahe proposed a geo-heliocentric model, in which the earth remained at the centre and was orbited by the stars, moon and sun, which itself was then orbited by the five other planets. Published in 1588,

Brahe's new system attracted more attention than Copernicus's had nearly fifty years earlier. Europe's scientific community was now attuned to these questions, and geo-heliocentrism appeared to solve certain problems – in particular the absence of the stellar parallax – that dogged heliocentrism. But it also provoked similar questions.

Brahe sent a copy of his new book to Christoph Rothmann, a German astronomer with whom he had already corresponded, and who would visit him at Hven two years later. Rothmann was at the time one of the vanishingly small number of fully paid-up Copernicans in Europe and, as such, he had had to face up to scriptural questions that earlier Protestant astronomers had been able to avoid. Both Rothmann and Brahe were 'accommodationists'. They recognised that the Bible was not written for philosophers or astronomers but for common people, and thus in a common language. The Bible 'speaks after their capacity of understanding', Rothmann explained.[14] In this way, he insisted, it could accommodate geo, helio-geo or helio-centrism.

Brahe wanted more than this, however. In particular, he was concerned that by recognising the accommodation of language, imagery and ideas to the understanding of the common man, 'philosophers' were in danger of undermining the scriptures or, worse, imputing untruths to them. Granted, physics and astronomy had to 'adjust themselves very well to the level of the crowd'. But where did one draw the line? Was there *no* truth in the prophets' conception of the physical world? Was there not a danger of throwing out theological truths when acquiescing to common, 'accommodated' language? 'Far be it from us to decide that they speak in so vulgar a way that they do not also seem to set out truths,' he said of the prophets.

This was not an overriding concern, even for Brahe, who took the authority of the scriptures very seriously. Indeed, aside from his correspondence with Rothmann, he rarely raised the issue, having greater concerns about the scientific difficulties of Copernicanism. Nor, self-evidently, was it an issue for Rothmann himself. Sensing the charge of impiety in Brahe's letters, Rothmann asked indignantly, 'Where have I downplayed the authority of the Holy Spirit in any way?' Rothmann took the Bible as seriously as Brahe. He simply had a more flexible (and less anxious) approach to its interpretation and – crucially – a more confident approach to intellectual and professional boundaries. 'Unless this question is decided by us,' he wrote to Brahe, 'us' here meaning astronomers and mathematicians, 'it will not be decided by anyone, whether theologian or [philosopher].'

Nevertheless, for all of Rothmann's confident accommodation and drawing of disciplinary boundaries, the question of how much of the literal reading of Holy Books should be accommodated to the new discoveries of science would dog later thinkers, in particular centuries later, when the discussion widened from the skies to incorporate questions of morality, ethnicity and humanity.

'A martyr for magic': the death of Bruno

Brahe's theory of geo-heliocentrism was not the only significant book on astronomy published in 1588. In the same year, Nicolaus Reymers Baer, commonly known as Ursus, published his *Fundamentum astronomicum*. Ursus was court astronomer and mathematician to Holy Roman Emperor Rudolf II, and his book argued that the medium of the heavens was not aether, as everyone believed on Aristotle's authority, but air, the density of which gradually decreased the greater the distance from the earth. At the same time, Giordano Bruno published his *Camoeracensis acrotismus* (*The Pleasure of Dispute*), which also argued against solid celestial spheres but went a great deal further. Thereafter, the final decade of the century burned bright with new ideas about the cosmos.

Born in 1548 near Naples, Bruno entered the Dominican Order in 1565. Brilliant, blessed with a sharp and enquiring mind and an excellent memory, he was also provocative, with a practised ability to lose friends and alienate people. He left his order in 1576 – half-expelled, half-escaped – and travelled around Europe, describing himself as an 'academician of no academy', before finding patronage and protection with Henry III of France. While serving as royal lecturer in Paris, he wrote extensively against Aristotelian natural philosophy and published a book on the art of memory in which he first voiced support for heliocentrism. In 1583, he travelled to England as a guest of the French ambassador, and lectured on Copernicanism at Oxford. His lectures did not impress the future archbishop of Canterbury, George Abbot, who later wrote how 'in truth it was his own head which rather did run round, and his brains did not stand still.' Nor did it impress the rest of his Oxford audience, who ran him out of town after three lectures, though on the grounds of plagiarism rather than Copernicanism.

Bruno was then run out of England altogether when, two years later, a mob attacked the French embassy in London. He took to travelling again, first through France, then Germany and finally, fatally, Italy. It was in the midst of this final bout of travelling, while in Germany, that he outlined the full extent of his heterodox astronomical views, in the hope of getting support from astronomers, conscious as he was that his views were primarily philosophical ideas, without much (indeed any) mathematical or observational support.

Bruno stood on Copernicus's shoulders but thought he could see vastly further than his Polish master. He argued that the universe was not limited but infinite and eternal. He said that in such a cosmos there was, by definition, no centre, whether earth or sun, or indeed any absolute motion. The stars that were scattered throughout were, by Bruno's reckoning, all just suns, and our own sun was just another star. The visible planets shone solely because they reflected the sun's light. The universe was populated with innumerable other planets, many of which were inhabited by other beings, as were the stars, which were not necessarily hot all over. It was an astonishingly 'modern' vision of the universe, though one born of Bruno's uniquely fecund imagination rather than any new observation or calculation.

If Bruno's imagination was uniquely generative, his ideas were not unique. The late medieval mind was, as we have noted, adept at conjuring provocative ideas, some about a decentred universe, some about the existence of other worlds. These were 'mere' thought experiments, although the philosopher Nicholas of Cusa had argued more seriously in his book *On Learned Ignorance* that only an infinite universe would be suitable for an infinite God, and had also suggested that the earth was a heavenly body that would shine like a star from a distance. Before Bruno, both the ninth-century Arabic astronomer Al-Battani, who was referenced by Copernicus, and the thirteenth-century Polish natural philosopher Witelo had argued that the planets shone from reflected sunlight. More recently, Rothmann and Brahe had debated the liquidity of celestial spheres, and Thomas

Digges had proposed an infinite universe. In short, Bruno's ideas were impressive, heterodox, evidentially weak, but not unprecedented.

Bruno deployed the same scriptural justification for this as did Rothmann and Rheticus before him. Thus, for example, he argued, had the author of the book of Ecclesiastes written 'the earth turns round to the east, leaving behind the sun which sets', instead of 'the sun riseth and goeth down', his listeners 'would have rightly accounted him a madman'. This was already a popular and well-worn argument but it rang hollow coming from Bruno, who was as well known for his unusual religious views as his astronomical ones.

It is not entirely clear what religious beliefs Bruno held, partly because he was accused of much that he denied and partly because the documents outlining the charges brought against him were lost during the Napoleonic Wars. He was accused of doubting the incarnation, the divinity of Christ, the virgin birth and the transubstantiation of the Mass, as well as of holding heretical views on the Trinity, the Holy Spirit, the immortality of the soul, prayer to saints and the efficacy of relics. It seems that he was a kind of pantheist, seeing God and matter as co-dependent and eternal. He tangled in magic, mixing in the circle of the great Elizabethan magician John Dee in London and producing several books on the topic. And he was friends with, and dedicated books to, the militantly Protestant Sir Philip Sidney. Whatever his exact beliefs and however skilfully he defended himself, some things were unforgivable.

When he returned to Italy in 1592, he first failed to secure the Chair in Mathematics at Padua University – it went to Galileo the following year – and then accepted an invitation to tutor in Venice, at the time one of Italy's most independent-minded and liberal republics. His ability to alienate people resurfaced, however, and when, two months later, he tried to leave, his host denounced him to the Inquisition. For the next eight years he was imprisoned in Rome, during which time the Inquisitors accused him of pretty much everything.

In reality, the Inquisition was no more provoked by Bruno's speculations on the size and shape of the cosmos than the Council of Trent had been by Copernicus's new ideas. Denying the Trinity, the virginity of Mary, the incarnation and transubstantiation, as he was accused of doing, were far more serious crimes, as was dabbling in divination. However, the two areas were not wholly unconnected. The question of the size and potential infinity of the universe was a scientific one, but if the universe were infinite, and if it were populated with an infinity of worlds, and if those worlds were inhabited, where did that leave the humans on this particular planet? Were they in any way special or different? Did God care for them in particular? Did he die for them specially? Could they still be saved?

These were not insuperable questions or even, necessarily, forbidden ones. In the religiously more settled times of earlier centuries, they were precisely the issues that theologians had theorised about. But Europe in the 1590s was not religiously settled, and Bruno was not merely theorising. His heretical views were evidence enough; his scientific ones were ancillary to the charge. Although he was reported to have defended himself ably, eight years of imprisonment, interrogation and ultimately torture broke him. Or perhaps not quite broke: one gloating witness claimed that, at his final appearance before his judges, Bruno made a threatening gesture and said, 'Perhaps you who pronounce my sentence are in greater fear than I who receive it.'[15] He was taken to the Campo de' Fiori in Rome on 17 February 1600, where he was hung naked, upside down and then burned at the stake. Bruno is sometimes presented as a martyr for science, much like Hypatia. 'Within ten years after the martyrdom of Bruno,' Andrew Dickson White would write knowingly, 'the truth of the doctrine of Copernicus was established by the telescope of Galileo.'[16] Bruno was 'science's first martyr', according to Michael White's more recent biography.[17] In reality, for all his brilliance, Bruno was not a scientist, even by the standards of the time. He died, as did many in the fevered, fearful atmosphere of the late Reformation, for his theology, his blasphemy and his interest in the forbidden arts. As John Gribbin has remarked, 'rather than being a martyr for science . . . [Bruno] was actually a martyr for magic'.[18]

PART 2

GENESIS

Galileo Galilei in prison. It was said that he would
rather lose a friend than an argument, a talent
that would help turn a cautiously positive response
to his ideas into a full-blown conflict.

SIX

GALILEO GALILEI

'A prisoner to the Inquisition': on meeting Galileo

In 1638, the young English poet John Milton, travelling through northern Italy, called on Europe's most famous prisoner. 'There it was that I found and visited the famous Galileo, grown old, a prisoner to the Inquisition,' Milton would write six years later in *Areopagitica*, his defence of liberty. Milton had enjoyed Catholic hospitality and literary culture on his travels. He had delighted in 'men endowed with both learning and wit'.[1] He had been welcomed by, among others, Cardinal Francesco Barberini, the pope's nephew and one-time friend and protector of Galileo. But Galileo's fate nonetheless confirmed for this self-consciously Protestant poet what he knew in his heart.

The scientist was under comfortable house arrest outside Florence for, as Milton put it, 'thinking in astronomy otherwise than the Franciscan and Dominican licensers thought'. Unlike intellectuals such as himself, who were born into 'philosophic freedom', Galileo's fate confirmed 'the servile condition into which learning [among Catholics] was brought'.

Six years earlier, the great astronomer had been broken by the Congregation of the Holy Office. Threatened with torture – probably 'of the rope', in which the victim's wrists were tied behind his back, and he was hauled up by a pulley, dropped and then suddenly yanked upwards, thereby wrenching the shoulders from their sockets – Galileo denied his life's work. He was sixty-nine years old. Kneeling before judges and witnesses, with a lit candle in his hand, he read out a statement, swearing that 'with a sincere heart and unfeigned faith' he abjured, cursed and detested all Copernican errors and heresies, and promised never again to assert anything that might arouse suspicions about him. Satisfied, though still fuming, the pope had ordered him into an unusually severe house arrest. 'Perpetual incarceration' by the Inquisition usually meant three to eight years. For Galileo, life meant life.

In reality, Milton may not have actually met the ageing scientist. He was certainly fascinated by his science, which he would weave into *Paradise Lost* thirty years later. 'The Glass/ Of Galileo', he wrote in Book 5, observed 'Imagind Lands and Regions in the Moon'.[2] Satan's shield 'hung on his shoulders like the Moon, whose Orb,/ Through Optic Glass the Tuscan Artist views'.[3] The fiend, when he lands upon the sun in Book 3, is in 'a spot like which perhaps/ Astronomer in the Sun's lucent Orbe/ Through his glaz'd Optic Tube yet never saw', referencing Galileo's lesser-known work on sunspots.

Poetic fascination does not biography make, however, and historians have cast doubt over the meeting. Galileo's house arrest was tight. Visitors, let alone heretical ones like Milton, were not encouraged. There is no other record of or evidence for the meeting, although Milton definitely did meet Galileo's illegitimate son Vincenzo. Milton fails to mention Galileo in the account of his Italian journey in his *Second Defence of the English People*.

Above all, *Areopagitica* was a rhetorical masterpiece, not documentary history. Milton argued that the 'project' of licensing the press 'crept out of the Inquisition'. 'The Popes of Rome', ever greedy for political power, spread their dominion 'over men's eyes, as they had before over their judgements, burning and prohibiting to be read, what they fancied not'. England must not become like 'other Countries, where this kind of inquisition tyrannizes'. Exhibit A: Galileo Galilei. Ultimately, whether Milton did in fact meet Galileo hardly matters. Galileo ascended to myth very soon after he died, and Galileo the myth has always exerted a more powerful influence on history than Galileo the man.

'Have faith, Galileo, and go forth': navigating a new world

Some people are born great. Some achieve greatness. Some have greatness thrust upon them. And some, like Galileo Galilei, manage to combine all three.

Born in Pisa in 1564, Galileo's brilliance was visible from early days. He studied first at the Benedictine monastery of Vallombrosa which, to his father's horror, he joined as a novice. Duly removed, he took to medicine at the University of Pisa although, never one to do as he was told, he soon veered off course, attended lectures on Euclid and was converted to mathematics.

Although history remembers him primarily as a pioneering astronomer, it was his talents as a mathematician and an engineer that shone first. He was a prodigious experimenter in an age before the experiment, although the story of him dropping weights from Pisa's tower to show they fell with uniform acceleration is probably his biographer's rather than his. He was a fine musician, inheriting his father's talents, and according to his biographer he was good enough to compete with the best lutenists in Tuscany. He was also a serious Renaissance man of letters, able to recite large tracts of Petrarch and Dante, and to discuss them with his poetic peers. And again according to his biographer, he was artistic enough to advise painters on matters of taste.

Endowed with so much talent, Galileo lacked only judgement, tact and modesty. Later biographers have tended less to hagiography than his first, and most now note how he never outgrew the habit of scoring points off people he judged wrong or stupider than himself, which was more or less everyone. By the reckoning of one contemporary, he preferred to lose a friend than an argument. It was a habit that helped him alienate, anger and arm his enemies. 'I have always had more love for him than he has for me,' complained Father Grassi, a young Jesuit mathematician who was called to give evidence at Galileo's first trial in 1616 and ended up a bitter adversary. 'I took the utmost care to allay minds harshly disposed toward him and to render them open to conviction of the strength of his arguments . . . But he has ruined himself by being so much in love with his own genius, and by having no respect for others. One should not wonder that everybody conspires to damn him.'[4]

Born with greatness, it was not until he turned his telescope to the sky in 1609 that Galileo first achieved it. Like Darwin, with whom he is eternally linked in the history of science and religion, he found scientific glory, rapidly followed by religious controversy, only in his late forties.

Experimenting with lenses the previous year, Dutch spectacle makers had created a device that magnified distant objects. At first, it was only by a factor of three but the device's accuracy improved and news of it spread. By the summer of 1609, Paolo Sarpi, monk, polymath, friend of Galileo and eternal thorn in the flesh of the Roman authorities, had obtained a device which he described to his friend.

By August, Galileo had constructed his own instrument with a nine-fold magnification and by early December he had managed a 30x device which he then trained on the moon. He saw, with unprecedented clarity, that the surface was pockmarked with what he interpreted as mountains and craters, though he rather exaggerated their size in his subsequent book in order to make his case more compelling. Either way, it looked nothing like the clear and perfect sphere Aristotelian physics said it was. He then turned the instrument on the Milky Way and observed it comprising innumerable, seemingly distinct stars and not, as Aristotle had claimed, a uniform fiery exhalation in the sublunary atmosphere. The following month he studied Jupiter and found a number of tiny stars orbiting the planet, which turned out to be the planet's moons.

Galileo was not always the first person to see these things. Four months before him, the English astronomer Thomas Harriot had surveyed the moon and noticed its rugged surface, but he had used a less powerful device and had left the maps he'd drawn scattered through his disordered papers. In contrast with Harriot and indeed his 'master' Copernicus, Galileo went straight into print with his *Sidereus nuncius*, or *Starry Messenger*, which came out in March 1610. Publish or perish.

There was more to follow. In the summer of 1610, Galileo began to observe Saturn, and mistook what would come to be known as its rings as two immobile satellite companions that never shifted from its side. Saturn, it appeared, was 'three-bodied'. He also studied Venus and found evidence for a piece of the jigsaw that had eluded Copernicans for seventy years, namely moon-like phases of the planet, which added further support for heliocentrism.

In early 1612, he was prompted to start looking at the sun by the Jesuit astronomer, mathematician and professor of Hebrew Christoph Scheiner. Scheiner had detected spots that appeared to be moving across the surface of the disc and posited that they were tiny stars orbiting the sun. Galileo was impressed by Scheiner's observations though not his conclusion, and suggested that the marks were clouds and that, more radically, the sun itself rotated as Copernicus had argued the planets did.

He published again, this time *Observations concerning Sunspots*, in 1612, now moving from observation and description to demonstration of the Copernican character of the cosmos. It was rocky terrain. The debate about Copernicanism had hardly moved forward in a generation and the same objections remained at large. Scientific: Why was no stellar parallax detectable? Physical: Why did bodies fall directly down on a moving planet? Philosophical: Where did all this leave the Aristotelian distinction between corruptible earth and the incorruptible heavens? Scriptural: What about Joshua's solar miracle? Theological: How, now, was the earth or were humans in any way special? Neither astronomy nor mathematics, among the humblest of the sciences, could be expected to move these mountains.

In as far as there had been movement, it was in the spread of Tycho Brahe's geo-heliocentric system, which had none of the elegance of the Ptolemaic nor the Copernican, but was demonstrably more accurate than the former, and demonstrably less demanding than the latter. In the last year of his life, Brahe had worked with a young German astronomer who was also slowly changing the script.

Johannes Kepler had originally had other plans for his life. 'I wanted to become a theologian,' he wrote to Michael Maestlin, his professor at Tubingen. Maestlin had recognised his brilliance and redirected his career heavenwards but he remained intensely devout, perhaps more so now he was set on another course. 'Behold how through my effort God is being celebrated in astronomy,' he wrote to Maestlin.[5] Studying the book of nature was, for Kepler, a form of prayer, a way of honouring the harmony, beauty and rationality of creator and creation. Made in God's image, humans were intended to know God from his 'corporeal works', from stars to snowflakes, on which he wrote a short pamphlet. 'It is my intention in this small treatise,' he began the preface of his first book, *Mysterium Cosmographicum* or *The Sacred Mystery of the Cosmos*, 'to show that the almighty and infinitely merciful God, when he created our moving world and determined the order of the celestial bodies, took as the basis for his construction the five regular solids [i.e. meaning the octahedron, icosahedron, dodecahedron, tetrahedron and the cube] ... and that he co-ordinated in accordance with their properties the number and proportion of the celestial bodies.'

Kepler's God was a cosmic mathematician, whose being could be read from (or into) the geometrical perfection of the cosmos. Astronomy was no exercise in hubris, he wrote in his *Astronomia nova* (1609) but a calling to faithful human beings. The end of all astronomical research, he reminded readers, is the 'knowledge, admiration and worship of the omniscient God'.

Kepler sent two copies of the *Mysterium* to Italy where one ended up, via an intermediary, in Galileo's hands. Galileo responded, thanking the author for being 'an ally in the search for truth'. He had, he confessed to his new friend, been a secret Copernican for some time – it is not clear since when – but he also admitted his unwillingness to come out 'as I've been deterred by the fate of our master Copernicus' having been 'ridiculed and derided'.[6]

Kepler was delighted at having found an ally. He was convinced that the movement that he and the still tiny band of heliocentric apostles were engaged upon was not only world-transforming but divinely ordained. 'God Himself has waited for 6,000 years for someone to study Him,' he reasoned, in an appealing narrative that Galileo would himself adopt. He tried to strengthen Galileo's resolve. 'Have faith, Galileo, and go forth.'

Galileo would not, at least not yet. Another supernova, in 1604, this one visible for eighteen months, picked yet another hole in the Aristotelian cosmos. Five years later, Kepler published *Astronomia nova*, which presented ever stronger mathematical arguments for heliocentrism, posited elliptical (as opposed to circular) orbits for the planets and offered theological arguments against the familiar scriptural objections. Still Galileo would not go forth.

This was neither cowardice nor premonition. The territory on to which Kepler urged him was precarious. However much universities may have once encouraged exotic thought experiments, the academic climate at the end of the sixteenth century was highly sensitive. Deep confessional lines now sliced through Europe, making academic culture defensive and conservative. The statutes of the University of Padua, where Galileo taught between 1592 and 1610, stated bluntly that 'all doctors under penalty of losing their lectureships are obliged to and must read and explain clearly and demonstrate the authors they are obliged to teach, word for word.'[7] Originality was not prized, let alone experimentation.

Then there was the Bible. Over and above the now well-voiced threat that heliocentrism seemed to pose to a handful of scriptural verses, there lurked the more subtle but more important question of whose right and responsibility it was to interpret the scriptures. Protestants claimed that the Vulgate, the fourth-century Latin translation that had long been authoritative within the Church, was error-strewn. The Catholic response, put forward at the Council of Trent, was that the original Hebrew text, which St Jerome had drawn on for his translation, and which was now lost, had been superior to the current Hebrew text, which had been corrupted over the centuries. The Council of Trent had stated baldly that in matters of faith and morals, 'no-one, relying on his own judgement . . . shall dare to interpret [the scriptures] according to his own conceptions.' The questions of textual reliability, compositional process, authorship and interpretative authority were intensely policed battle lines in the late sixteenth century. The idea that natural philosophers might be the ones to get it right, or to reinterpret seemingly self-evident texts in novel and creative ways, was as unthinkable as it was absurd.

In addition to all this, there was the challenge Copernicanism posed to the Aristotelian worldview to which the Catholic Church had self-consciously allied itself at the Council of Trent. The fact that observation was slowly undermining the model of the cosmos that Aristotle had drawn was tolerable. Indeed, the manner in which faithful Catholic astronomers shifted towards Brahe's system around the time demonstrated that Aristotle's physics was negotiable. The Greek, as leading Catholic thinkers increasingly recognised, had not known everything. But Aristotelian thought did provide the underlying metaphysical principles for much Catholic theology. In the Mass, for example, it was the substance of the bread and wine that was transformed into Christ's own body, while their accident – their physical nature and appearance – remained the same. The logic was Aristotelian. Pulling out a cosmological block here or there might not seem much in itself, but it could just send the whole structure to the ground.

Protestant kingdoms would have sung psalms of praise at such a prospect, which hardly helped matters. All threats to Catholic doctrine were potential Reformations: intellectual and possibly political invasions that merited an appropriately severe response. By the time Kepler was nudging Galileo out into the open, the Inquisition was a fact of life for everyone in northern Italy, 'a low level background terrorism' in John Heilbron's winning phrase.[8] The authorities spent much energy reprimanding, dismissing and sometimes imprisoning people who crossed theological lines, often mixing personal animosities in with doctrinal queries, in a merry-go-round of score settling.

Given these countless, submerged, hazards, it would have taken diplomatic craft of the most exquisite and agile sensitivity to navigate the waters, with a heliocentric cargo, and come out unscathed. Galileo Galilei was not a diplomat. Considering how delicate and complex everything was, it was remarkable that things didn't blow up sooner or more violently.

'Why stand ye gazing up into heaven?': rising star

Single observations should not change the world. 'I cannot understand why knowledgeable people want to make the [entire] heavens corruptible in order to be able to pronounce the nova a star,' reasoned Galileo's friend, the astronomer Guidobaldo del Monte, in the wake of 1604.[9] This wasn't simply obtuseness: philosophers of science still wrestle with the questions of how much, how significant, how replicable and how reliable empirical evidence needs to be to overturn an entire 'paradigm'. More than a single nova, del Monte thought. Supernovas were, after all, singular, often controversial and obviously unrepeatable events: once in a generation if you were lucky. A mountainous moon, Jovian satellites, three-bodied Saturn, phases of Venus and a rotating sun, all coming in a few years, were not so easily ignored.

Galileo was fêted. He was the sky's Columbus, discovering new worlds armed with no more than an 'Optic Tube'. He was a prophet of a new age, as his very name implied. Were not the disciples asked at the start of the book of Acts, 'Men of Galilee, why stand ye gazing to the heavens?' as they searched for their resurrected and now ascended Lord, before they went on to spread the good news. Kepler celebrated the verbal coincidence. Others would turn it to more critical ends.

Galileo received accolades for his *Sidereus nuncius*, and a folder of such praise – containing forty hexameters, four distiches and two epigrams, all in Latin, all by Jesuits – still survives.[10] Cardinal Francesco del Monte wrote to Grand Duke Cosimo II, 'Were we living in the ancient Roman Republic, I have no doubt that a statue would be erected in the Campidoglio in honour of his outstanding merit.'[11]

Flushed with such praise, Galileo arrived in Rome in Holy Week 1611 to curry favour with the people that mattered. Cardinal Bellarmine, who had effectively turned down the papacy in 1605 but remained a hugely influential figure, had commissioned a report from four eminent professors from the Jesuit College – Christopher Clavius, Christopher Grienberger, Odo van Maelcote and Gio Paolo Lembo – which affirmed all of Galileo's discoveries. The Milky Way was indeed made up of a vast number of stars, Saturn was oval in shape, Venus had phases, Jupiter had satellites. Their one small reservation, voiced by Clavius, was that while the moon certainly had a visibly irregular surface, they thought this might be on account of differences in density rather than height.

The report did not commit to any wider interpretations of these phenomena but nonetheless Galileo was riding high. Maelcote went on to deliver an oration praising him later in the year. Cardinal Bellarmine looked approvingly through his glass and was willing to concede that Aristotle had not discovered everything there was to discover in the heavens. Galileo was invited to join the Accademia dei Lincei – Academy of the Lynx – which was, in effect, the world's first learned society dedicated to natural philosophy. He was lauded by Pope Paul V and he befriended Cardinal Maffeo Barberini. Although he chafed at Clavius's lunar reservations – what was so special about sphericity after all? – he recognised the visit as a great success and crowed, in private correspondence, about how the Jesuits were now, finally, beginning to confess the truth.

Not everyone saw things his way. Within three months of the *Starry Messenger*'s publication, Martin Horky, an assistant to Giovanni Magini, a Bolognese professor of mathematics, had published *A Very Short Excursion Against the Starry Messenger*, which itself then earned responses from Galileo's allies. There was undoubtedly a conflict here, but that was simply because this was a new scientific idea. The process and progress of science, even from these earliest moments, was inherently conflictual.

In reality, Horky's objections had more to do with his myopia and the very real problems there were in getting the earliest telescopes to work as they should. A number of critics claimed that Galileo's observations were simply optical illusions, and that his lenses merely refracted, distorted and misled. Galileo tried to address these objections by enlisting the support of Cosimo de Medici to send new telescopes to the cardinals, courts and kings of Europe, so they could validate his discoveries. The gesture is less important for what it achieved than for what it signified – that the truth could now be ascertained by observation, by anyone, anywhere – and what this threatened to do to established structures of authority.

Other more substantial scientific objections remained – supremely the stubborn absence of a stellar parallax – alongside the scriptural, theological and philosophical objections, and these intensified as the implications of Galileo's observations became harder to avoid. Some problems were self-inflicted. The Lynxes offered to publish Galileo's letter on sunspots in 1612, for which they chose as an epigraph the scriptural text 'the kingdom of heaven suffers violence, and men of violence take it by force'. It was a breathtakingly stupid verse to use, calculated to provoke opponents who predictably complained that the Lynxes and their pet astronomer wanted to overthrow the theologians violently. The book encountered problems with the censors. Galileo had originally written how 'divine goodness' had directed him to go public with Copernicanism (the censors changed it to 'favourable winds'). He had pronounced that the immutable heavens were 'not only false, but erroneous and repugnant to the indubitable truths of the Scriptures'. The censors wondered how this layman mathematician could pronounce so authoritatively on the matter of biblical interpretation.

In truth, Galileo's published *Letters on Sunspots* concealed far more of his contempt than it revealed. His private correspondence shows how he frequently judged his opponents as incapable of following even the simplest argument. His Aristotelian critics were 'superficial and vulgar writers'. They 'simulated religious zeal'. Galileo, being Galileo, found it hard to keep such contempt private. The Jesuits, initially if tentatively supportive, were alienated by his gratuitously scathing attacks on some of their leading mathematicians in *The Assayer* in 1623. When writing his letter to the Grand Duchess Christina of Lorraine, who was then aged fifty, setting out his scriptural defence, he approvingly quotes the Church Father St Jerome on:

the chatty old woman, the doting old man, and the wordy sophist . . . [who] read [the Scriptures] in pieces and teach them before they have learned them . . . [and others who] philosophize concerning the sacred writings among weak women [while] others – I blush to say it – learn from women what they are to teach to men.

This was not the most tactful of lines to adopt with a duchess, although it did at least show that Galileo, like his critics, was indignant when the wrong people tried to interpret the Bible.

Galileo came to write to Christina in 1615 because, two years earlier, the formidable woman had found herself next to Benedetto Castelli, Galileo's pupil, friend and supporter, at breakfast where she had proceeded to interrogate him about these 'celestial novelties'. Calling him back to her rooms after the meal, now accompanied by the grand duke and four other eminences, she took him to task over his understanding of the scriptures. 'I behaved like a champion,' Castelli reassured his friend, 'despite the fact that the majesty of their Highnesses was enough to frighten me.'

Galileo wrote back a week later, encouraging his friend and articulating his own views. 'Though the Scripture cannot err,' he wrote, 'nevertheless some of its interpreters and expositors can.' Both scripture and nature were from God, he believed, but because the former said 'many things which are different from absolute truth, in appearance and in regard to the meaning of words', whereas the latter 'is inexorable and immutable', it was essential that, at least when it came to natural phenomena, the former be read in the light of the latter. 'Two truths can never contradict each other', he opined, but in this instance that meant 'the task of wise interpreters is to strive to find the true meanings of scriptural passages agreeing with those physical conclusions.'

Reasonable, sincere and emollient – at least by his standards – Galileo's letter still threatened to turn the world upside down. That scripture had to be interpreted had rarely been doubted, and certainly not by Catholics. Few thought, as Galileo put it, that God had 'feet, hands and eyes' or that he exhibited 'human feelings like anger, regret, [and] hate', or that he was 'forgetful of things past or ignorant of future ones'. Interpretation was not a problem; the question was by whom, according to what criteria and to what ends. Galileo's answer pivoted on the new things he had seen through his glass, around which all reality, including the scriptures, now had to orbit. Scripture was to be interpreted according to nature, which meant that theologians should, at least when thinking about the nature of the cosmos, take their cue from natural philosophers. As it happens, and as Galileo himself would point out at length a couple of years later, not even this was a wholly new idea. But it was still an unnerving one, coming after decades in which the question of who should interpret the Bible and on what grounds had run like a fault line through Christendom.

It was this that was liable to upset the authorities, more than any new astronomical observations or the implicit undermining of Aristotle, who was, despite his greatness, still only a pagan. As early as December 1611, a friend had written to Galileo informing him of a group of malicious opponents who were plotting against him at the house of the archbishop of Florence. Galileo had allies aplenty, including some in high ecclesiastical office. But a theological cloud over his work only grew after the publication of his *Letters on Sunspots*. 'Your actions are observed minutely,' warned a friend.[12] Public denouncements were still rare but then, on 21 December 1614, at the church of Santa Maria Novella in Florence, the star preacher Friar Tommaso Caccini delivered a scathing sermon accusing Galileo, and other mathematicians, of heresy for contradicting the scriptures and the Church Fathers. 'Ye men of Galilee, why stand ye gazing up into heaven?' he asked pointedly, with none of Kepler's admiration.

'How one goes to heaven and not how heaven goes': coming out writing

Caccini's sermon caused a scandal, to the extent that the Dominican General even wrote to Galileo to apologise. Nonetheless, the climate was cooling and Galileo began to worry that his letter to Castelli would get him in trouble. His letters, in particular the important ones, were often copied and circulated, and his letter to Castelli made it into the hands of a Dominican friar, Niccolò Lorini, who passed it to the Inquisition in Rome a few months after Caccini's sermon.

Panicked, Galileo had Castelli return the original to him and on 16 February 1615 he wrote to a friend in Rome, the minor Vatican official Piero Dini, complaining of the 'wickedness and ignorance' of his enemies, and suggesting that Lorini had doctored his letter. He enclosed a new version, which he claimed was the correct one, and he asked Dini to pass it on to the Inquisition.

Copies of this were also made but Galileo's original letter to Castelli was lost, and it subsequently became impossible for historians to know who was telling the truth, what was the original text and who had doctored what and to what effect. The truth emerged only in 2018 when Salvatore Ricciano, a postgraduate student from the University of Bergamo, was searching through the Royal Society archives in London and, remarkably, came across the original letter, complete with edits, which had somehow found its way to London but been misdated when archived.

The letter showed that Galileo had lied to Dini and softened his own version for fear that its tone would indict him. Thus, where he had written that certain propositions in the Bible were 'false if one goes by the literal meaning of the words', he replaced 'false' with 'look different from the truth'. Elsewhere, he had edited a reference to the scriptures as 'concealing' its most basic dogmas, replacing this with the more emollient 'veiling'. Galileo was forced into self-censorship.

Ironically, in this instance, he need not have been quite so anxious. The Inquisition examined Lorini's copy of the letter and found little to object to. Dini reported back that Bellarmine had informed him that he could not believe Copernicus's book was going to be prohibited and that the worst that might happen was that a note would be added to the effect that 'its doctrine is put forward to save appearances', rather than being a state-ment of demonstrable truth.

This frustrated Galileo, just as it had Copernicus when Osiander had suggested it seventy years earlier. Annoyed by Clavius's doubts about the moon, animated by the Grand Duchess Christina's grilling of Castelli and now worried about the turning tide of ecclesiastical opinion, he recog-nised the need to put forward his case in the best theological light possible.

Galileo solicited advice from theologians and drew on the work of the respected Carmelite and professor of theology Paolo Foscarini. Foscarini had just written a treatise arguing that Copernicanism 'agrees with, and is reconciled with the passages of Sacred Scripture which are commonly addressed against it', (to quote its subtitle), which he had sent to Bellarmine.

Bellarmine had replied cautiously, remarking that as a theory – 'speaking suppositionally' – Copernicanism was acceptable but that, as an affirmation of reality, it was 'a very dangerous thing', which was against the consensus of the Church Fathers and liable to 'irritate all scholastic philosophers and theologians'. However, he added that:

> If there were a true demonstration that the sun is at the centre of the world and the earth in the third sphere, and that the sun does not revolve around the earth but the earth against the sun, then we would have to use great care in explaining those passages of Scripture that seem contrary.[13]

The door was still ajar, just.

Galileo's resulting work was written as a letter (to avoid censors) and addressed to the same duchess whose interrogation of Castelli had provoked him. His arguments were substantially those of his letter to Castelli, only this time, crucially, justified with reference to an impressive range of theological sources, both historical – Tertullian, Jerome, Dionysius, Peter Lombard, Aquinas – and contemporary. Above all, his letter was almost a commentary on the most authoritative Church Father of them all, St Augustine.

Galileo references or quotes (usually at length) Augustine fourteen times, making particular use of his *Commentary on Genesis*, which we encountered in chapter 1. Taking aim at those who opposed him – professors who 'showed greater affection for their own opinions than for the true ones' and whose writings were 'full of useless discussions sprinkled with quotations from the Holy Scriptures, from passages which they did not understand' – he advised that they would have done well to listen to Augustine.[14] The Church Father had written, 'with greater prudence', that the authors of holy scripture had more or less ignored natural philosophy.[15] These were not unimportant issues, Augustine had explained – 'these things should be examined with very subtle and demanding arguments' – but they were of 'no use for eternal life' and of 'little benefit' to the Holy Church. Augustine recognised that even pagan philosophers could acquire 'conclusive reasons or observations' about the world. He pointed out that the scriptures were open to differing interpretations. The Bible was accommodated to the understanding of the 'common people', and should not always be read literally. He reasoned that when philosophers could 'truly demonstrate something about natural phenomena', it was incumbent upon theologians to show it as compatible with the scriptures, rather than the other way round. And he lamented how shameful it was when pagans heard Christians holding forth on matters ignorantly, thereby undermining anything else they had to say about the faith. In short, this most venerable of Church Fathers supported Galileo's key contention that the interpretation of the scriptures must be open to what science has to say about the nature of the world. By his reckoning, the two were compatible with one another. As Galileo famously put it, in fact referencing the eminent Cardinal Baronio, 'the intention of the Holy Spirit is to teach us how one goes to heaven and not how heaven goes.'[16]

Contrary to the myth that would later pit him against the Church and so see little more than subterfuge and self-preservation in such sentiments, Galileo was both faithful and sincere in all this. He endured smears aplenty against his orthodoxy, even before he achieved greatness. Friendships with people like Paolo Sarpi and Cesare Cremonini, a Paduan professor often in trouble with the Inquisition, spread whispers. In 1604, one Silvestro Pagnoni, who had worked as a copyist for Galileo for eighteen months, claimed that he had seen him go only once to Mass and never to confession.[17] The authorities investigated and dismissed the accusations as 'frivolous' and essentially spiteful.

Galileo was not, it seems, especially pious but he was no sceptic, let alone a heretic. 'I know him well', wrote his patron Cosimo to a cardinal, and 'he is a good man, very observant and pious in religion.'[18] Well, perhaps not 'very' but pious enough and with a genuine and motivating faith. I thank God 'from the bottom of my heart', he wrote after his first lunar discoveries, 'that he has pleased to make me the sole initial observer of so many astounding things, concealed for all these ages.'[19] The sentiment exemplifies how powerfully Galileo believed, both in God and in Galileo.

That being so, he recognised, as Augustine had done twelve hundred years earlier, that if the Church were publicly to denounce that which increasingly seemed to be demonstrable and inarguable, it would do no more than provoke laughter among true sceptics. He constructed his argument to the grand duchess not only to save his skin – though that was no doubt a major consideration – but because, knowing Copernicanism to be true, he wanted to save the Church from an act of humiliating self-harm.

In the end, he decided not to send the letter. It was probably a wise move. For all that he had managed to deploy the Augustinian big guns in his favour, his framing rhetorical device – of treating Christina as the primary audience to whom he complained about his ignorant and hypocritical critics – inevitably meant corralling them into an enemy camp in which not all belonged. Not everyone who opposed Galileo was as ignorant or hypocritical – or, indeed, as opposed – as he assumed.

One year and fifteen thousand words of unpublished epistolary prose later, however, and he was still left with the same problem as when Caccini had preached against him. Worse, in fact. Caccini had given a formal deposition to the Inquisition in March 1615, accusing Galileo of heresy, and another Dominican, Ferdinando Ximenes, claimed before the Inquisition in November that Galileo and his followers did indeed hold certain heretical ideas. Eventually, the Inquisition decided to examine the *Letters on Sunspots*. In December 1615, after a long illness, Galileo travelled to Rome to try to clear his name and prevent the Inquisition from condemning Copernicanism. The visit was somewhat less successful than his first.

On 19 February 1616, the Inquisition commissioned a committee of eleven to assess two Copernican propositions. Five days later, it reported back, unanimously. The idea that 'the sun is the centre of the world and completely devoid of local motion' was, the committee judged, 'foolish and absurd in philosophy' and 'formally heretical' in as far as it contradicted the literal and familiar meaning of the scriptures. The attendant idea, that 'the earth is not the centre of the world, nor motionless, but it moves as a whole' was similarly 'absurd', philosophically speaking, although only 'at least erroneous in faith'. The following day, 25 February, Pope Paul instructed Bellarmine to warn Galileo to abandon his views. Should he refuse to do so, he was to be issued with an injunction 'to abstain completely from teaching or defending this doctrine and opinion or from discussing it'. Should he reject this, he was to be imprisoned. Bellarmine complied the following day. Galileo straightaway acquiesced.

At the same time, the cardinal members of the Inquisition met to decide whether and which books should be placed on the Index Librorum

Prohibitorum, the list of works prohibited by the Church. Two of them, one being Galileo's friend from 1611, Maffeo Barberini, managed to secure agreement for a weaker censure. In light of its mathematical brilliance, Copernicus's book was suspended 'until corrected', rather than banned. Diego de Zuniga's *Commentary on Job* suffered the same fate. Foscarini's book was banned altogether but Galileo's publications were not mentioned. Had he ever published his letter to Christina, they would have been.

Galileo wrote the Tuscan Secretary of State a surprisingly upbeat letter. Caccini's attempt to label him a heretic had failed. Copernicus's book was to be edited only slightly. 'I am not mentioned', he wrote, in the Index. Indeed, he would not have been involved at all, he went on, had his enemies 'who have not refrained from any machination, calumny and diabolical suggestion' not 'dragged me into it'.[20] A week later, he sounded similarly positive, recounting a 'warm audience' he had been granted with the pope the previous day, the two having 'strolled and reasoned' together for three-quarters of an hour. Galileo had complained about the maliciousness of his enemies. The pope reassured him that he was 'aware of my integrity and sincerity'. Galileo had fretted that he would heretofore always be on the watch list. The pope 'consoled' him that he should live with his mind in peace and that, as pontiff, he 'was very ready at every occasion to show me with actions his strong inclination to favour me'. Such warm sentiments were not just wishful thinking on Galileo's part. When, in the spring, he heard rumours to the effect that he had personally been put on trial and condemned by the Inquisition, he asked Bellarmine to intervene personally and write a declaration on his behalf, denying them. The cardinal consented and gave Galileo what he mistakenly thought was a get-out-of-jail-free card.

The events of 1616 were not, then, a personal catastrophe for Galileo but they did mark an ominous step for his Church, all the more tragic for being unnecessary. The misfortune, according to one modern biographer, was that Pope Paul, Bellarmine and the Inquisition believed that they had had to act. 'Galileo's attempt to set up an independent school of cosmology and biblical criticism looked like the budding of a new head of the Protestant hydra.'[21] To have remained silent, not least when Caccini and others were whipping up fear of new heresies, risked further harm to an already wounded body. The Church's official opposition to Copernicanism in 1616 remained what it had always been – philosophical and theological – and although it had shown signs of softening in the light of Galileo's new discoveries, the resistance remained principled. Its actions, however, were pragmatic and political.

'Who would there be to settle our controversies if Aristotle were deposed?': turning tides

The events of 1616 had a freezing effect on any realist interpretations of Copernicanism in Catholic lands. When, the following year, a young Barnabite astronomer, Redento Baranzano, took it upon himself to defend Copernicanism, he was rebuked by his Order and compelled to deny his views in a subsequent pamphlet. This was a line not to be crossed.

It did not, however, mean the end of scientific study, either for Galileo or other Catholic philosophers. A comet in 1618 fuelled further speculation and caused the leading Jesuit mathematician Orazio Grassi, having observed the parallax, to place the comet beyond the moon without hesitation. This did not impress Galileo, who believed comets were sublunar phenomena, mere atmospheric illusions, and in his *Discourse on Comets*, Galileo ridiculed Grassi, insulted his fellow Jesuit mathematician Christopher Scheiner and mocked the geo-heliocentric model of Tycho Brahe, about whom Galileo was almost as rude.

The argument, and the style, continued in his ensuing and more substantial *The Assayer*, in which Galileo further attacked Grassi, giving him the pseudonym Lotharia Sarsi. 'Sarsi' had implied that Galileo's arguments were thinly veiled Copernicanism, an allegation Galileo could not let stand. His tone of sarcastic superiority was at its most practised. He cited how Sarsi had argued, on ancient textual authority, that the Babylonians had cooked eggs by whirling them in slingshots and had used this, as in his case, against the earth's motion. Galileo was not persuaded. He had tried to repeat the endeavour but with no success. Something must be missing. 'Now we do not lack eggs, nor slings, nor sturdy fellows to whirl them', he reasoned, and 'yet our eggs do not cook'. There could be only one conclusion. 'Since nothing is lacking to us except being Babylonians, then being Babylonians is the cause of the hardening of eggs.'[22]

Beneath the mockery was a serious point. 'I cannot but be astonished that Sarsi should persist in trying to prove by means of witnesses something that I may see for myself at any time by means of experiment,' Galileo lamented. 'How can he prefer to believe things related by other men as having happened two thousand years ago in Babylon rather than present events which he himself experiences?' Experience – experiment – was a better guide to truth about the natural world than textual authority.

Such an approach was strengthened by the argument for which *The Assayer* is most famous, namely that the universe, continually open to our gaze, is written in the 'language of mathematics'. Its characters are 'triangles, circles, and other geometric figures'. Only those expert in this tongue can read and understand it. Once again, authority slid away from the Word.

By the time *The Assayer* was published in 1623, Galileo was feeling emboldened. Paul V had died in 1621 and his successor, the first Jesuit pope, Gregory XV, occupied the throne of Peter for only two years. In his place, the conclave elected Maffeo Barberini, who took the name Urban VIII. Barberini was 'one of them'. Although not a member of the Lynxes, he was considered a fellow traveller. He had been a friend of Galileo's for over a decade and a moderating influence through the events of 1616. He was a keen poet, although more applauded than talented, his poems being considered 'desolately empty of any poetic inspiration' by one modern critic.[23] His 1620 volume included verses on Mary Magdalene, Saint Louis IX and, improbably, Galileo, to whom he sent the book with a warm inscription. Galileo reciprocated by dedicating *The Assayer* to Urban, and presenting him with a copy, which the pope read with pleasure, having no great love of Jesuits.

Things boded well. Galileo travelled to Rome again, a year after Urban's election, where he was welcomed and given six papal audiences, two medals and the promise of a pension for his son. He also befriended the German cardinal Frederic Zollern, who took the Copernican case to Urban, arguing that German Protestants held it true and mocking Catholics for their lack of understanding. Urban replied, to the letter of the law, that Copernicanism had never been judged heretical and would not be so, and that there was no reason to assume that a proof would never be forthcoming.

Urban did not give permission for Galileo to reopen the Copernican question but the climate was clearly thawing. Galileo returned home and began work on a book that would, eight years later, become his *Dialogue Concerning the Two Chief World Systems*. Shortly after he returned, there was an anonymous attack on *The Assayer*, accusing it of promoting an 'atomism' that erased the distinction between 'substance' and 'accident' and therefore being incompatible with the doctrine of the Mass. The Inquisition investigated the book and found it to be unobjectionable. Its author was fully exonerated. More promising signs. In 1630, Galileo's old friend Castelli wrote to him reporting a story from Father Campanella, a Dominican philosopher and keen Copernican, who had written a defence of Galileo in 1616 but withheld its publication until 1622. Campanella informed Urban about an opportunity he had recently had to convert some German Protestants to the true faith. Success beckoned until the Germans heard about the prohibition on Copernicanism. They were 'scandalised' and unable to proceed. 'It was never our intention,' Urban responded, 'and if it had been up to us that decree would not have been issued.'

By this time Galileo had completed his book. Originally specifically about tidal motions, a subject with important ramifications for the movement of the earth, Galileo prefixed a lengthy 'dialogue' (technically a trialogue) about 'the' two world systems, the Copernican heliocentric and the Ptolemaic/Aristotelian geocentric. It was a sleight of hand. The discoveries of the last twenty years had clawed holes in the latter system and ever fewer people adhered to it in its pure form. More popular was the compromise geo-heliocentric system of Tycho Brahe, which had now formally been adopted by the Jesuits, and the rather more accurate elliptical heliocentric system of Johannes Kepler. Galileo overlooked both of these, by being disrespectful to Brahe in the *Dialogue* (for which Kepler berated him) and ignoring Kepler altogether, even though his calculations now offered the best case for heliocentrism.

The book's discussion took place between three philosophical figures. Salviati was named after a friend and patron, and was Galileo's alter ego, a clear but cautious advocate for Copernicanism. Sagredo was named after another friend and was similarly intelligent but as yet undecided on the matter. Simplicio was supposedly named after the sixth-century philosopher Simplicius of Cilicia, but the name essentially just meant 'simpleton', and the figure is really an amalgam of contemporary Aristotelian philo-sophical arguments, some sounding dangerously like Pope Urban.

Simplicio did not come out well. 'Who would there be to settle our controversies if Aristotle were to be deposed?' he asks timorously on Day 1. To be sure, he was not ridiculed as relentlessly as Sarsi in *The Assayer* – his companions even exempting him from the general condemnation of wordy philosophers on Day 3 – but he was at best simply a foil, and at worst a figure of fun. Galileo couldn't help himself.

He travelled to Rome in spring 1630 to secure permission to publish and enjoyed a long audience with Urban in mid-May. Urban was a friend but also a politician under pressure. Temperamentally vain and unstable, he was paranoid about his position, and particularly about ominous astro-logical predictions. (He issued a bull in 1631 against divination, particu-larly those that foretold the death of popes.) For all that he had apparently said to Father Campanella, he still had reservations aplenty with Copernicanism.

As it happened, the friends' meeting went well. Galileo's manuscript was received positively and provoked no major objections. He left Rome feeling esteemed and honoured. He had failed to obtain any formal impri-matur but that summer he received a written endorsement from the Vatican secretary and chief censor. If certain conditions were met – pertaining to the book's title, content, preface and ending – the book would receive papal assent. The conditions effectively rebalanced the dialogue, softened its defence of Copernicanism and made it clear that this was all still in the realm of theory. 'It must be made clear that the purpose of the work is solely to demonstrate that the Catholic Church knows all the reasons that bear on the question, and that it was not for not knowing them that the Copernican opinion was banned in Rome.'[24] Perhaps most importantly, the conditions included the request to recog-nise, in a postscript, that God, in his omnipotence, could treat the world in any way he chose and was not constrained by what humans observed or calculated. It was the thirteenth-century debate about Aristotelian science all over again, only this time with a new scientific paradigm in view.

All this was a big ask – or rather a big demand – but Galileo was willing to be flexible now that the victory of a reopened discussion on Copernicus appeared to be in sight. He returned to Florence and revised the text.

There were delays. In the summer of 1630, an outbreak of the plague paralysed communication in Italy. There were wrangles over the edits and formal Vatican approval. Eventually, frustrated, Galileo negotiated to have the volume published in Florence. Castelli advised him that this would be acceptable providing he could get Urban's approval, but warned that the pope was fickle at the time. Within a fortnight, however, the long-awaited permission arrived, in spite of the fact that no censor had approved the final text. 'The book will not encounter any obstacle here in Rome,' he was assured by the chief censor in Rome, providing he had followed the instructions given.[25]

The book went to print in June 1631 and became available in February the following year. Galileo had met the conditions, changing the title and explaining in the preface how he was writing to defend the Church from the charge of ignorance of heliocentrism. He also added the requisite postscript in which he said it was wrong to limit or compel divine power and wisdom, and that God could give the tides their motion (a topic that the book's protagonists had discussed at length) 'in many ways, some of them even inconceivable to our intellect'.

Astonishingly, however, he put this sentiment – to which Urban was so fiercely attached – in the mouth of the Simpleton and then, in case the association was lost on readers, had him remark: 'I once learned [this] from a man of great knowledge and eminence.' Galileo could simply not allow his own Salviati to utter the words. He would insist, in his trial the following year, that he had done nothing wrong. To the brittle, volatile and vengeful Urban, however, it read like a deliberate public humiliation.

The book arrived in Rome in May 1632 and immediately provoked rumours. Galileo had many enemies who had long waited for him to stumble. Urban was furious at having been misled, as he saw it, by those who should have recognised the book's provocation and censored it, and he was incandescent at the betrayal by his, by now former, friend. When the Inquisition unearthed the document from sixteen years earlier, forbidding Galileo from holding, teaching or defending Copernicanism, the die was cast. He was summoned. Old, ill and now afraid, Galileo asked that the trial be switched to Florence or that he might offer written testimony. The Inquisition, over which Urban personally presided, refused. Three Florentine physicians signed a statement confirming that Galileo was too ill to travel. The Inquisition dismissed their objection. In January 1633, Galileo made his will and set out for Rome.

'I resorted to the natural gratification everyone feels for his own subtleties': endgame

Galileo finally had greatness thrust upon him in 1633. He lived at the Tuscan embassy in Rome, in isolation, for two months after he arrived, while a panel of three theologians read and passed judgement on the *Dialogue*. Not surprisingly, none had any doubt that Galileo held and defended Copernicanism in the book. On 12 April, he appeared before the Inquisition, accused of contravening the order of 1616.

Galileo produced Cardinal Bellarmine's certificate from 1616 defending him, and then wriggled before his inquisitors. Cardinal Bellarmine 'told me that Copernicus' opinion could be held suppositionally, as Copernicus himself had'. He was asked about the formal prohibition. 'It may be that I was given an injunction not to hold or defend the said opinion, but I do not recall it.' The Inquisitors presented him with the written injunction. It helped jog his memory. Galileo did now recall that it forbade him to hold, defend or teach the doctrine, but 'I do not recall . . . that there was the phrase *in any way whatever*, but maybe there was.' He was asked whether he had obtained official permission to write his book in the light of this ban. 'I did not . . . because I do not think that by writing this book I was contradicting . . . the injunction.' The book had, however, been licensed by the Master of the Sacred Palace. Had he informed the Master about the 1616 injunction? 'I did not . . . because I did not judge it necessary.' Had the Master reviewed the final text? Frustratingly not; the plague had intervened. Galileo had not sent it. The manuscript 'certainly would have been damaged, washed out, or burned, such was the strictness of the borders'. It had, however, been reviewed by the relevant authorities in Florence. And with that, his first appearance before the Inquisition ended and he was – unusually – detained in the luxury of the chief prosecutors' rooms, with his own servant, for the rest of the month.

The Chair of the Inquisition asked for permission to deal with Galileo 'extrajudicially' in order to extract a confession. It was, it appeared, unnecessary. At the end of the month, Galileo made a second deposition.

Reflecting on his earlier exchange with the Inquisition, he admitted he had gone too far with the book. The arguments, especially about sunspots and tides, were unduly strong. 'I resorted . . . to the natural gratification everyone feels for his own subtleties and for showing himself to be cleverer than the average man.' He recanted. 'I neither did hold nor do hold as true the condemned opinion of the earth's motion and the sun's stability.'

Ten days later he made a further statement in his defence. He returned to his poor recollection of the initial injunction and to Bellarmine's certificate as a way of explaining his failure to notify the Master of the Sacred Palace as he should have done. From this, he pleaded, 'I think I can firmly [discredit] the idea of my having knowingly and willingly disobeyed the orders given me.' He acted 'not . . . through the cunning of an insincere intention but rather through vain ambition.' I implore you, Most Eminent Lordships, he ended, 'to consider the pitiable state of ill health to which I am reduced'.

The Lordships did indeed mention this in their final report to Urban in early June but it hardly mattered in a document that, by wrapping up the interrogations with other evidence, such as from the *Letter to Christina* that had somehow made it into their hands, made it look as if Galileo had been violating doctrine and authority for twenty years. Urban ordered him to be interrogated a final time and, if he answered satisfactorily, he was to abjure before the Congregation of the Holy Office, clearing himself from the 'vehement suspicion of heresy'. 'For my part I conclude,' Galileo ended his fourth deposition, 'that I do not hold, and after the determination of the authorities, I have not held the condemned opinion.' It was enough. No further physical persuasion was required. Galileo was sentenced on 22 June. Kneeling down before his judges and witnesses, holding a lit candle in his hand, he read out a pre-prepared statement and then left, without another word.

The events of 1633 shook Galileo but did not break him, and he continued his scientific work. He stayed with the archbishop of Siena for six months, and began a new work on mechanics. The story that he uttered the words *eppur si muove* – 'and still it moves' – as he left his Roman inquisitors seems to originate with this archbishop and a picture, representing the scene of his humiliation erroneously in a dungeon, which displayed the famous words. The words and setting suggest that the mythologising began immediately.

Milton was the first to make a martyr of the man but his disciples in Italy shared his sentiments. His death in 1642 received little public attention, and although the Duke of Tuscany had wanted to bury him in the Church of Santa Croce, with an elaborate statue in his honour, the adamant and vindictive Urban VIII refused, on the basis that he had been condemned on the charge of 'vehement suspicion of heresy'. His final resting place was in a side chapel.

Or not quite final. The body was reburied in the main body of the church in 1737. By that time, his reputation was recovering. His disciple, Vincenzo Viviani, had begun Galileo's biography-cum-hagiography in 1654, but this was not published until 1717. It left the reader in no doubt of his greatness – 'Nature chose Galileo as one who should reveal part of those secrets' – or of the epochal battle he fought: 'Galileo showed himself to be more adherent to the Copernican hypothesis which had been condemned by the Holy Church as repugnant to the Divine Scripture.'[26]

Viviani also pursued the Duke of Tuscany for a commemorative statue, and was eventually rewarded, after his death, with a bust above Galileo's house in 1693, and a statue over his new grave in 1737. But Galileo's work continued to draw less attention than his life. Newton praised him in his *Principia Mathematica* but his own mathematics and astronomy clearly surpassed those of the man on whose shoulders he stood. By contrast, the emerging anti-clericalism of the early Enlightenment found in Galileo a wondrous icon. 'Every inquisitor ought to be overwhelmed by a feeling of shame in the deepest recesses of his soul at the very sight of one of the spheres of Copernicus.' So thought Voltaire, seventeen years after Galileo's reburial, and so would think a host of anti-clerical, deist, sceptical, rationalist and eventually atheist thinkers who would follow in his wake.

Florin Périer measuring mercury in a barometer at the top of the Puy de Dôme in 1648. Given Catholicism's later anti-science reputation, it is ironic that the world's first formal experiment was based in the gardens of a monastery, drew on the help of its monks, was conducted by the Catholic Florin Périer, requested by the devout Blaise Pascal, and suggested (at least according to his own claims) by René Descartes.

SEVEN

THE MANY BIRTHS OF SCIENCE

'Experimental Christians': 'Protestant' science

Science is revelatory. It delights to lay itself open to enquiry. It is not satisfied with its conclusions. Its whole aim is to illuminate every dark recess, to improve everything on rational principles. But it is not easy or natural. The Greeks achieved it in some measure or, at least, Archimedes did. But then it was eclipsed for nearly eighteen centuries 'till Galileo in Italy and Bacon in England at once dispelled the darkness.[1]

Such, at least, was the view of Sir John Herschel, astronomer and polymath, in his *Preliminary Discourse on the Study of Natural Philosophy*. Published in 1831, the book was a milestone in the philosophy of science, setting out formally the ideas of observation and theory, the lawfulness of nature and the significance of mathematics, and above all the principle and methods of inductive reasoning that were essential to the practice of science. It proved influential. Twenty-eight years later, Charles Darwin had his publisher send an advance copy of *The Origin of Species* to Herschel, to whom he wrote: 'I cannot resist the temptation of showing . . . my respect, & the deep obligation, which I owe to your *Introduction to Natural Philosophy*. Scarcely anything in my life made so deep an impression on me.'[2]

But the book was also important for the light it shone backwards, to the origins of science, as well as forwards, to its progress. For Herschel, as for so many of his countrymen, Francis Bacon stood alongside Galileo at the threshold of science. 'It is to our immortal countryman Bacon that we owe the broad announcement of this grand and fertile principle [of induction]', he began chapter 3. Bacon will be 'justly looked upon in all future ages as one of the great reformers of philosophy'.[3] Indeed, before his *Novum Organum*, 'natural philosophy . . . could hardly be said to exist'.[4]

Herschel acknowledged that Bacon's scientific achievements were nugatory – 'his own actual contributions to the stock of physical truths were small' – but his followers 'ransacked all nature for new and surprising facts'.[5] Or rather the 'followers of Bacon and Galileo': for Herschel, like many of his (Protestant) countrymen, saw in Bacon a Protestant legitimation of the entire scientific enterprise, just as he saw in the fate of Galileo a (Catholic) delegitimation of the same endeavour. Science had deep theological roots – or, as Protestant historians would have it, deep Protestant roots – and that meant it was caught up in the rhetoric of religious warfare from its earliest days.

Bacon was no Galileo. He was not a scientist. He was not even a natural philosopher. A lawyer and a long-serving Member of Parliament who rose to the heights of Lord Chancellor, he remained financially and politically vulnerable throughout his life: imprisoned for debt, charged with corruption and finally dragged from office by parliamentary intrigue in 1621. Natural philosophy was for him a personal intellectual passion, a refuge from political storms. 'I have vast contemplative ends,' he confessed to his uncle shortly after he turned thirty. 'I have taken all knowledge to be my province.'[6]

There was plentiful theological justification for such a passion, and Bacon's writings were saturated with biblical quotations and theological arguments. For all that some Church Fathers had counselled against the study of nature, the idea that knowing God's works was a way of knowing Him Himself had an even longer history. In his letter to the Romans, St Paul had baldly stated that God's 'invisible' nature could be discerned through 'the things that have been made'. It was an idea that Bacon shared. 'The Psalms and other Scriptures do often invite us to consider and magnify the great and wonderful works of God,' he reasoned in *The Advancement of Learning*, but they did more than that. 'If we should rest only in the contemplation of the exterior of them,' Bacon reasoned, 'we should do a like injury unto the majesty of God as if we should judge or construe of the store of some excellent jeweller by that only which is set out toward the street in his shop.' George Orwell once described the

Christian idea of heaven as choir practice in a jeweller's shop. For Bacon, it was the earth that was the jeweller's shop, its true brilliance comprehensible only from within.

There were good reasons to believe the shop was an ordered one. 'The Lord by wisdom hath founded the earth; by understanding hath he established the heavens,' proclaimed Proverbs 3:19. There was an underlying structure to the natural world, in spite of its confusing diversity. Bacon concurred. 'In the work of the creation we see a double emanation of virtue from God,' he wrote. God's power was responsible for the very existence of things, and his wisdom for the beauty of their form. Rather than a confusion of 'mass and matter', creation was orderly. Chaos was overcome through God's six methodical days of creation. The resulting heavens and earth bore 'the style . . . of a law, decree, or counsel'.[7]

Furthermore, there was reason to believe that human beings might – at least in theory – be able to comprehend this order. Mankind was the climax and glory of God's creation and shared something of God's wisdom and knowledge. After all, mankind was given the privilege of naming the creatures in Genesis 2: 'naming' implying knowledge implying understanding. 'God hath framed the mind of man as a mirror or glass, capable of the image of the universal world,' Bacon reckoned; humans 'not only delight[ing] in beholding the variety of things', but were 'raised to find out and discern the ordinances and decrees' of creation.

There was, therefore, a range of powerful theological arguments that could be deployed to catalyse and justify the study of nature. But they were not uniquely Baconian or even Protestant, and were heard just as commonly in Catholic countries. Indeed, the conviction that human reason was capable of accessing the truths of creation was *stronger* in Catholic thought than in Protestant. Why then, other than national chauvinism, did Herschel pin the honours for leading humanity from the darkness of ignorance so firmly on Bacon? Why was Bacon judged the father of the Scientific Revolution?

One has to be careful when answering this question. Some historians have cast doubt on whether there was such a thing as a Scientific Revolution. Steven Shapin famously began a book on the topic with the line, 'There was no such thing as the Scientific Revolution, and this is a book about it.' Others of a less sceptical bent have wondered whether we should talk about scientific revolutions rather than a, still less *the*, Scientific Revolution. Still others have recognised the reality of such a revolution but explained it through wholly non-religious ideas, such as the discovery of the New World or the growth of commerce.

Even were one determined to award scientific laurels for Christian theology, the mere existence of Galileo – not to mention Pierre Gassendi, Marin Mersenne, René Descartes, Blaise Pascal, Nicolaus Steno and others we shall encounter in this chapter and the next – advises against placing them on only Protestant foreheads. Catholics did science too, and some of them did it extremely well. In short, we should not claim (Protestant) theology as the sole or sufficient engine for the Scientific Revolution.

Acknowledging this, however, is not the same as saying that theological, indeed quintessentially Protestant, ideas played no significant role in 'the' Scientific Revolution. They did, and three in particular are worth highlighting.[8]

First, in place of the twin-track Christianity of the Middle Ages, in which the noblest way anyone could honour God was through the contemplative or religious life, the Protestant reformers placed a new emphasis on the ability of all believers to honour their creator through their daily activities, however manual or menial. Work was newly sanctified. This idea was famously picked up as an explanation for the origins of capitalism by the German sociologist Max Weber and he, in turn, inspired the American historian Robert Merton, whose ground-breaking doctoral thesis in 1936 proposed a link between Protestantism and the development of experimental science. Although the thesis was subsequently critiqued for claiming too much for reformation theology, it still has something to recommend it. The secrets of creation were not best discovered in a monastery's scriptorium. The acquisition of *scientia* could be a practical affair. Mankind required, in Bacon's words, 'inquiry *and* invention'. 'In a society where a sharp line was drawn between gentlemen, who had soft hands, and craftsmen and labourers, whose hands were hard,' writes the historian David Wootton, 'Bacon insisted that effective knowledge would require co-operation between gentlemen and craftsmen, between book-learning and workshop experience' – and this co-operation was made easier once the work of the workshop had, in effect, been baptised.[9]

This was not to claim that libraries or books were redundant. The ancient metaphor of God's two books – scripture and nature – remained a controlling one for much scientific work in the early modern period. Bacon used it powerfully. 'Our Saviour saith, "You err, not knowing the Scriptures, nor the power of God",' he noted in *The Advancement of Learning*. In so doing, Jesus was laying before the people 'two books or volumes to study . . . the Scriptures, revealing the will of God, and then the creatures expressing His power'.[10]

The characteristically literal Protestant approach to the text was to change the impact of this 'two books' metaphor. From at least the third century, the Bible had been read on a number of distinct levels. The words could be read literally and they were – almost no one doubted the historical reliability of the scriptures. However, this was often the least important or interesting approach to the Bible. More meaningful, more profound, more spiritual was a moral reading, which told you how you should act, or an allegorical reading, which told you what you should believe, or an anagogical reading, which told you what you should hope for.

As with the book of scripture, so with the book of nature, where the objects also assumed complex moral, allegorical and anagogical meanings. 'This whole sensible world,' wrote Hugh of St Victor in the twelfth century, 'is like a kind of book written by the finger of God . . . and each particular creature is somewhat like a figure . . . instituted by the divine will to manifest the invisible things of God's wisdom.' None of this proscribed natural philosophy, as we saw in Part 1, but it did mean that nature, like scripture, was best and most fully understood symbolically.

Protestant reformers rebelled against this entire fourfold structure, the second change that led to the Scientific Revolution. They declared it not only inadequate and misleading, but downright iniquitous, reading into the pure, clear words of scripture all manner of false Catholic doctrines. They insisted not only on *sola scriptura* – scripture alone – but also on a solely literal reading of scripture (at least in theory; they soon discovered a certain flexibility when it became necessary).

In time, such a literal approach to one book would filter through to the other. Nature was stripped of the rich, layered garments of symbolic resonance and meaning that it had acquired in medieval theology, and stood naked before the gaze of philosophers. Those who ventured into its jewellery shop were mandated to find only jewels, not heavenly lights. 'Let no man . . . think . . . that a man can search too far, or be too well studied in the book of God's word, or in the book of God's works,' Bacon wrote in one of his most ringing endorsements of science. 'Let men endeavour an endless progress or proficiency in both.' But there was a proviso. 'Only let men beware . . . that they do not unwisely mingle or confound these learnings together.'[11] God's book of works, like his book of words, should be understood on its own terms. In effect, a literal reading of scripture helped generate a natural reading of nature.

Third, and most important of all, Protestant influence on the formation of science lay in the return to Adam. Humans were created to know. Made in God's image, and entitled to name the rest of creation, Adam's senses were acute, his knowledge was encyclopedic, his understanding complete and his command over nature total. Adam was mankind as he was meant to be, in mind as well as morals. This had all been disrupted by the Fall, the marks of which were as much epistemological as they were spiritual. Adam's capacity for knowledge was weakened and his control over nature was lost.

Opinions differed on how far. The Church Fathers, and the Orthodox tradition in their wake, took a relatively lenient view of the effect of Adam's transgression. The Fall was a matter of regret rather than despair. Latin Christendom was generally more severe, taking from Augustine the idea of human corruption and depravity. This was tempered somewhat in the later Middle Ages, when scholastics argued that humanity had lost its supernatural rather than its natural gifts at the Fall, and our capacity for reasoning remained more or less still intact. But the reformers re-appropriated Augustine's bleak assessment of human nature and the effects of the Fall. Just as humans could do nothing to effect their own redemption but had to rely entirely and without question on the grace of God, so their capacity for reasoning their way to the truth was extinguished. Living in the long shadow of Adam's fall, neither our senses nor our minds were sufficiently reliable. Natural reason was snuffed out every bit as completely as supernatural.

There were some Protestant preachers who, as a result of this, denounced all efforts at natural knowledge as a futile distraction or, worse, a perilous abandonment of spiritual health. Others, however, made a virtue out of this painful necessity. Bacon was unambiguous about the severity of the problem. 'The root cause of nearly all evils in sciences', he claimed, is that humans 'falsely admire and extol the powers of the human mind'.[12] This, however, induced modesty and inquisitiveness in him, rather than despair and resignation. 'Those given to effusive praise of human nature . . . rest content in admiring what they already possess and regard the science cultivated among them as perfect and complete', he reasoned. By contrast, those who 'exhibit a more just and modest sense of mind' are 'perpetually driven to new industry and novel discoveries'.[13] In effect, recognising human limitations bred the kind of intellectual humility that is essential to science.

Bacon had sympathy, he wrote in *Novum Organum*, with those who 'denied that certainty could be attained at all'.[14] He was not quite so down-beat, however, or at least he drew different conclusions. While they 'go on to destroy the authority of the senses . . . I proceed to devise and supply helps for the same'. It was precisely because unaided human reason and sense were not to be trusted, that they needed to be aided.

This might mean literal aids to the senses, such as were offered by spec-tacles, telescopes and microscopes, but the latter two were new, controver-sial and not always reliable devices. An alternative 'aid' lay in the experi-ment. Again, we have to be careful here. Experiment in the early seventeenth century did not mean what it means today. The word shared a root with 'experience', with which it was treated synonymously in the Middle Ages. Thus, Adam was said to have gained 'experimental knowl-edge' of the animals that came before him in need of a name in Genesis 2. It was also a heavily loaded theological term, used by those who believed that the direct and unmediated experience of God was essential to a fully redeemed life. Puritan preachers spoke of 'experimental Christians' or 'experimental prayer' in direct contrast to the kind of spiritual life that was mediated by priests or written authorities (other than the scriptures).[15]

Such 'experimental' Christian life shed light on experiments with the natural world. If our surest knowledge of God was to be found in our experience of him, so, perhaps, might we acquire true knowledge of nature. The 'experimental' life might be both spiritual and natural. There was more. The very obscurity of nature's nature meant that mankind was warranted in using artifice to obtain her truths. Bacon became famous as the man who recommended that humans 'torture nature' or 'put nature to the rack' to extract her – and it was *her* – secrets. In actual fact, as recent scholarship has shown, Bacon neither used those famous phrases nor advocated torture as the rightful approach to nature.[16] Nevertheless, the opacity and recalcitrance of fallen nature did legitimise a more aggressive form of interrogation than had been the case. Experiencing the world directly – through observation that, where possible, was aided by new devices and new techniques – became the best way humans could hope to regain Adam's knowledge.

And not only his knowledge. Because, if 'knowledge is power' as Bacon nearly said (his exact phrase, from the Latin, was 'knowledge itself is power'), patiently acquiring knowledge via observation and experiment also meant recapturing something of Adam's control over nature. The curse that God laid on Adam and Eve didn't make the creation 'a complete outlaw for ever' he explained towards the end of *Novum Organum*. Rather, it meant that 'by various labours' – and importantly not by the kind of 'disputations or empty magical ceremonies' that had heretofore characterised natural philosophy and alchemy – 'man will in due course and to some extent compel the created world to provide him with bread.'[17]

'As pious as he is capable': 'Catholic' science

Neither intellectual humility nor observation nor experimental science were the preserve of Francis Bacon or his fellow Protestants. Vesalius had observed. Galileo had experimented. Montaigne was sceptical, and he did not require Protestant theological reasons to be so. One did not need Augustine to doubt human nature or our capacity to know things clearly.

Catholic lands boasted some of Europe's most impressive scientific minds in the early seventeenth century. At their heart stood Marin Mersenne, a friar in one of the most ascetic orders of the Church, who, from the 1620s, became the principal correspondent within Europe's most extensive network of scientific thinkers. Mersenne's own scientific motivations were not simply theological, but fervently orthodox. Like Galileo, with whom he corresponded, Mersenne wanted to prevent heretics from shaming his Church. Suspicious, at least at first, of Galileo's anti-Aristotelianism, he nonetheless championed the Italian's science, introducing him to a French audience and imitating his work by making observations and calculations with balls, pulleys, planes and ballistics as a part of his wider project to defend the Church's intellectual reputation.

His contemporary, correspondent and lifelong friend Pierre Gassendi was more empirically minded still. Gassendi was a theologian, philosopher, priest and professor of mathematics and yet, in spite of this plethora of reasons for certainty, he was troubled by the revival of classical scepticism. Critical of Aristotle, he posited a mechanistic and atomistic basis of nature that would ground (and necessitate) experimental science. This was dangerous territory – novel, pagan in inspiration, incompatible with Aristotelianism, hard to reconcile with Catholic dogma – but Gassendi argued forcefully that it was not only theologically sound but a more coherent alternative to the reigning scholasticism. Sceptical, empirical, mechanistic, atomistic, Gassendi's approach constituted, in the words of historian Richard Popkin, 'the formulation, for the first time, of what may be called the "scientific outlook" '.[18]

Younger than both and more complex than either was another of Mersenne's correspondents, Blaise Pascal. Pascal is impossible to categorise. Unusually, he received no formal teaching in theology or philosophy; indeed, technically speaking he received no formal training at all, as he was home taught by his mathematician father. He was introduced to Mersenne's circle as a mathematician – he is still best known for his 'wager' – and he corresponded with, among others, Pierre Fermat, of 'last theorem' fame, about probability and gambling.[19] Beyond mathematics, he maintained an interest in the new experimental science, inventing calculating machines and thinking about vacuums, and in 1647 he travelled to Paris where he met Descartes and talked about the possibility of an ambitious experiment that he himself was too ill to conduct. He was also, however, an intense and ascetic Jansenist, belonging to a seventeenth-century movement within the Catholic Church that was characterised by a powerful sense of human sin and divine grace. He abandoned his mathematical and scientific work after an ecstatic experience on the night of 23 November 1654, the record of which he had sewn into his coat and carried around with him until his premature death eight years later.

Brilliant as Gassendi, Pascal, Mersenne and others in their circle were, however, they would be eclipsed by another of their tribe. René Descartes was a mathematician of the first rank, rivalling Pascal in his originality, but he also wrote on geometry, meteorology, physiology, physics and metaphysics. The ascetic Pascal would doubt the seriousness of Descartes's faith, and in particular his attitude to specific Christian doctrines like sin. He would not be the last to do so. Descartes sought to separate the worlds of faith and philosophy, and there were few traces of distinctively Christian, as opposed to generically theistic, ideas in the latter. He preferred to naturalise miracles, as they offended against his idea of an omniscient creator. His rejection of the scholastic division of substance and accident had worrying implications for the doctrine of the Mass and would lead to Cartesianism, the philosophy that took his name, being proscribed by the French crown and then the Holy Office. It is easy to see why Descartes became an icon of reason and science's attack on religion.

Yet, for all that Descartes could, like Galileo, leave the orthodox uneasy, he was both sincere and honest in his Catholicism. He ostentatiously deferred to the Church in matters of faith, and dedicated his *Meditations* to the faculty of theology at the Sorbonne as a demonstration of the existence of God and the immateriality of the soul. It was at a meeting at the residence of the papal nuncio, Cardinal Pierre de Bérulle, in Paris in 1628, at which Mersenne was also present, that Descartes disclosed his ambition to found the sciences on a new method, a mission in which he was encouraged by de Bérulle. When Queen Christina of Sweden, his host in the last year of his life, renounced her throne and converted to Catholicism in 1655, five years after Descartes's death, she credited the philosopher, whose whole philosophical system pivoted on God. God's perfection and immutability were, in Descartes's mind, the only guarantor of reason and indeed existence. The same could be said for his approach to natural philosophy. 'Man cannot achieve correct knowledge of natural things so long as he does not know God,' he wrote.[20] It was God's reliability that underpinned the physics and mechanics of creation.

Although Pascal criticised Descartes's faith, the two were friends, and met and talked in Paris about mathematics, the existence of a vacuum and recent experiments with mercury. Pascal had worked with mercury barometers in 1647, and he discussed the results with Descartes, before writing to his brother-in-law, Florin Périer, asking him to conduct another more physically demanding experiment on his behalf. On Saturday 19

September the following year, Périer climbed the Puy de Dôme in central France, in the company of a number of clergymen and friars from the city of Clermont. Earlier in the morning, in the gardens of a nearby Minim monastery, he had placed a four-foot-long glass tube, sealed at one end but open at the other, in a bowl of mercury, and measured how far the mercury had risen. 'I repeated this experiment twice at the same spot, in the same tubes, with the same quicksilver, and in the same vessel,' he later wrote, finding the same results each time. He then set out for the summit, asking Father Chastin, one of the brothers of the house, 'a man as pious as he is capable', to observe any changes through the day. When Périer and his party reached the summit, around 3,000 feet higher, they repeated the experiment, and found that the mercury rose three and a half inches less. To ensure his accuracy, Périer repeated the experiment in five different places at the summit, and then did so again twice on the way down, and once again in the monastery garden. The results were progressively less down the mountain, and back to the original level on the morning when they returned to the monastery. 'This was the final confirmation of the success of our experiment,' Périer concluded.[21]

Fifteen years later, Robert Boyle would claim the events of that day constituted the moment that the 'New Physics' was validated. It was more than that: it marked the world's first experiment – in the sense of a planned, organised, hypothesised, designed, observed, measured, repeated and verified procedure, which was soon written up, disseminated and replicated. Centuries later, Catholicism, then Christianity and then religion would acquire the reputation for being anti-science. It is somewhat ironic, therefore, that the world's first formal experiment was based in the gardens of a monastery, drew on the help of its monks, was conducted by the Catholic Florin Périer, requested by the Jansenist Blaise Pascal, suggested (at least according to his own claims) by the Catholic Descartes and beat an attempt to conduct a similar experiment by the Friar Mersenne.

All of which invites the question as to why Herschel did not find himself writing about the *Catholic* origins of science. After all, Bacon and Descartes shared much more than their philosophy textbook stereotypes – the archetypal empiricist vs. the archetypal rationalist – allow. Both trained as lawyers (though unlike Bacon, Descartes did not practise). Both were hostile to scholasticism, the marriage of Aristotle and Christian theology dominant in Catholic Europe. Both rejected the idea of final causes: the idea that things happened because of the ends or conclusions to which they were drawn. Both disregarded the idea that the purposes of God could be read off nature. Both wanted to separate the worlds of theology and philosophy. Both sought to reform the entire basis of knowledge, Descartes's ambitions to do so rooted, by his own account, in a dream he had while on campaign as a soldier in November 1619. Moreover, Descartes was a significantly better mathematician and a more successful natural philosopher; he did more than Bacon, indeed more than anyone else, to popularise the idea that objects obeyed laws of nature rather than having intrinsic qualities themselves; and he articulated a more comprehensive mathematical and mechanical worldview than any of his peers, including Bacon. Why, then, was it to his 'immortal countryman' Bacon that Herschel owed so much, rather than to the immortal work of Descartes, or indeed Mersenne, Pascal, Gassendi or Florin Périer?

The effect of the Galileo affair is the obvious but only partial answer. The trial of Galileo had the impact of an earthquake, with an epicentre, peripheral tremors and then, beyond, plentiful rumours. Mathematicians and astronomers within the reach of Rome were very careful to censor both themselves and any copies of Copernicus or Galileo in their possession. Few dared engage with heliocentrism and loyalists dutifully toed the official line.

Beyond Italy, however, the shock waves were less destructive. Descartes, on hearing of Galileo's fate, suppressed his work, *The World*, on account of its heliocentrism; the book was published in full only a quarter of a century after his death. But Galileo's ideas were still widely discussed in Catholic countries north of the Alps, and some thinkers, like Gassendi, even published apologies for heliocentrism.

Among those who had taken religious orders, scientific activity continued. Mersenne was, as we have seen, a Minim friar, as was the physicist Emmanuel Maignan and the mathematician Jean-François Niceron. Jean-Baptiste Du Hamel and Nicolas Malebranche were both Oratorians. Bonaventura Cavalieri and Stefano degli Angeli were mathematicians and Jesuati, a medieval religious order not to be confused with the Jesuits. The cartographer Vincenzo Coronelli was a Minorite. Benedetto Castelli was a Benedictine. Pierre Gassendi was a priest, as was the pioneer geologist Nicolas Steno, who converted from Lutheranism and ended up as a bishop.

The Jesuits had emerged from the Galileo affair as arch-villains, at least according to the Protestants and a good number of Galileo's followers. However, their science was pre-eminent in the seventeenth century. Not only did they educate some of the finest scientific minds, including Mersenne and Descartes, but they boasted numerous eminent scientists in their own right, including the astronomers Christoph Scheiner and Giovanni Battista Riccioli, the physicist Francesco Grimaldi and the polymath Athanasius Kircher, who wrote on, among other things, music, magnetism, astronomy, sinology, Egyptology, cryptology, geology, numerology and biblical studies. Jesuits engaged in experimental science. They had a keen sense of the need for measurement and precision in their work. And they drew on a truly global network of information and sources.

It was not, then, that the Galileo affair suffocated all serious scientific activity in Catholic countries or by Catholic scientists, even those fiercely loyal to Rome. The real challenge to nascent science came in the nature of the authority, both institutional and intellectual, that presided over that activity.

The Galileo affair showed that the papacy could close off entire areas of scientific enquiry if it felt the need to do so. Heliocentrism wasn't the only casualty. Although far less advanced at the time, and indeed hardly distinguishable from alchemy, the nascent discipline of chemistry became a primarily Protestant affair in the seventeenth century, and this, combined with its association with the distinctly heterodox sixteenth-century Paracelsus, meant that proto-chemistry was effectively off-limits to Catholic thinkers. Denounced as diabolical magic, its texts were commonly placed on the Index.

More seriously, Cartesianism was judged a significant threat to Church teaching. Descartes's *Meditations* was put on the Index in 1663. His teaching was banned in several universities in the following decade, and thereafter widely denounced by religious orders. To be clear, many of the elements of Cartesianism – its scepticism, its atomism, its mechanisation of the world – worried Protestants just as much as they did Catholics. Bacon's pioneering work had helped catalyse the cause of science in England and would, as we shall see in the following chapter, help turn the nation into one of the leading centres of natural philosophy, but that did not mean the English were receptive to all new ideas.

The philosopher Henry More, who introduced Cartesianism into Cambridge University, came to denounce it as fundamentally irreligious, and Isaac Newton declared that it was 'made on purpose to be the foundations of infidelity'.[22] The difference was that the Catholic Church could act effectively on such fears. In the words of one historian of early modern Catholic science, 'the threat of censorship and inquisitorial proceedings was always in the air . . . dampening the spirit of enquiry'.[23] Natural philosophy was always a potential threat. When the potential became real, the Catholic Church could do something about it.

And not only could, but needed to. Aristotelianism had provided the foundation for European science since the mid-thirteenth century. Yet many parts of that foundation were now proving inadequate or downright misleading. To remove them, however, would require enormous labour, not to mention a clear and defensible idea of what might replace them, which was not immediately forthcoming.

More significant even than this, however, was the fact that Aristotelianism had been adapted and adopted as the intellectual foundation for certain key theological dogmas. In effect, not only had the science of the early modern period been built on Aristotelian foundations but so had much of its religion. Replacing those foundations thus could feel like a religious threat rather than a scientific one.

It was not that Catholic thinkers could not question or challenge Aristotle. None other than the future Cardinal Bellarmine had departed from Aristotelian orthodoxy in his lectures on astronomy at the University of Louvain in the 1570s. Mersenne, although educated by Jesuits, would come to rebel against scholasticism. Galileo held it in contempt. Pascal, as a Jansenist, inherited the Augustinian notion of human sin, and with it a scepticism towards human reason that radically undermined scholasticism. Nor was this the province solely of clever individuals. Aristotelian ideas could be challenged by orthodox orders, such as the Jesuits, as evidenced by the way in which, by the 1620s, they had firmly rejected the Aristotelian/Ptolemaic version of the cosmos in favour of Tycho Brahe's geo-heliocentric hybrid model. Still, it was a hard, risky and commonly distrusted work.

And here the contrast with Protestantism *is* important. Bacon is, again, a good example. Luther had once lamented, with characteristic venom, that in the universities, 'the blind, heathen master Aristotle rules alone.'[24] The Cambridge at which the thirteen-year-old Bacon arrived in 1573 was a good example. Bacon, however, had already absorbed a distaste for Aristotle with his mother's milk. Anne Bacon was a woman of strong Calvinist convictions (her father having tutored England's first Protestant king, Edward VI), and she provided Francis and his brother with suitably Reformed tutors in their youth.

Bacon's encounter with Aristotelian thought at Cambridge, including the philosopher's collected work on logic, *Organon*, had done nothing to improve his opinion, and it cemented his aversion to its 'frivolous disputations, confutations, and verbosities'. Although he retained a certain respect for Aristotle himself, his scholastic interpreters were treated with undisguised contempt. They established nothing new, merely played with words and preferred sterile disputation to discovery or actual learning.

Bacon sought to replace Aristotle's *Organon* with his own *Novum Organum*. This was no more straightforward a task for him than it was for his Catholic contemporaries. There was no more self-evident scientific replacement available to Bacon than there would be to Descartes, thirty years later. The difference was that for Bacon, challenging Aristotle was less burdened by weighty theological implications. Indeed, the heavy anti-Aristotelian lifting had already been done by the reformers generations ago. To be Protestant was, in effect, already to have significant reservations about Aristotle.

We should not exaggerate the extent of Protestant intellectual freedom in this period, nor the speed with which Aristotelianism was dethroned in Protestant universities. Protestant regimes could be every bit as intellectually authoritarian as Rome. For all that Bacon thought of Aristotle's *Organon* in the 1570s, it was still an essential part of the curriculum when Newton went to study at the same college, nearly ninety years later. Nevertheless, dislodging an intellectual paradigm that was clearly now more of a hindrance than a help to science was made easier in a culture that had already challenged its authority and no longer based its core theological commitments on it. In effect, the retardation of Catholic science in the seventeenth century, in spite of the pioneering brilliance of many Catholic scientists, was down primarily to the wholehearted way in which the Church had endorsed the previous, longstanding and seemingly secure scientific paradigm. The problem was not too much hostility to (new) science but too much trust in (old) science.

'Not to intermeddle in Spiritual things': 'secular' science

In 1630, Marin Mersenne travelled to Holland. There he met a number of Protestant thinkers, including the Huguenot André Rivet and the Socinian Martin Ruar. Huguenots were French Protestants. Socinianism stood even further from Catholic orthodoxy, rejecting the doctrine of the Trinity altogether. Ruar arguably was even further away, advocating religious tolerance and the complete separation of theology from natural philosophy. The vigilantly orthodox Mersenne, who had once condemned such heretics, befriended Rivet and Ruar and integrated them into his expanding network of correspondents. Shared scientific interests were more important than theological differences.

Mersenne's network was not the only one in Europe. The Jesuit Athanasius Kircher had his own web of correspondents that rivalled Mersenne's in Paris. Not only was Kircher's brilliance recognised across the European religious divide – he corresponded with Protestant scientific luminaries like Christiaan Huygens and Gottfried Leibniz, and was called 'Germanus Incredibilis' by the Lutheran poet Georg Harsdörffer – but his network really was worldwide, drawing on the Jesuits' now global reach. Kircher himself had wanted to be a missionary in China and retained a lifelong fascination with the country.

In England, the German-born Henry Oldenburg, having travelled across Europe, settled in London, and picked up and developed the correspondence network of Samuel Hartlib, another German-born, London-settled scientific man of letters. Less extensive than Kircher's, Oldenburg's network would, in time, incorporate correspondents from Italy to Denmark, including Europe's most 'dangerous' thinker, Baruch Spinoza. In addition to Mersenne, Kircher and Oldenburg, the French astronomers Nicolas-Claude Fabri de Peiresc and Ismaël Bullialdus had similar networks, all of which ultimately overlapped in the transnational Republic of Letters. While kings campaigned and armies fought in the Thirty Years

War over theologically infused territorial claims, scientists and philosophers wrote to one another.

In November 1639, Descartes wrote to Mersenne to tell him that he intended to send his new work, the *Meditations*, to 'the twenty or thirty most learned theologians' for their comments before it was published. Mersenne's network was put to work and in the end Descartes collected seven responses, including from Gassendi; the Catholic theologian Antoine Arnauld; Thomas Hobbes, who like Spinoza would be condemned for alleged atheism; and Mersenne himself. Descartes published them alongside the book and a reply to his critics. Mersenne's network had been deployed in effect to produce the world's first peer-reviewed piece of scholarship. Oldenburg would develop and begin to systematise a similar process years later when he began to edit the *Philosophical Transactions of the Royal Society*.

By the time Mersenne received Descartes's letter, his network numbered in the hundreds. A few years after his Dutch visit, now settled in Paris, he established a slightly more formal 'Academia Parisiensis', which met weekly to discuss primarily mathematical problems, the group including Fermat and, for a time, Pascal. It was not the first such 'academy'. As we saw in the previous chapter, Galileo was invited to be a member of the Accademia dei Lincei in Rome, which was founded by the Italian aristocrat and scientist Federico Cesi in 1603. The Accademia survived only twenty years after Cesi's death, but Mersenne's more informal academy, and his wider correspondence network, would outlive his death in 1648 and form the kernel of what, eighteen years later, would become the Académie des sciences in Paris. Mersenne's hopes of bringing together scientific scholars from across a continent, presently being wasted by the Thirty Years War, were realised.

Francis Bacon was not among Mersenne's correspondents but he shared, indeed outdid, the Minim friar's ambitions. One of Bacon's last works, written about three years before his death but only published posthumously, set out his vision for 'the society of Salomon's House'. *The New Atlantis* tells the story of European sailors who stumble upon the island of Bensalem en route from Peru to China. Bensalem is a utopian country (though that word is not used in the book), 'free from all pollution or foulness', at the centre of which stands Salomon's House, or 'the College of the

Six Days Works'. The college is the 'very eye of [the] kingdom', dedicated to the study of the works and creatures of God, and run by thirty-six 'fellows' with different 'offices'. Twelve venture abroad to fetch 'books, and abstracts, and patterns of experiments', three 'collect the experiments which are in all books'. Three collect 'the experiments of all mechanical arts', three 'try new experiments' and three more 'draw the experiments of the former . . . into titles and tables, to give the better light for the drawing of observations and axioms out of them'. Three 'look into the experiments of their fellows, and cast about how to draw out of them things of use and practise for man's life, and knowledge', three 'direct new experiments, of a higher light', three 'execute the experiments so directed, and report them', and a final three 'raise the former discoveries by experiments into greater observations, axioms, and aphorisms'. Experimentation, and the number three, were evidently important in Salomon's House.

Bacon's was a colourful depiction of a kind of scientific academy given the reins of power, or more precisely one in which 'the knowledge of causes, and secret motions of things' is turned to 'enlarging of the bounds of human empire [and] the effecting of all things possible'. Salomon's House is a place where knowledge facilitates power. That knowledge, however, was *secular*.

The word 'secular' is apt to mislead us today, so it is worth pausing to clarify what it means in this context. The word comes from the Latin 'saeculum', meaning 'generation' or 'age', and was adopted in early Christian writings to mean 'this age' or, more precisely, 'confined to this present age that is passing away'. In the Middle Ages, it came to be used of 'the world', in the sense that 'secular clergy' were those priests who lived 'in the world' among their flock, as opposed to those who were in 'religious' orders and able to direct their attention undisturbed to the eternal. By the early modern period, the word 'secularise' had been coined to indicate a move from the religious to the worldly, and was used to denote, for example, converting an ecclesiastical institution or its property to civil possession and use, or simply removing something from clerical control.

In this sense, Bacon wanted to secularise knowledge, in effect by separating the intellectual domains. The study of nature could illuminate God's power and his glory although it had its limits. Inquisitive humans should not presume 'by the contemplation of Nature to attain to the mysteries of

God'. That was the role of revelation and personal, spiritual 'experiment'. By the same reckoning, 'the scope or purpose of the Spirit of God is not to express matters of nature in the Scriptures.' If you wanted to understand how the heavens moved or the creatures of the earth behaved, you should turn to observation rather than the Bible. The disciplines were, in effect, separate and discrete: non, or perhaps barely, overlapping magisteria, as the evolutionary biologist Stephen Jay Gould would phrase it centuries later.

Secularising knowledge entailed secularising knowers. Those engaged in the study of nature would do well to keep their focus on natural philosophy rather than getting entangled in theological disputes which – while important in themselves – were a distraction from the specific task in hand. In this way, Bacon was responsible, in the words of one modern biographer, for 'the first systematic comprehensive attempt to transform the early modern philosopher from someone whose primary concern is with how to live morally into someone whose primary concern is with the understanding of and reshaping of natural processes'.[25]

And just as Bacon became a model for 'secular' natural philosophy, so did the scientific institution that would become associated with his name: the Royal Society. Universities may well be self-governing institutions but their loyalties were thoroughly caught up in Europe's bitter denominational divisions, and their curricula captured by inadequate Aristotelian science. In contrast with them, scientific academies, like Bacon's fictional Salomon's House and the actual institutions it inspired, were free from clerical control and uninterested in doctrinal dispute, offering (greater) freedom of enquiry and the genuine opportunity to learn something new.

And yet – to underline how specific was the sense of 'secular' in all this – Salomon's House could hardly have had deeper religious foundations. New Atlantis was a pious place. When the sailors first spoke with the governor of the island, they asked him how the nation was converted to Christianity, automatically winning the governor's respect in doing so: 'Ye knit my heart to you, by asking this question in the first place; for it sheweth that you first seek the kingdom of heaven.' New Atlantis's scientific house was named after the Old Testament king renowned for his wisdom, knowledge and intellectual humility. Solomon was, according to 1 Kings 4, the king who described 'trees, from the cedar tree that is in

Lebanon even unto the hyssop that springeth out of the wall [and] . . .
beasts, and of fowl, and of creeping things, and of fishes', and the people of
New Atlantis claimed to have precisely those writings since lost to the
world. The fellows of Salomon's House, dedicated to experiments as they
were, 'have certain hymns and services, which we say daily, of Lord and
thanks to God for his marvellous works'. In this way, Bacon's secularised
scientific institution was, at the same time, a thoroughly religious one.

The same seemingly (to us) paradoxical religious secularism underlay
the foundation of formal academies from the middle of the century,
perhaps the best known of which was England's Royal Society. Hopes for
an actual Salomon's House foundered in the turmoil of the 1640s, although
some informal discussion groups, such as Robert Boyle's 'invisible college',
remained operating. At the Restoration in 1660, a number of natural
philosophers and physicians came together with the intention of founding
a college for discussion, experiment and 'Physico-Mathematical Learning'.
Two years later they were formally incorporated as the Royal Society.

Made in Bacon's image, the Royal Society was an excellent example of the
'secular' nature of the new science. An early memorandum states how the
Society's fellows scrupulously avoided 'meddling with Divinity, Metaphysics,
Moralls'. The institution's first biographer – writing, improbably, only five
years after its incorporation – wrote how the Society was 'abundantly cautious
not to intermeddle in Spiritual things' and that its fellows 'meddle no otherwise with divine things'.[26] Sir Robert Moray, who, in his sixty-five years, was
soldier, spy, diplomat, natural philosopher and one of the Society's original
founding committee, said that its journal, *Philosophical Transactions*, would
not concern itself with 'theological matters'.

And yet, as in Bacon's image, this was also a conspicuously religious
form of secularism. The Society's 1663 Charter declares that the activities
of the Society shall be devoted 'to the glory of God the Creator'. Its early
biographer Thomas Sprat was to become dean of Westminster and then
bishop of Rochester. His history – indeed 'the weightiest and most solemn
part of it' – was intended to defend the new institution and its new experimental learning 'in respect of the Christian faith'. He argued, in good
Baconian style, that the activities of the Society were intended to regain
Adam's lost knowledge and control, and that, through his experiments,
the natural philosopher was 'well furnish'd with Arguments to adore [the

Deity]', finding 'before his eyes the beauty, contrivance, and order of God's Works'.

In the 1930s, when Robert Merton and Charles Webster were first suggesting the Protestant origins of modern science, the American historian Dorothy Stimson analysed the composition of the early Royal Society and its informal predecessors and found that, of those whose religious affiliations could be identified, a disproportionate number were Puritans. Robert Merton confirmed this, and indeed the fact that it was also true among comparable academies on the Continent, and concluded that 'the originative spirits of the Society were markedly influenced by Puritan conceptions', although he latterly conceded that the term 'Puritan' in effect meant Protestant.

The early Royal Society, then, was a secular institution, in so far as it sought to free itself from clerical control, the universities' hidebound Aristotelianism and theological debates, which had, of late, been politicised and militarised to devastating effect. But it was not secular, in the sense of being critical of or hostile to religious piety. In effect, it was the next step on from Mersenne's own scientific network and his informal academy. No one could have accused the Minim friar of hostility or even indifference to religious concerns. But the fact that his network grew to incorporate people from across the religious spectrum, on the basis of their scientific interest and acumen rather than their theological loyalties, was an important step on the road to the secularisation of science. As with the Royal Society, the authority was scientific rather than theological.

Bacon would have approved. Of course, by this time he had been dead for forty years. But like Galileo, his reputation went before him. 'Bacon, like Moses, led us forth at last;/ The barren wilderness he past,/ Did on the very border stand/ Of the blest promised land' ran the poem by Abraham Cowley, with which Sprat's biography began. He was applauded beyond English shores, used, albeit selectively, by Descartes, Mersenne, Gassendi and others in France, and influenced the French Académie des sciences, which coalesced around Mersenne's correspondents. Florence's more informal and short-lived Accademia del Cimento (Academy of Experiment) was similarly modelled on Baconian principles, and in the Netherlands his works were reprinted forty-five times before the end of the century.

His death, alas, was less glorious than his post-mortem reputation, at least according to John Aubrey, the natural philosopher and gossipy biographer. From what Thomas Hobbes told Aubrey, the elderly Bacon was taking the air in a carriage near Highgate one winter when he began to wonder whether snow might prove as effective a preservative as salt. He got out of the coach, bought a hen off a poor woman who lived nearby, had her gut it and then stuffed its body with snow. The chicken was not the only flesh to be chilled, and Bacon fell seriously ill. Retiring to the Earl of Arundel's house nearby, he went to bed with a warming pan. The bed, alas, was damp and unused and Bacon's cold worsened. 'As I remember he [Hobbes] told me, he died of suffocation.' He was an experimenter to his last.

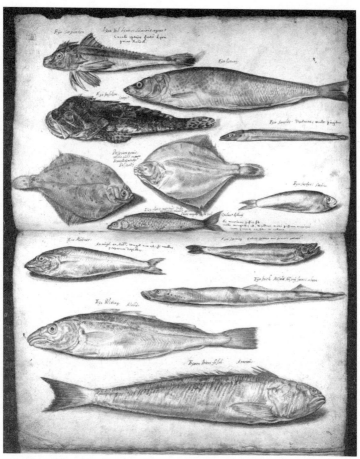

John Ray's *De Historia Piscium* (*A History of Fishes*) published in 1686. Despite its reputation for splitting the two, the period of the Enlightenment, particularly in England, was the period of greatest *harmony* between science and religion. Arguably, there was too much harmony, preparing the ground for the tensions of the nineteenth century.

EIGHT

THE PERILS OF PERFECT HARMONY

'This most elegant system': religious enlightenment

Every culture has its creation myth. The Babylonians recounted how, in the story known to us as *Enuma Elish*, the young god Marduk fought and killed the ancient ocean goddess, Tiamat, 'split[ting] her in two, like a fish for drying', and then assembled the cosmos from the halves of her carcass. The Hebrews played on and subverted this (and other) ancient myths, to tell various creation stories, including the one that was iconically placed at the head of the canon, in which YHWH formed the heavens and the earth in a peaceful and ordered way, without needing to slay any gods or combat any chaos. Modern Westerners, or the more self-consciously secular and rationalist among them, tell a still more inspiring tale of how, from the bloodshed of religious warfare and the darkness of superstition, armed with no more than observation, reason and witty coffee house conversation, the Enlightenment constructed a world of toleration, rational enquiry, political liberty, progress and the pursuit of happiness. In this narrative, the Enlightenment begat reason and reason begat science and science did unto religion what Marduk did unto Tiamat.

Thus, according to Harvard psychologist Steven Pinker, the Enlightenment was responsible for the reason, science, humanism and progress that shape modernity, or at least would be if religion would let it. At a more sophisticated level, Anthony Pagden sees in *The Enlightenment: And Why It Still Matters* the source of today's 'broadly secular, experimental, individualistic, and progressive intellectual world', while Jonathan Israel, in his series of monumental(ly long) books on the subject, traces all that is admirable in modernity, not just to the 'moderate' Enlightenment, but also to the naturalistic 'radical' anti-religious Enlightenment of the Dutch philosopher Baruch Spinoza.

There are dissenting voices, of course. Scholars have long recognised that different Enlightenments – French, American, British – adopted different attitudes to religion. In his book *The Enlightenment: The Pursuit of Happiness*, Ritchie Robertson points out that, however anti-clerical it may have been, the Enlightenment was not really anti-religious. The scholar Ulrich Lehner has even written a history of the 'Catholic Enlightenment', which he rightly calls a 'forgotten history'. Nevertheless, dissenters aside, the popular vision of the Enlightenment is of a period of European and American history, lasting from around the last quarter of the seventeenth century to the end of the eighteenth, in which science began to overwhelm and overthrow religion.

And yet, there is Isaac Newton. Even those who trace the *real* Enlightenment to other figures, whether Spinoza, Descartes, Bacon or David Hume, recognise that the eighteenth century near worshipped Newton. From the 1720s onwards, from France to Russia, theologians and sceptics alike celebrated him for revealing the secrets of the universe. Yet Newton was fiercely religious, treating the subject with the utmost seriousness. The overriding concern in his life, according to his modern biographer Robert Iliffe, 'was to show in what ways humans had been created in the image of God'.[1]

His commitment was personal. He berated himself for infringing the Sabbath by making a feather hat on the sacred day, and also a mousetrap, and a clock, and pies and for eating an apple in church. And he also berated others, including the classical scholar Richard Bentley and the astronomer Edmund Halley, for speaking 'ludicrously' about religion.[2] But that commitment was also 'professional'. He 'rejoiced' at the pious potential of his work and, unusually for him, said so loudly and publicly. 'This most elegant system of the sun, planets, and comets could not have arisen without the design and dominion of an intelligent and powerful being,' he reasoned towards the end of his *Principia*.[3]

His devout writings, which included pages upon pages on biblical exegesis, theology, prophecy and Church history, not to mention alchemy, were something of an embarrassment to later admirers, many of whom tried to excuse them as a mere pastime, occupying the later years of his life, when his true intellectual powers had dried up. Alas, it was not so. The range and intensity of Newton's theological interests have only recently come fully to light and they show, unequivocally, that not only was his piety lifelong (we always knew that) but that his intense study of the Bible and Church history ran in parallel with his mathematical and scientific work throughout his entire adult life, and was undertaken with every bit as much seriousness. Rather than a symbol of the Enlightenment's alleged antagonism between science and religion, Newton was an icon for their harmony.

Newton was extreme in this, as he was in so much else, but far from alone. Take Robert Boyle. Boyle was a pioneering experimenter and chemist, his *Sceptical Chymist* becoming that discipline's foundational text. He was an early Fellow of the Royal Society, and is remembered today primarily as the discoverer of Boyle's law, the inverse relationship between the pressure and volume of a gas at constant temperature. He was perhaps the only English-speaking scientist of his time who could compare to Newton. He was also every bit as pious as his younger Cambridge contemporary.

Like Newton, the devout Boyle also refused Holy Orders. Like him he studied the scriptures in forensic detail, reading the New Testament in Greek and studying Hebrew, which he discussed with Jewish scholars. Like him, he wrote early and passionately on religion, in books that were either not published or published many years later. Like him, he considered science to glorify God; indeed for Boyle, it was an activity worthy even for the Sabbath. And like him, he denounced the impiety and cultural Christianity of his peers, although Boyle was more generous and ecumenically minded than his mathematical peer. In the words of his biographer Michael Hunter, 'the central fact of Boyle's life from his adolescence onwards was his deep piety, and it is impossible to understand him without doing justice to this.'[4] Similar points could be made about the other great scientific figures of his age, such as Gottfried Leibniz, John Ray, Carl Linnaeus, Christiaan Huygens, Antonie van Leeuwenhoek and Jan Swammerdam, as well as a host of lesser scientific lights. All testified in their work and writing to the perfect harmony of science and religion. And yet, to avert the danger of replacing the familiar Enlightenment narrative of simplistic opposition with an unfamiliar Enlightenment narrative of simplistic harmony, we should note that Newton is also an icon of discord. He was one of a vanishingly small number of scholars at the time who refused to take Holy Orders and only managed to stay at Trinity College, Cambridge, as a layman on account of special dispensation from Charles II. His intense biblical studies led him to the conviction that the early Church was both wrong and deceitful in its belief that Jesus was co-eternal with the Father, and although he never denied Christ's divinity, it was a subordinate divinity. Newton was adamant that the Father and the Son were not co-equal, and Christ was not uncreated and of the same 'substance' as God the Father. He was, in effect, an Arian, advocating a heresy that had been defeated and banned by the Church in the fourth century, and was as heretical in Newton's time as it had been then. Newton, ironically given his alma mater, was not a Trinitarian. Private and suspicious at the best of times, he naturally kept this conclusion to himself. Had he not done so, he would have lost his academic position and might now be numbered among the radical Enlighteners who tried to pull down Christianity one dogmatic brick at a time.

Or take Boyle again. The chemist's faith was deep, personal and sincere but set about by uncertainties. 'Of my own Private, & generally unheeded doubts,' he wrote in his twenties, 'I could exhibit no short Catalogue.' In the hands of those who trace a narrative of conflict between science and religion in the Enlightenment, such a confession is revealing indeed. Doubt creeps in under the cloak of scientific research. In reality, in Boyle's case the doubts were rooted in the intense introspection of his Protestantism rather than his scientific work. 'He whose Faith never Doubted, may justly doubt of his Faith,' he wrote. Nevertheless, his longstanding concern to answer 'infidels [and] atheists' intimates how deep those doubts went. In short, the examples of Newton and Boyle, and indeed many others who fill the story of science and religion during the Enlightenment, remind us the story is no more straightforward in this iconic period than it is in any other.

'On the wonderful secrets of nature': physico-theology

Theology had helped incubate science in the early decades of the seventeenth century. Science then returned the favour. For many years, indeed for many centuries, theological ideas and biblical texts – about the wonders of God's creation, its order and lawfulness, about Adam's lost knowledge and Solomon's great wisdom – had encouraged and legitimised the study of nature. The creator pointed approvingly towards the creation. That relationship would persist. But from the third quarter of the seventeenth century onwards, the balance began to shift and early scientists deployed their new discoveries in the service of religion. Creation was judged, with ever-growing certainty, to point towards the creator.

This was not an entirely new phenomenon. 'Natural theology' long predated the Scientific Revolution, being in effect natural philosophy's metaphysical twin. For centuries, thinkers, by no means all of them Christian, had sought to understand and define the divine through observing, and reasoning about, nature. The discipline gained new impetus and urgency in the sixteenth century. The European discovery of peoples who had no (recognisable) knowledge of divine revelation; the resurgence of classical learning, and in particular the revival of scepticism; the doubt cast by Catholics on Protestant interpretations of scripture, and by Protestants on Catholic foundations of authority; and the desire to bypass entirely the increasingly violent disagreements about revealed theology: all of this contributed to a desire to ground knowledge of God in something that was more universal and less contentious than revelation. Something like nature.

Natural theology, however, morphed into physico-theology around the middle of the century, when the nature, depth and accuracy of observation changed.[5] Galileo had shown that there was more to the skies than met the naked eye and although 'microscopes' were being used within a few years of his discoveries, it was not until the 1650s (when the English word was coined) that they were deployed successfully to observe objects too small for unaided vision.

The Italian physician Marcello Malpighi – a Catholic who was elected as a Fellow of the Royal Society and invited by Pope Innocent XII to become the papal physician: another example of the transnational and 'secular' nature of science – studied blood circulation in small animals and discovered capillaries, as well as learning more about skin pigmentation and embryology. Robert Hooke improved the device, published his elaborately illustrated *Micrographia* – the first major book on microscopy, the charm of which kept Samuel Pepys up at night – and coined the word 'cell' (after the cells in which monks dwelt) as a basic structure within biology. Antonie van Leeuwenhoek improved the device still further, independently confirmed Malpighi's discoveries about blood flow and became the first person in history to see the multitudinous micro-organisms that throng within water. The world was all of a sudden much more complex, intricate and abundant than had heretofore been imagined. Glory be to God.

And they did give glory to God, scientists and divines alike, volubly and frequently, in innumerable sermons and books published over the next century, encompassing every conceivable dimension of the new world that scientific observation was revealing. Seminal texts, like William Derham's, spoke simply of *Physico-Theology* but there were books entitled or about Astro-theology (space), Helio-theology (the sun), Planeto-theology (planets), Cometo-theology (comets), Pyro-theology (fire), Hydro-theology (water), Chiono-theology (snow), Seismo-theology (earthquakes), Bronto-theology (thunder), Phyto-theology (plants), Chorto-theology (grass), Petino-theology (birds), Ichthyo-theology (fish), Insecto-theology (insects), Akrido-theology (grasshoppers), Bombyco-theology (silkworms), Lokusta-theology (locusts), Melitto-theology (bees) and Testaceo-theology (mussels and snails). Time and again, authors offered 'a demonstration of the Being and Attributes of God from' a more or less significant aspect of the natural world, newly noticed and understood by science.

Many of these authors were clerical naturalists, that seminal hybrid role that so characterised the English Enlightenment, to which we shall return, but physico-theology was not a solely English affair. Indeed, the apparently harmonious marriage of science and religion in the Enlightenment was a thoroughly pan-European – or, at least a north European – phenomenon.

The tradition ran particularly deep in the Netherlands. Article 2 of the Belgic Confession of Reformed churches, on 'How God Makes Himself Known to Us', claimed that we know Him first 'by the creation, preservation, and government of the universe; which is before our eyes as a most beautiful book, wherein all creatures, great and small, are as so many letters leading us to perceive clearly the invisible things of God, namely, His eternal power and deity'. In this way, the Dutch Reformed Church became the only one in Europe to bake natural theology into its constitution.

That, however, did not stop the Netherlands becoming a beacon for thinkers who strayed beyond its borders of orthodoxy, as the country became the most intellectually open territory in Europe. Descartes had lived most of his life in the Dutch republic. The lawyer Isaac La Peyrère published his *Prae-Adamitae*, a polygenetic book about the existence of humans before Adam, in Amsterdam in 1655. Europe's most notorious sceptic, Baruch Spinoza, lived, worked, thought and published in the Dutch republic. Physico-theology found fertile soil in a country that needed to defend its good name.

Early in 1669, the Dutch biologist Jan Swammerdam published a general history of insects, which made good use of the microscope to outline in precise detail the anatomy of these tiny – once pestilential, now apparently pleasing – creatures. Swammerdam's justification for his work was explicitly theological: 'it is the duty of all men, through all creatures, no matter how small, to climb to God the Creator himself.' And so were his results, which revealed an order, detail, complexity and organisation that was so awe-inspiring it could not but be due to divine craftsmanship. Never had the popular saying attributed to Pliny the Elder's *Natural History* – 'Maxima in minimis': God's greatness is best seen in small things – been better illustrated.

A generation later, in his posthumously published *Cosmotheoros*, the Dutch physicist and astronomer Christiaan Huygens reasoned similarly. Huygens had invented the pendulum clock, solved the mystery of Saturn's rings, discovered its moon Titan, described the wave theory of light and written the first book on probability theory. He was, by any reckoning, one of the scientific giants of the century. But in his posthumously published *Cosmotheoros*, in which he speculated on the existence of extra-terrestrials, it was life that impressed him. 'Nobody will deny but that there's somewhat more of Contrivance . . . in the production and growth of Plants and Animals, than in lifeless heaps of inanimate Bodies,' he argued. 'The finger of God, and the Wisdom of Divine Providence, is in them much more clearly manifested than in the other . . . everything in them is so exactly adapted to some design, every part of them so fitted to its proper life, that they manifest an Infinite Wisdom, and exquisite Knowledge in the Laws of Nature and Geometry.'

There was a similar fascination in German lands. Luther's theology (if not necessarily Lutheran theology) offered a clear imprimatur for the study of God's creation, and from the late sixteenth century books were published along the lines of Jakob Horst's endearingly entitled *On the wonderful secrets of nature, and on the fruitful contemplation of them which are not only useful but also pleasant to read about*. As those wonderful secrets were probed and exposed, the genre gained in popularity and for a century, from the 1670s, German scholars eagerly debated the theological implications of nature's secrets. 'What purposes of the Creator can be found in the figure of a flower?' asked the Leipzig-based literary society, the Deutsche Gesellschaft, in a popular essay prize question. Usefulness and beauty, argued Georg Friedrich Baermann in his winning entry. Flowers were beautiful, of course, but the question could equally be asked of frogs, as in Friedrich Menz's *Rana-Theologie* of 1724, or of stones, in Friedrich Christian Lesser's *Lithotheologie*, or of insects, as in his *Insecto-theologie*, whose subtitle explained 'how men can attain a vital knowledge of, and admiration for, the omnipotence, wisdom, goodness and justice of the great God through an intensive observation of those normally neglected insects'.[6]

Other Reformed lands told similar tales. Sweden had a less developed scientific community in the early Enlightenment – it had to wait until 1739 for its own Royal Academy of Sciences – but it did boast the eighteenth century's – indeed history's – most influential taxonomist in the form of Carl Linnaeus. Linnaeus was an obsessive systematiser by nature and needed little justification for his work; work that would lead him into muddy waters about the age of the earth and the place of humans within it. However, he was nonetheless firmly rooted in a physico-theological tradition in which cataloguing and analysis was understood as justifying the ways of God to man. 'The maker has furnished the globe, like a museum, with the most admirable proofs of his wisdom and power', he wrote in his *Reflections on the Study of Nature*, and 'man . . . is alone capable of considering the wonderful economy of the whole.'[7] It thus followed that 'man is made for the purpose of studying the Creator's works that he may observe in them the evident marks of divine wisdom.' Kepler could not have put it better.[8]

Swiss scientists felt the same. Johann Jakob Scheuchzer studied mathematics and medicine in Germany and the Dutch Republic before becoming chief medical officer at the Foundling Hospital in Zurich, one of the most important Reformed dominions in the Swiss Confederation. Élie Bertrand was a geologist and theologian, who served as special counsellor to the king of Poland, after which he returned to the Pays de Vaud in Berne, another Reformed territory. Despite their different paths there, both were captured by the landscape of the Alps, into which they journeyed, and both helped, through their studies, to shift the understanding of the mountains from being terrifying and essentially pointless – 'a heap of Stones and rubbish', in the words of an early geologist, Thomas Burnet – to an awe-ful and useful dimension of God's creation. 'Come . . . into the mountains that the divine power raised up with so much majesty and resourcefulness,' Bertrand wrote. 'Admire the masterpieces of the beneficent hand . . . the beauty, and the necessity, the usefulness and the purposes of the mountains.'[9]

The Dutch republic, German kingdoms, Sweden, the Reformed dominions within the Swiss Federation: physico-theology appears to have been a quintessentially Protestant affair. It wasn't, quite. Jakob Horst's book was in fact a translation (and Protestant-isation) of a Catholic text by a Dutch physician and priest, who studied under Vesalius, Levinus Lemnius. In France, the abbé Noël-Antoine Pluche published his multi-volume *Le Spectacle de la nature* over eighteen years from 1732. It went through over fifty editions in France, was translated into English, German, Dutch, Italian and Spanish and became one of Enlightened Europe's most popular scientific texts. It may even have been the most widely read physico-theological book in the Enlightenment, as unapologetic in its scientific details as it was apologetic in its religious intent.

In Italy, Laura Bassi became the first woman in Europe to receive a university chair, in Bologna for experimental physics, while her contemporary, Maria Gaetana Agnesi, received a lectureship for her mathematics, also at Bologna, both insisting, according to Agnesi's posthumous book

The Mystic Heaven, that the rational contemplation of nature led the mind up to the mystical contemplation of God. In Spain, the Benedictine monk Benito Jerónimo Feijóo y Montenegro wrote on a wide range of scientific and natural historical topics, remarking on how they reflected the intelligence of the creator, such as the dog's heart which, when dissected, caused him to remark that neither he nor his fellow monks had ever 'contemplated anything which gave such a clear idea of the power and wisdom of the supreme Craftsman'.[10]

These were more exceptions than rules, however, and neither natural theology nor physico-theology found as fertile a soil in Catholic lands as they did in Protestant countries. In spite of the popular myth, this had very little to do with the Catholic Church's attitude to science in the wake of the Galileo affair, the freezing effect of which was, as we have seen, more localised and less damaging than the myth would have it. Rather, it was down to the fact that the sources of intellectual authority were more intact in Catholic lands. Aristotelian science still reigned, in spite of the hit it had taken over heliocentricity. Moreover, to be a Catholic was precisely to accept the Church's authoritative teaching on the omnipotence, wisdom, goodness and justice of God. There was less *need* to turn to nature. There was as much curiosity about the workings of the natural world in southern Catholic Europe as there was in Northern Reformed Europe, and just as much theological justification for its study, but the twin authorities of Aristotle and Rome meant there was less *need* for physico-theology there.

'Our employment in heaven': science in the English Eden

If the marriage of science and religion was harmonious across much of Europe in the Enlightenment, it was positively blissful in England. Dutch, German, Swedish, Swiss, French, Spanish and Italian scientists wrote about how insects, frogs, planets and stars demonstrated the truth of Christianity, but it was from their English counterparts that they took their cue. English texts sold very well abroad. John Ray's *Wisdom of God Manifested in the Works of Creation*, first published in 1691, was translated into French and German and republished seventeen times. William Derham's *Physico-Theology* did better still, and was translated into Italian,

French, Dutch, German and Swedish. Voltaire made his name popularising Newton. The Anglomania that spread across the Continent in the second quarter of the eighteenth century was testimony to this new, harmonious understanding of creation, so enthusiastically baptised by English theology.

The process began in England much earlier. The civil wars of the 1640s and the temporary collapse of the licensing system led to a proliferation of pamphleteering. Sectarians released a bewildering variety of religious and metaphysical beliefs on a febrile public, many of the ideas veering so far from Christian orthodoxy that they were (seen as) little better than thinly disguised atheistic materialism. Such heterodoxy catalysed English divines to reassert The Truth. However, because the problem was down to over-zealous, personalised interpretations of the scriptures, a *scriptural* reassertion of orthodoxy risked simply being dragged back into the theological undergrowth. The book of nature offered a clearer, less contentious and more secure foundation for true belief.

The natural philosopher Walter Charleton wrote *The Darkness of Atheism Dispelled by the Light of Nature: a Physico-Theologicall-Treatise* in 1652, in the process becoming the first person to use that hyphenated term. His stated context was 'the swarms of Atheisticall monsters' that had descended of late, nibbling at the authority of the scriptures and shredding that of the Established Church. Charleton's stated desire was to demonstrate 'the Existence of God, by beams universally de-radiated from that *Catholick* Criterion, the Light of Nature'. The word is highly instructive: for Charleton, it was *nature* that was catholic, meaning universal, not the Church. The following year, the Cambridge don Henry More wrote his *Antidote against Atheism*, in which he, like Charleton, elected not to build his argument 'upon any Sacred History' because, as he lamented, 'I knew the Atheist will boggle more at whatever is fetch'd from established Religion.'[11] In other words, More's case for religion was best made without recourse to religion.

If this particular impetus to physico-theology faded after the Restoration in 1660, its wider appeal did not. Robert Boyle lived through the civil wars and interregnum as a young man and he feared for their impact on Christianity. He continued to draw religious lessons from his observations and experiments for many years after the sectarian threat had disappeared. Science, and what it revealed of nature, was a stimulus to piety, irrespective of any sectarian or materialist threat, because the more people discerned 'how Congruous the Means [of nature] are to the Ends to be obtain'd by them, the more Plainly we Discern the Admirable Wisdom of the Omniscient Author of Things'.[12]

Indeed, physico-theology offered more than that. Not only did the discoveries of science illuminate God's work and stimulate piety, but its practice even instantiated godly virtues. Thomas Sprat, in his pre-emptive history of the Royal Society, ingeniously claimed that the experimental philosopher was, in effect, practising Christian virtues – of repentance, humility, self-examination. 'The doubtful, the scrupulous, the diligent Observer of Nature, is nearer to make a modest, a severe, a meek, an humble Christian, than the man of Speculative Science', who was puffed up with 'better thoughts of himself and his own Knowledge'.[13] Scientists were godly by definition.

Boyle went even further. Scientists were the new priests, representing creation and worshipping God through their ritual of observation and experiment in the temple that was nature. Just as priests in the line of Aaron offered sacrifice to God on behalf of the people of Israel, so scientific reason and knowledge 'can confer a Priesthood without Unction or Imposition of Hands'. Science was 'an act of Piety to offer up for the Creatures the Sacrifice of Praise to the Creator'. That was why this most pious of chemists could countenance science on the Sabbath. It was, in the most specific sense, a genuinely *religious* activity.

In this way, physico-theology found fertile soil in England in the last third of the seventeenth century. Indeed, almost literally so. It was, after all, easier to discern harmony, peace, order and beauty in a country that was abundant in rich, fertile, lowland agricultural soil, and comparatively free from inhospitable and apparently useless mountains and swamps; in a country that was not generally troubled by earthquakes or volcanic eruptions; in a country in which your likelihood of being eaten by a wild animal was lower than being knocked over by a stagecoach; and in a country that was largely free from poisonous snakes and spiders and where insects might bite or irritate but were distinctly unlikely to devour your crops and starve your village. For sure, the weather was not always clement and bad harvests remained a real and present danger, especially during what we now call the Little Ice Age. However, the English landscape, at least in the eyes of those who studied nature rather than worked it, had more in common with Eden than the post-lapsarian wasteland into which Adam and Eve were cast.

The harmony was not just agricultural. English physico-theology came of age in the final years of the century. The country had appeared to be sliding towards another civil war as the Catholic James II ascended to the throne, secured an heir, suspended Parliament and tried to rule by personal decree. In a remarkably short and bloodless reversal, William of Orange invaded by invitation, James fled and the English found themselves enjoying a stable constitutional monarchy, a bill of rights, parliamentary rule, comparative religious toleration and growing economic prosperity. Order and harmony reigned.

And so did physico-theology. The intricate, congenial order of the natural world was seemingly paralleled by the just, balanced order of the political one. And both were underpinned by God. In 1690, the now eminent Robert Boyle published *The Christian Virtuoso*, which showed, according to its subtitle, 'that by being addicted to experimental philosophy, a man is rather assisted than indisposed to be a good Christian'. He died the following year but left in his will a bequest to found an annual series of lectures in London 'for proving the Christian religion, against notorious Infidels, viz. Atheists, Theists, Pagans, Jews and Mahometans'. As with the Royal Society, the Boyle lectures were to leave theological differences well alone and concentrate instead on how science might prove Christian orthodoxy. The first lecture was delivered the following year by Richard Bentley, who spoke on *The Confutation of Atheism*, and the series continued annually for another forty years.

A couple of years after Bentley delivered his 'Boyle lecture', as it came to be known, John Ray published his *Physico-Theological Discourses*. These had been drafted and delivered many years earlier, as Cambridge sermons on Psalm 104 verse 24 – 'How manifold are thy works, O Lord? In Wisdom hast thou made them all' – but were now revised, updated and augmented with sections on creation and the biblical flood. Foetal formation, the structure of the ear, buoyancy of fish, male nipples: nothing was beyond Ray's studious gaze. Like Boyle, he exhorted readers to experience nature directly: 'Let it not suffice us to be book-learned . . .

but let us ourselves examine things as we have opportunity, and converse with Nature as well as books.'[14] And like Boyle, he considered science important enough to do on the Sabbath day. Indeed, as he said in his *Wisdom of God*, 'the contemplation and consideration of the words of God may probably be some part of our employment in heaven.' Science was literally a heavenly activity.

Impressive and influential as Boyle and Ray (and Derham and Bentley and others) were, however, by the mid-1690s they were all in Newton's shadow. In advance of delivering his Boyle lectures, Bentley had written to his academic friend and master for advice. Newton wrote back four detailed letters that were subsequently published, the first of which began: 'When I wrote my treatise about our Systeme I had an eye upon such Principles as might work with considering men for the beleife of a Deity & nothing can rejoyce me more then to find it usefull for that purpose.' If the greatest scientist in Europe thought this, who was anyone to say otherwise?

Newton's 'treatise about our System' – his *Philosophiae Naturalis Principia Mathematica* or *Mathematical Principles of Natural Philosophy* – had been published five years earlier, in 1687, but its origins go back more than two decades. Newton had emerged from the Aristotelian shadow in which Cambridge still lay within a few years of his arrival at Trinity in 1661, and by 1664 he was reading and critiquing Descartes. He spent the following two years back at home, in Woolsthorpe, after Cambridge closed on account of the plague. It turned out to be a rather profitable sabbatical. Having effectively taught himself mathematics, he invented calculus (or 'the method of series and fluxions'), the way of calculating continuous change by means of summing infinitesimal differences; split out the constituent elements of white light; made initial association of planetary motion with terrestrial gravity; and nearly blinded himself by conducting experiments with a bodkin in his eye.

He returned to Cambridge in 1667 as a Fellow of Trinity and then two years later became the second Lucasian Professor of Mathematics. Despite taking the chair and being elected to the Royal Society three years later, he remained reluctant to publish anything. Suspicious to the point of being paranoid, thin-skinned, short-tempered, secretive and mercurial, he did not take criticism well. When his 'Theory about Light and Colours' was published in 1672 in the Royal Society's *Philosophical Transactions*, only to be criticised by its Curator of Experiments, Robert Hooke, the fall-out lasted a lifetime. Newton's genius for making discoveries was matched only by his genius for making enemies. He theatrically withdrew from scientific intercourse and focused on (what he considered to be) the more important business of alchemy, biblical studies and Church history.

In reality, he did remain scientifically engaged – corresponding with the first Astronomer Royal, John Flamsteed, about comets, and with Edmund Halley about gravity – but it was not until the mid-1680s that Halley, with whom Newton did get on, persuaded him to re-enter the fray. Although the sheer size and difficulty of the *Principia* meant that it took some time for its true significance to be realised, it left few in doubt of Newton's pre-eminent position. Not only was heliocentrism now, finally, proved beyond any possible doubt, but the *Principia* also showed the remarkable unity of things in heaven and earth, hammering the final nail into Aristotle's cosmic coffin, *and* established how, as Galileo had mused, nature was indeed written in the language of mathematics. It was a remarkable achievement for a man who expended more intellectual energy on working out the true nature of the divine than of the heavens.

For all his almost puritanical piety and theological fascination, Newton mentioned God just once in the first edition of the *Principia*. It was only later, in the letters he wrote to Bentley and in the 'General Scholium' or commentary that was appended to the second (1713) and third (1726) editions of the *Principia*, that the theological reverberations of his general system were clearly heard. In a similar way, theological reflections made only a late appearance in the 'Queries' to his 1704 book, *Opticks*.

God most commonly appeared in early modern natural philosophy either at the beginning, as a premise, or at the end, as a conclusion. Of these two, Newton preferred the second. 'Such a wonderful Uniformity in the Planetary System [and] the uniformity in the Bodies of Animals,' he wrote at the end of *Opticks*, 'can be the effect of nothing else than the Wisdom and Skill of a powerful ever-living Agent.'[15] That recognised, Newton's God was more than just a conclusion. Troubled by the apparent instability of his solar system, which might, it seemed, collapse in on itself or scatter into eternity over the long run, Newton concluded that 'a continual miracle is needed to prevent the Sun and the fixed stars from rushing together through gravity.'[16] Ironically, given the way in which the Newtonian system came to represent the archetypal mechanistic or clockwork universe that needed no divine intervention save an initial winding up, Newton's system was predicated on divine involvement.

A similar point could be made for gravity. Newton was unclear what caused gravity and so wondered, hesitantly, whether this too might be divine action. 'It is inconceivable, that inanimate, brute Matter should, without the Mediation of something else, which is not material, operate upon, and affect other Matter without mutual Contact,' he wrote in one of his letters to Bentley. 'Gravity must [therefore] be caused by an Agent acting constantly according to certain Laws,' he reasoned, before concluding, 'whether this Agent be material or immaterial, I have left to the Consideration of my Readers.'[17] His readers had fewer doubts and many an early Newtonian stated conclusively that gravity was God's action in the universe.

An agent acting constantly 'according to certain Laws' does not really constitute divine *intervention*, however, and 'continual miracles' are not really miracles at all, at least in the widespread understanding of that word. (The parallels with the occasionalism of al-Ghazâlî come to mind.) For all intents and purposes, the phrases meant 'a lawful cause that is as yet undefined' rather than 'a contravention of the laws of nature'. In that regard, Newton's God was not a God of the gaps, because his presence and activity were lawful and everywhere. As Newton himself remarked, 'if [miracles] should happen constantly according to certain laws impressed upon the nature of things, they would no longer be called miracles, but might be considered in philosophy as part of the phenomena of nature notwithstanding that the cause of their causes might be unknown to us.'[18]

Not everyone was impressed with this reasoning. Gottfried Leibniz, whose relationship with Newton was already toxic on account of a dispute over who invented calculus (it seems they both did, independently), thought his system absurd. By his reckoning, it treated God like a shoddy workman who needed continually to intervene to prevent his creation from disintegrating, while also effectively abolishing the traditional idea of miracles. Leibniz's objections were theological rather than scientific and Newton fought back, as did his loyal supporters; the mathematician

Roger Cotes, in his preface to the second edition of *Principia Mathematica*, memorably calling Leibniz a 'miserable reptile'.

For all that physico-theological tracts, Boyle lectures and Newtonian mechanics took aim at atheism, however, there were precious few atheists in early modern Europe. The Church courts record innumerable armchair (or more usually tavern) sceptics, hauled before the bench for saying, as John Deryner of Great Bedwyn in Wiltshire had done, that 'there was no God and no resurrection, and that men died a death like beasts'.[19] But these were just the rough, ill-educated edges of everyday Christendom.

In reality, the label 'atheism' was less about whether you believed in God than about which kind of God you believed in and, more importantly, what you did about it. When in the early seventeenth century the Jesuit François Garasse identified five different types of atheism in France, he labelled them 'furious and enraged atheism', 'atheism of libertinage and corruption of manners', 'atheism of profanation', 'wavering or unbelieving atheism' and 'brutal, lazy, melancholy atheism'. Atheism was a moral crime rather than a metaphysical one.

As a result, it wasn't enough for physico-theology to demonstrate the existence of God. It had to reveal his nature, and his providence, and the truth of his divine history and the authority of his moral commands. And it was so. God, nature confirmed, was a master craftsman, awesome in his scale and intricate in his detail. He was also wise and omnipotent and caring and good. Nature worked. It fitted together, from the bodies of beetles to the orbit of planets. It was beautiful and ordered and fruitful and productive, as, therefore, was its maker.

God was self-evidently attentive to human needs, for what better way could there be of appreciating the value of creation than in recognising its utility. Aesthetic appeal was important, of course, but usefulness was surely the truest measure. It was, Ray reasoned, highly providential that every kind of plant produced seed not only for its own 'continuation and propagation' but also for 'gratifying man's art, industry and necessitates'. Derham concurred. 'It is a very remarkable act of divine providence,' he noted, 'that useful creatures are produced in great plenty and others in less.' Nature showed God to be a God of providential care, justice and mercy.

This was all good, but the study of nature offered still more. Divine history, or at least snippets thereof, could be gleaned from it. The early Danish geologist, and Protestant-turned-Catholic, Nicolas Steno cast doubt on the traditional belief that fossils grew in the earth, and also laid the foundations for modern stratigraphy in his recognition that different rock strata were laid down upon one another. The implication – that fossils, often found miles inland or even up mountains, were once living creatures, and that the earth had undergone a time of watery inundation – clearly pointed to the Genesis flood. A handful of texts even dealt with parthenogenesis (virgin birth) and resurrection, although Boyle, for one, admitted in a short tract on the subject that resurrection 'is not to be brought to pass according to the common course of Nature'. Not everything could be demonstrated through physico-theology.

Virtue and moral duty could be, though. The tradition of seeing virtues in other creatures was as old as civilisation, and had a venerable tradition within the Bible. 'Go to the ant, thou sluggard; consider her ways, and be wise,' advised Solomon in the book of Proverbs.[20] Which is what scientists did. The behaviour of ants, and other creatures, was now emerging more clearly into the light, offering ever more profitable moral examples. To be sure, an existential chasm still separated humans from animals. As Ray said, 'understanding', that 'supreme faculty of the soul', was the chief difference between us and 'brute beasts', making us 'capable of virtue and vice' in a way animals were not. However, the said beasts, even the least among them, could still offer ethical instruction to wayward humans. Picking up on Solomon's injunction, William Derham expanded on the 'great wisdom of this little creature', by drawing readers' attention to their 'unparalleled tenderness, sagacity and diligence about their young':

> Tis very diverting, as well as admirable to see with what affection and care they carry about their young in their mouths, how they expose themselves to the greatest dangers rather than leave their young exposed; [and] how they remove them from place to place in their little hills . . . for the benefit of convenient warmth, and proper moisture.

Human parents take note.

So it was that nature's moral laws were as foundational as scripture's. Good and bad were written into creation at the deepest level. In an example of proto-utilitarian, naturalistic ethics, John Ray reasoned, in his *Wisdom of God*, that pain 'provokes us to seek ease and relief when we labour under it, but also makes us careful to avoid for the future all such things as are productive of it'. Fortunately, as he went on to say, those things that are 'hurtful to our bodies, and destructive of the health and well-being of them . . . also are for the most part prohibited of God, and so sinful and injurious of our souls'. There was a happy, indeed near-perfect and providential, coincidence of pain, pleasure, ethics and God's moral law, woven together into the very fabric of nature.

'Too much to deny and too little to be sure': underneath the harmony

Both parties did well out of this harmonious marriage of science and religion but one did rather better, in the long run, than the other. By the time Boyle, Newton, Ray and the others were writing, science had inspired much but achieved little. At the time, science was a child bride, without dowry, and dependent on her aged partner for protection and care.

She needed it. When, Robert South, Public Orator at Oxford, presided at the opening of the Sheldonian Theatre in 1669, less than a decade after the Royal Society had been founded, he rubbished its fellows by saying that they 'can admire nothing except fleas, lice and themselves'. His remarks were obtuse but not senseless. Observations of insects served no obvious purpose, and experiments with air seemed positively pointless, not least for the animals involved. One of Robert Boyle's colleagues, we don't know who, composed a poem about an early experiment on the effects of a vacuum on a cat: 'Out of the glasse the Ayre being screwed,/ Puss dyed and ne'er so much as mewed.'[21] The cat got off comparatively lightly. Robert Hooke recounts, in Sprat's institutional biography, an experiment on a dog, 'in prosecution of some enquiries into the nature of respiration'. The dog was dissected but kept alive, 'a certain pipe thrust into the windpipe'. The heart, the fellows observed with curiosity, 'continued beating for a very long while after all the thorax and the belly had been opened, nay after the diaphragm had been in great part cut away, and the Pericardium removed from the heart'. The observations were fascinating; the achievement less obvious.

This was a problem in the 1660s and remained one for decades. When Jonathan Swift's Lemuel Gulliver, on the third of his travels, beginning in 1706, visited the Grand Academy of Lagado, he was informed how the institution had been founded, forty years earlier, by people who had returned from the flying island, Laputa, 'with a very little smattering in mathematics, but full of volatile spirits acquired in that airy region'. Disliking the management of 'everything below', they 'fell into schemes of putting all arts, sciences, languages, and mechanics, upon a new foot' and procured a royal patent for erecting an academy of 'projectors', who were now running a range of projects in order to improve on life and nature. Gulliver was shown around the Academy. He met one projector who 'has been eight years upon a project for extracting sunbeams out of cucumbers . . . to warm the air in raw inclement summers'. A second was engaged in an 'operation to reduce human excrement to its original food'. A third was learning how to plough fields by means of pigs and strategically placed acorns, while a fourth was working out how to get silk from spiders. A 'most ingenious architect' was contriving a new method for building houses 'by beginning at the roof, and working downward to the foundation', while a group of blind men were being taught to distinguish between colours of paint by feel and smell. Reflecting on the Academy's great ambitions in his matchlessly deadpan tone, Gulliver laments that 'the only inconvenience is, that none of these projects are yet brought to perfection; and [that] in the mean time, the whole country lies miserably waste, the houses in ruins, and the people without food or clothes.'[22]

The new science clearly needed intellectual legitimacy. Religion offered it. Even if it hadn't paid much of a return on its promises, and was seemingly absurd at some times and cruel at others, the new science did at least lift the veil on the divine, revealing the wonders and glories of his creation in heretofore unprecedented detail. It was, as Boyle and others stressed, a religious activity in itself, something that was underlined by the fact that so many scientists were either transparently pious, like Boyle or Newton, or actually ordained.

The clerical (or clergyman or parson) naturalist was a predominantly, if not exclusively, English figure, going back to William Turner in the sixteenth century. From the later 1600s, the role became firmly established, with the parson who split his time between pastoral duties and natural philosophy (though not necessarily equally) well recognised. In addition to John Ray, there was Seth Ward (mathematician, bishop of Salisbury), William Derham (natural philosopher, rector of Upminster), Steven Hales (botanist, physiologist, curate of Teddington), John Morton (naturalist, rector of Great Oxendon, Northamptonshire, author of *Natural History of Northamptonshire*), Matthew Dodsworth (expert on ferns, rector of Sessay, Yorkshire), William Stonestreet (expert on shells, rector of St Stephen Walbrook, London), Adam Buddle (botanist, rector of North Fambridge, Essex), John Lightfoot (biologist, rector of Gotham), Spencer Cowper (meteorologist, dean of Durham), Gilbert White (ecologist, ornithologist, author of *Natural History and Antiquities of Selborne*) and William Paley (archdeacon of Carlyle and author of *Natural Theology or Evidences of the Existence and Attributes of the Deity*). In his own way, Darwin was the last in this series, training for Anglican ordination, observing the natural world and living in a rectory, lacking only the dog collar and the Christian faith.

Religion's legitimisation of scientific activities also served to catalyse specific disciplines. Nicolas Steno is perhaps the first person we can legitimately call a geologist in modern Europe. His observations were, with hindsight, inaccurate, misleading and hampered by a biblical chronology, but he did generate genuinely new and valuable ideas (such as the existence of rock strata and the nature of fossils) and, more importantly, legitimised the study of the earth in naturalistic terms.

Chemistry can also trace its roots to this period, and in particular to the work of Robert Boyle. Likewise entomology: Jan Swammerdam published his *General History of Insects* in 1669, John Ray his *Historia insectorum* in 1710 and Eleazar Albin his beautifully illustrated *Natural History of English Insects* in 1720. All of them were grounded in the same piety. You could see the 'Omnipotent Finger of God', Swammerdam ejaculated, in the anatomy of these smallest and most despised of creatures, all of which showed 'miracle heaped on miracle'. He was talking about a louse in this instance, but he could equally have been speaking of ants, bees, beetles, silkworms, mayflies, cheese mites or even wasps. Entomology was truly a field of divine revelation.

Science, then, did well from the marriage. And so, at first, did religion. Science was religion's first defence against atheism. 'I appeare now in the plane shape of a mere Naturalist, that I might vanquish Atheisme,' wrote Henry More valiantly in his *Antidote against Atheism*.[23] In reality, the threat of atheism was never what the divines claimed. The world was not 'miserably over-run with Scepticisme and Infidelity' as Anglican bishop and natural philosopher John Wilkins lamented. Nevertheless, physico-theology was a powerful ally in this phoney war.

Science also helped heal religion's civil wars. Many of England's pioneer scientists lived through decades of religious and political animosity. They found in natural philosophy a ministry of reconciliation. Thomas Sprat spoke for many when he argued that experimental philosophy was the best antidote to our 'civil differences and religious distractions'. In experiments, he wrote hopefully, 'there can be no cause of mutual Exasperations', for 'in them [men] may agree, or dissent, without faction, or fierceness'. That wasn't quite true, even then, as the relationship between Newton and Hooke, or Newton and Leibniz, or Newton and Flamsteed showed. But it was true enough. Science saved religion from itself.

But at what cost? Not even the most enthusiastic natural philosopher claimed that observation and experiment could replace revelation altogether. Many did, however, claim that physico-theology actively confirmed or, at the very least, paved the way for Christian doctrine. In the Netherlands, Bernard Nieuwentijt, a physician who conducted air pump experiments that were good enough to be reported by the Royal Society, claimed, in his bestselling 1715 book *The True Use of Contemplating the World*, that not only did scientific discoveries prove God's existence and reveal much of his nature, but they also confirmed the literal reading of the Bible. Not all physico-theologians were quite so bold – Nieuwentijt was living under and boxing against Spinoza's dark shadow, and so felt a certain pressure to his work – but most would have broadly concurred.

The argument, however, was not a safe one. Science simply could not sustain the enormous weight that was being placed upon it. Some small cracks began to appear. It was all very well to see in fossils the evidence for the universal flood but the fact that some fossils were completely different to any known living species, and others were fish, rather weakened this argument. Halley, for one, while accepting that there had been a catastrophic event such as that described in Genesis, doubted that it could have happened within the last six thousand years. Linnaeus found it hard to believe that the 5,600 species he named could all have been crammed into the Ark. Even John Ray found it hard to completely reconcile the biblical account of the flood with the variety and distribution of insects, birds, fish and plants. Nothing science revealed, at this stage, incontestably contradicted the biblical narrative. But more sceptical souls wondered whether the new disciplines were as supportive as the theologians claimed.

Small cracks gave way to larger ones. Physico-theology found order and harmony in creation, in which it saw the perfection of the creator. This in itself was a remarkable shift of emphasis. Bacon had set out on the scientific road on the basis of the Fall: the methods of science were justified on the grounds that things were wrong; humans and nature were out of joint, characterised by imperfect knowledge, imperfect senses, imperfect reason and by suffering. And yet the world science seemed to discover was, in William Paley's later, iconic words, 'a happy world, after all' – perfect, intricate, beautiful, useful. It was a move that would have momentous consequences in the long run. From having believed in God because things were wrong (and needed righting), people were now encouraged to believe in him because things were already, self-evidently, right. It left an enormous hostage to fortune.

The problem was evidenced early on, during the moments when the apparent harmony broke down. Long before the poor citizens of Lisbon were shaken by an earthquake, crushed by their cathedral's falling masonry and then drowned by a tsunami on the quayside where they had taken refuge, reports of natural disasters abounded. Shortly before Ray's *Physico-Theological Discourses* were due to be published, news arrived of an earthquake that had destroyed the English colony at Port Royal in Jamaica. Ray paused, reflected, and included a description of the disaster in his book, in which he observed, 'the people of this plantation being generally so ungodly and debauched in their lives, this Earthquake may well be esteemed . . . a judgement of God upon them.' This was biblically defensible, of course, but ill-fitted the perfect and harmonious picture he had drawn.

Not all were persuaded by his argument and some sought simpler routes out of the conundrum. The problem was that such routes led away from Christianity. Perhaps such disasters were part of the natural plan, after all. Or perhaps God never intervened at all. Newton's God may have been intimately involved in his system, but the God of the Newtonians was not. Voltaire found in Newton's science precisely the denuded God he sought – stripped of illogical, inconsistent, immoral ecclesiastical accretions – with which he could then attack Christianity. Theology begat physico-theology begat deism, in which God fired the starting pistol and then retired to the stands to watch the race play out.

Sometimes the cracks were so wide they threatened to bring down the whole physico-theological edifice. The argument from design, it seemed, however much aided by observation and experiment, went vastly further than was merited. It might just about be possible to support the 'simple proposition' that the 'causes of order in the universe probably bear some remote analogy to human intelligence', as Hume's Philo put it at the very end of his *Dialogues Concerning Natural Religion*, the Scottish Enlightenment's most successful critique of natural theology, but it could *prove* or *demonstrate* nothing, and certainly nothing like as much as the physico-theologians claimed.[24] Immanuel Kant would agree.

We remember Hume's *Dialogues* and Kant's *Critiques* as epitomising *the* Enlightenment: the mythical, combative, unitary intellectual revolution with which we began this chapter. But the *Dialogues* were only published (posthumously) in 1770, Kant's *Critique of Pure Reason* in 1781 and his *Metaphysical Foundations of Natural Science* in 1786. In other words, for all their power, they were latecomers to an intellectual movement in which the harmony of science and religion remained dominant.

Nor were they entirely new, at least in their conclusions. From a very different position, a century earlier, Blaise Pascal had made more or less the same point about physico-theology. In as far as God has left signs of himself in his creation, Pascal argued, they lay within the human heart, mind and imagination, rather than material objects. 'Nature presents to me nothing which is not matter of doubt and concern,' he lamented. 'If I saw nothing there which revealed a Divinity, I would come to a negative conclusion; if I saw everywhere the signs of a Creator, I would remain peacefully in faith. But, seeing too much to deny and too little to be sure,

I am in a state to be pitied.' Such knowledge as nature presented of the divine was, at best, partial and ambiguous. Experience was a better guide to God than experiment. In any case, Pascal wrote, arguments from nature, even at their best, could only take you so far. 'Proofs can only carry us to a speculative knowledge of God,' he said, adding, 'to know him in this manner is not to know him at all.'[25]

Margaret Cavendish was one of the most daring thinkers of the late seventeenth century, one of the few women invited to participate in meetings of the Royal Society. Her ideas could sound as if they came from Darwin's notebooks two centuries later: 'Neither can I perceive that man is a Monopoler of all Reason, or Animals of all Sense, but that Sense and Reason are in other Creatures as well as in Man and Animals.'

NINE

MECHANISING THE SOUL

'The mechanism of the human mind': machine learning

In 1749, a 44-year-old English physician, David Hartley, published a long book entitled *Observations on Man, his Frame, his Duty, and his Expectations*. Hartley was the son of a poor Anglican clergyman. Both of his parents died before he reached sixteen but he made it to Cambridge, where he studied under Nicholas Saunderson, the brilliant, blind professor of mathematics. Expected to follow in his father's pastoral footsteps, Hartley found he could not sign the 39 Articles, the foundational statements of the Church of England, and turned instead to medicine, first in London, then in Bath. He married twice, had children, excelled in his medical practice, became a Fellow of the Royal Society, promoted inoculation against smallpox and began to write. *Observations* was his masterwork.

In it, Hartley put forward a thoroughly material, empirical and physiological approach to human beings, and in particular their mental life. 'Since the human body is composed of the same matter as the external world,' he said, 'it is reasonable to expect that its component particles should be subjected to the same subtle laws.'[1] For Hartley, the mind was the brain, and ideas, memories and emotions were physical sensations.

His approach was Newtonian. Drawing on the 'Queries' in Newton's *Opticks*, Hartley posited that light vibrations strike the retina, thereby oscillating the nerves that lead to the brain. 'The tone of a musical string either rises or falls upon altering its tension,' he explained. 'Let us suppose something analogous to this to take place in the component molecules of the brain.' The repetition of such sensory vibrations left vestiges in the brain's 'medullary substance', which Hartley called 'vibratiuncles'. The resulting 'association' – of sensation with the residue 'vibratiuncles' – formed the material basis of memories and ideas. It was a thoroughly materialistic approach to mental life. Hartley wrote openly of 'the mechanism of the human mind' and defended the apparently deterministic conclusions of a theory which seemed to imply that 'each act results from the previous circumstances of body and mind in the same manner, and with the same certainty, as other effects do from their mechanical causes.'[2] It was strong stuff.

A year earlier, a French physician had published something remarkably similar. Julien Offray de La Mettrie was born in 1709, four years after Hartley. He began studying theology but soon turned to philosophy and the natural sciences, taking up medicine in Paris and Leiden. He married in 1739 but left his wife and children three years later when he gained employment as an army surgeon, an experience that shaped his views on what human beings were made of. A few years later, he caught a fever and, paying close attention to the effect it had on his thought processes, came to the conclusion that all human mental activity was merely an aspect of physical sensation.

He too turned to writing, and penned a series of books outlining his mechanistic understanding of human nature. The most famous – unambiguously entitled *L'homme Machine* or *Man Machine* – outlined an understanding of the human body and mind that would have resonated with Hartley. La Mettrie adopted the same approach to the composition of nature as did Hartley: 'Let us then conclude boldly that . . . in the whole universe there is but a single substance differently modified.' He claimed that mankind's spiritual nature was, at root, bodily sensation: 'the diverse states of the soul are always correlative with those of the body.' And he favoured a similarly physiological approach to mental activity. 'As a violin string or a harpsichord key vibrates and gives forth sound,' he wrote, using precisely the same analogy as Hartley would, 'so the cerebral fibres, struck by waves of sound, are stimulated to render or repeat the words that strike them.'[3]

So far, so similar. La Mettrie and Hartley sang from the same hymn book when it came to the physical basis of human thought and nature. Yet, they drew completely different religious conclusions from their work.

Hartley's early resistance to ordination was motivated by his religious faith rather than his lack of it. Pious and philanthropic, he was repelled by the idea of eternal torment that was, in theory, fundamental to Anglican theology at the time. He believed that God desired human happiness, as the 'paradisiacal state' of Eden and the concluding vision of Revelation showed. Humans were made to 'be partakers of the divine nature', in the words of 2 Peter 1, one of Hartley's favourite passages. The 'love of God, and of his creatures' was 'the only point at which man can rest'.[4]

Hartley's religion was not simply *compatible* with his science; it was central to it. Human happiness, morality and intelligence were not after-thoughts or attachments to an otherwise redundant, lifeless material creation; they were *part of* that material creation. The spiritual was not an alien implant into the material but a dimension of – or as we might say today, an emergent property of – the material. 'It does indeed follow from this theory,' he said with characteristic honesty, 'that matter, if it could be endued with the most simple kinds of sensation, might also arrive at all that intelligence of which the human mind is possessed.'[5] The idea shocked many of the orthodox, who were wedded to a kind of 'substance dualism', whereby the spiritual was inserted into otherwise inert matter before it returned to God, when that matter died. By Hartley's empirical reckoning, sensation begat intelligence and intelligence begat morality and spirituality. It was through patterns of sensation, vibration and association that humans learned what caused pain and pleasure, grief and happiness. And it was this that then formed the self, generating imagination, ambition, self-interest, sympathy, moral sense and theopathy, a Hartley coinage meaning, in effect, love of God. Humans were spiritual *and* material; indeed, they were spiritual because they were material. Hartley, in effect, produced a scheme that married material, mechanical, moral and spiritual in unprecedented harmony. 'Have you found leisure to look into Dr. Hartley's *Observations* . . . since your retirement to Monticello?' physician and Founding Father Benjamin Rush wrote to President Thomas Jefferson in 1811. 'Its illustrious author has established an indissoluble union between physiology, metaphysics and Christianity.'[6]

Not so La Mettrie, whose mechanical understanding of mankind grounded one of the most uncompromisingly atheistic visions of the eighteenth century. His scientific arguments were substantially those of Hartley. The idea that man was made of 'two distinct substances' was unsustainable. Everything must appear at the bar of observation and experience. The body was 'nothing but a clock', simply 'a machine which winds itself up'. There was no 'abrupt transition' between animals and men, although he acknowledged that 'even the best' animals do not know the difference between moral good and evil. 'Words, languages, laws, science and arts' emerged naturally, in the process polishing the 'rough diamond of our minds'.[7]

And yet his religious conclusions were radically different. La Mettrie kicked over his theological traces, with good reason. At one point, he recounted the disreputable opinions of 'a French friend of mine' who said 'the universe will never be happy, unless it is atheistic', going on to explain his reasons at length:

> If atheism, said he, were generally accepted, all the branches of religion would be destroyed and their roots cut off. Result: no more theological wars, no more soldiers of religion ... Nature, now infected by sacred poison, would regain its rights and its purity.[8]

At least, that was what this despicable 'wretch' claimed. Few were fooled. La Mettrie dismissed the 'verbiage' of physico-theologians, like Ray and Derham, as 'tiresome repetitions of zealous writers ... more likely to strengthen than to undermine the foundations of atheism'. However much haughty humans wished to 'exalt themselves', they were 'at bottom only animals and vertically crawling machines'. Man was 'originally nothing more than a worm', a 'veritable mole in the field of nature'. Why is the reason for man's existence 'not simply his existence itself'? What need was there for God? There was nothing more to matter than matter. The idea that humans had an eternal destiny could not be dismissed entirely but that was only because 'we know nothing on this subject'. In reality, we know no more of our fate than do caterpillars, not even the cleverest of which could ever have imagined 'that it was destined to become a butterfly'. Better surely that we 'submit to invincible ignorance' and – crucially – to the physical pleasure on which our happiness resides. 'Deaf to all other [i.e. religious] voices, peaceful mortals would only follow the spontaneous promptings of their own being . . . which alone can lead us to happiness through the pleasant paths of virtue.'[9]

For La Mettrie, those 'spontaneous promptings' might not necessarily be virtuous. The paths to happiness led through (or even to) pleasure rather than goodness. If humans were simply machines, and their 'lives' were moulded by pleasure and pain rather than by reason or revelation, they might just as readily (and justifiably) seek sensual pleasure as moral good. And if there were no afterlife to aspire to or immortal soul to protect, there was no need to regret this choice. Indeed, remorse was useless as it only made people suffer needlessly. As he wrote in his final book, *L'Art de Jouir* (*The Art of Pleasure*), nothing could gainsay pleasure, 'the sovereign master of men and gods, in front of whom everything vanishes, even reason itself'.[10]

So it was the physiological and psychological study of mankind which revealed what we were made of, how we worked and ultimately how we should live: morally, happily, piously or pleasurably, happily and godlessly. Much depended on who you were, or where you lived.

'Make the theologians swallow the poison': animal cruelty

The myth that Copernicanism destroyed human pride by decentring the earth is frustratingly persistent. In the early modern period, there were many humans and they had a great deal of pride, but relocating the sun did little to damage it. Rather, human vanity resided almost entirely on what the great intellectual authorities of the age said about who humans were, rather than where their planet was.

It was an article of faith that humans were different, marked by an indisputable superiority and authority over the rest of nature. Aristotle affirmed that although they shared the nutritive part of the soul with plants, and the sensitive part with animals, only humans were in possession of a rational soul. Genesis proclaimed that humans (alone) were made in the image of God. The Psalmist sang that God 'hast made [man] a little lower than the angels and crowned him with glory and honour'. Fleas, flowers and the orbit of planets might declare the glory of God, as the physico-theologians loved to claim, but nowhere did it shine more brightly than in the human apex.

There was disagreement, however, about how exactly that superiority manifested itself. What constituted human uniqueness? There was posture. Since classical times, the fact that humans stood vertical and looked up to heaven was contrasted favourably with those animals that could stare only down into the earth for which they were destined. This, though, was simply the outward form of the true, essential, inner distinction. That lay in the human capacity for speech, 'so peculiar to man', John Ray claimed, 'that no beast could ever attain it'. Or it lay in reason, the unique human intellect that generated our similarly unique capacity for numeracy, memory, imagination, aesthetic sense and curiosity. Or, it lay in religion, humans being the only creature on earth with a sense of time, an intimation of eternity, an appreciation of moral duty and a grasp of the divine.

Whichever option you chose, humans were different – though not necessarily absolutely or invariably so. People acted without reason all the time, most were apt to forget their moral responsibilities and, as the church courts showed, many were irreligious to boot. 'The busy mind of man [will] carry him to a brutality below the level of beasts when he quits his reason,' lamented John Locke in his *First Treatise on Government*.[11] Early modern historian Keith Thomas once observed that 'men attributed to animals the natural impulses they most feared in themselves – ferocity, gluttony, sexuality . . . it was as a comment on human nature that the concept of "animality" was devised'.[12] That was true, but there were innumerable preachers who liked to point out that humans often behaved *worse* than beasts. Animals did not lie, get drunk or engage in gratuitous or sadistic violence. As one Jacobean bishop preached sorrowfully, 'No one beast, be it ever so bad, can be matched to man.'[13]

For all their eternal destiny, humans were fleshly creatures that died like other animals, their bodies committed to the ground, earth returning to earth, dust to dust. And to complicate matters further, the Psalmist rejoiced that God preserved the life of all beasts, gave them their drink, provided them with their food and accepted their praise.[14] The line between man and beast could be frustratingly blurry. 'Even in that which we pretend our peculiar prerogative, ratiocination [reason], they seem to have a share,' the dean of Westminster preached in 1683.[15] Sharing characteristics with other animals was not thus necessarily a problem. It was whether there was anything truly distinct and peculiar about humans that mattered. 'Certainly man is of kin to the beasts by his body', Bacon wrote in his essay 'Of Atheism', but 'if he be not of kin to God by his spirit, he is a base and ignoble creature'.

The temptation to tidy up this vague, compromised and negotiable distinction between man and beast, and turn it into something hard and immutable, was irresistible, not least after those unimpeachable authorities of human nature – the Bible and Aristotle – came under sceptical scrutiny from the sixteenth century. Perhaps (natural) philosophy could pinpoint the difference and clarify the confusion. Unambiguously differentiating humans from animals was important, at least for humans.

Descartes is usually identified as the prime mover in this endeavour, although he was not the true first cause. Nearly a century before him, the Spanish physician Gómez Pereira had published a book which argued that animals were no more than automata. *Antoniana Margarita*, named after his parents, was written in response to a question that had been controversially live since the Italian philosopher Pietro Pomponazzi had published his *On the Immortality of the Soul* in 1516. Was it possible to *prove* the immortality of the soul (*contra* Aristotle) through human reason? Doing so demanded showing not only the soul's indivisibility but also its complete independence from the body, for otherwise it risked being tainted by the body's evident frailty and mortality.

Pereira rose to the challenge, demonstrating the soul's immortality and its detachment from bodily corruption. By his reckoning, the soul alone was responsible for human cognition *and* for sensory perception, which meant that animals, soulless as they were, not only lacked the ability to think but also the capacity to feel. Pereira recognised this went against the theological consensus, not to mention popular common sense. 'It has been accepted by both the learned and the unlearned that brute beasts and human beings alike have been endowed with the faculty of sensory perception.'[16] He was uncompromising in his conclusions, however. What we mistook for animal sensation and response was merely the natural reaction of instinctive tendencies to external stimuli, in the way that an organ pipe might resound if you struck it. There was no feeling there, let alone any thought.

Descartes put forward a strikingly similar view in his *Discourse on Method*, published in 1637. Those who wished to discredit him claimed he had simply plagiarised his ideas about animal mechanisation from Pereira. Descartes was indignant. 'I have not seen the *Antoniana Margarita*, nor do I believe that there is any great need for me to see it,' he protested in a letter to Marin Mersenne, though he added, 'I should like to know where it was printed, so that, if I should need it, I could find it.'[17] His critics' case was strengthened by the fact that, while reasoning on the soul's capacity for self-knowledge in *Antoniana Margarita*, Pereira had concluded, with a flourish, 'I know that I know something: whatever knows, exists: therefore I exist!'[18] It was a philosophical and rhetorical conclusion that bore more than a passing resemblance to Descartes's *Cogito ergo sum*.

However much Descartes did draw on Pereira – *Antoniana Margarita* was originally published more than eighty years before Descartes's *Discourse* and only reissued once, many years later in 1749 – his conceptualisation of man, soul, animals and automata was destined to be much more influential. Descartes's vision of humans was uncompromisingly dualist. Humans were in possession of a body and a soul, and although he was at pains to emphasise that the soul (being 'one and indivisible') was 'really joined to the *whole* body' (being itself 'a unity'), he also argued that 'there is a certain part of the body where [the soul] exercises its functions more particularly than in all the others.'[19]

This was the famous pineal gland, a small part 'situated in the middle of the brain's substance', by means of which the soul, as the seat of all human thought and volition, was able to control and direct the body. Descartes modified this slightly in his final work, *The Passions of the Soul*, suggesting that the causality ran two ways, and that the pineal gland 'can be pushed to one side by the soul and to the other side by the animal spirits', but his general picture of the body was of an entirely passive automaton. 'I suppose the body to be nothing but a statue or machine made of earth,' he said in his *Treatise on man*, written in the 1630s although only published some thirty years later.[20] Its operations were entirely mechanical, functioning without reference to the soul. Animals, bereft of a soul, were bodies, without capacity for thought, memory or will; in effect, just complex machines.

This scheme had the advantage of being physiologically grounded, philosophically sophisticated and, in the way it preserved human uniqueness and separation from animals, extremely clear and tidy. It had the disadvantage of being demonstrably wrong, even according to the physiology of the time. It also came at a significant cost, most obviously to animals, but ultimately to humans too.

The cost to animals was obvious. Although Descartes modified some of his more pitiless ideas about animal sensation, rowing back from the idea that they were completely incapable of feeling, many of his disciples showed less restraint. You could know a faithful Cartesian, it was said, because he had no problem with kicking his dog in public. One should not exaggerate Descartes's impact in all this. Cartesians did not invent animal cruelty, and scientific, philosophical, even theological discussion of this nature had only a limited effect on people's everyday treatment of animals. But it could *legitimise* that treatment. Erecting a wall of separation between humans and (other) animals allowed people to justify actions that might otherwise invite moral censure.

The human cost was subtler. Distinguishing the soul so absolutely from the body, and treating the latter as no more than a machine, was indeed to elevate humans well above all other creatures, but only by placing them on an extremely fragile pedestal. Descartes effectively bet the anthropic house on his 'substance dualism', which he then tried to prove through anatomy.

As that discipline developed in the seventeenth century, it became ever clearer that human and animal bodies bore very many similarities, and could both profitably be compared to machines. In the 1670s, Giovanni Borelli, a Catholic mathematician, astronomer and, more quietly, anatomist, wrote (his posthumously published) *On the Movement of Animals*, a pioneering explanation of how muscles worked, which likened them to a series of levers in the body. Important and influential, the book made no attempt to differentiate the workings of human and animal bodies.

At the end of the century, the English physician Edward Tyson dissected what was probably a chimpanzee that had been held at the docks in London, and published a book entitled *Orang-Outang, sive Homo Sylvestris: or, the Anatomy of a Pygmie Compared with that of a Monkey, an Ape, and a Man*. The very idea of comparing such species in this way was novel – the book is recognised as a founding text of comparative anatomy – and the way in which Tyson drew comparisons between the facial, skeletal and muscular characteristics of each was unnerving. But his conclusion that the species under examination resembled a human more than it did a monkey was rather more troubling. By the time Linnaeus was classifying species in the 1740s, he found himself compelled to write 'the fact is that as a natural historian I have yet to find any characteristics which enable man to be distinguished on scientific principles from an ape.' He knew all too well how explosive this potentially was, writing to a friend 'if I were to call man ape or vice versa, I should bring down all the theologians on my head', adding, 'but perhaps I should still do it according to the rules of science.'[21]

The only thing that prevented Descartes's acceptable 'beast machine' turning into an unacceptable 'man machine' was his immortal, immaterial soul and its anchor in the pineal gland. The problem was that the pineal gland wasn't really paying up. 'In the brain, which we term the seat of reason,' the physician Sir Thomas Browne wrote in his *Religio Medici – The Religion of a Doctor* – in 1643, 'there is not anything of moment more than I can discover in the crany of a beast.'[22] Twenty years later, the English physician and founding member of the Royal Society, Thomas Willis, was more focused in his criticism. Animals 'which seem to be almost quite destitute of Imagination, Memory, and other superior Powers of the Soul [also] have this [Pineal] Glandula . . . large and fair enough.'[23]

With so much riding on a soul that continued to elude anatomists, the path to full human mechanisation was opened up. La Mettrie obligingly took it. He paid a back-handed compliment to Descartes early on in *Man Machine*, honouring him for being 'the first to demonstrate that animals were mere machines', but lamenting that he would have been 'worthy of respect' had he 'known the value of experiment and observation, and the danger of cutting loose from them'. In other words, Descartes obscured the clear message of experimental science by his opaque and unsubstantiated philosophy – although, in his defence, La Mettrie claimed, Descartes's vague philosophising about two substances was really no more than 'a cunning device to make the theologians swallow the poison hidden behind an analogy that strikes everyone and that they alone cannot see.'[24] La Mettrie, physician that he was, was perfectly happy to administer that poison, killing off the soul, any lingering human difference and God's best line of defence too.

'Let us not limit nature's resources': polyp power

In 1740, a young Genevan called Abraham Trembley accepted a job in the Hague as tutor to the children of a Dutch politician, Count Willem Bentinck van Rhoon. Serious and scientifically minded, the Protestant Trembley took his two infant charges pond-dipping and the group examined their treasures under a microscope.

Trembley spotted among the multitude of creatures a type of polyp, now known as a *Hydra*. This had been classified forty years earlier as a plant, but Trembley noticed that it behaved rather like an animal. Not only did it move, but it was sensitive to movement: touch it and its tentacles recoiled. However, if this were an animal, it was no ordinary one, because it also displayed an enviable ability to regenerate. Cut it up and it simply regrew. 'From each portion of an animal cut in 2, 3, 4, 10, 20, 30, 40 parts, and, so to speak, chopped up, just as many complete animals are reborn,' wrote one contemporary journal.[25] Trembley sent fifty in jars to scientists across Europe – and then resent them as the wax seals had inadvertently killed the animals – and in so doing caused a sensation. The normally reserved *Memoires de l'Academie des Sciences* ejaculated 'the story of the Phoenix who is re-born from his ashes, as fabulous as it might be, offers nothing more marvellous.'[26]

In capturing his polyp, Trembley had waded into a treacherous scientific, philosophical and religious swamp. A mechanised and self-regulating universe, such as the one Newton had mathematically demonstrated – or, more accurately, such as his followers thought he had demonstrated – was one thing. It could be, and indeed was seen widely as, proof of the ordered, lawful magnificence of God. Mechanised and self-regulating *life* was another matter altogether, threatening to relegate life to the level of stuff. Just as, by some materialist reckonings, there was nothing special about being human, so, it seemed, there was nothing special, and certainly nothing spiritual, about being alive.

Trembley's polyp was disturbing. The manner in which it blurred and disrupted categories of nature cast doubts on the separate creation outlined in Genesis chapter 1; the map of nature no longer had clear lines drawn across it. Its impressive ability to regenerate put questions against the hierarchical ladder of creation that humans inferred from Genesis and knew from their own experience. If a mere polyp could repair itself in this way, what did that say about the supposed supremacy of humans who tended not to be able to regrow limbs, let alone multiply them?

Trembley's polyp even insinuated the possibility of change across species, or 'transformism' as the idea was then known. Was this a complex plant that had somehow transformed itself into a simple animal? Transformism was a particularly sensitive issue at the time, on account of the French diplomat and natural historian Benoît de Maillet's book *Telliamed, or conversations between an Indian philosopher and a French missionary on the diminution of the sea*. This book drew on the author's time in Egypt and sought to explain the existence of fossils in rocks a long way above sea level by proposing an ancient age for earth. It was a problem others had noticed. Linnaeus was similarly troubled and although he accepted the biblical account of the flood, he sought to expand its impact, time frame and indeed that of the whole earth. 'He who attributes all this to the Flood, which suddenly came and suddenly passed,' he wrote, 'is truly a stranger to science.'[27]

De Maillet built on Nicolas Steno's work with fossils to argue for a vastly older age of the earth than was imagined even by the boldest biblical revisionists. His guess of two billion years was not to be bettered until the twentieth century. He then went further to suggest that the earth had been shaped by chance rather than ordered by design, and that all animals, including humans, had developed from primitive sea creatures. Divine action and providence were edged from the scientific picture. *Telliamed* was the first published account of what would become the theory of evolution, a scientific idea whose religious, social and political repercussion would be incalculable. Trembley's polyp made de Maillet's ideas just that bit more credible.

Perhaps worse than any of this, however, the little polyp's seemingly miraculous ability to regenerate pointed, albeit vaguely, in the direction of self-creation, depriving God of one of his most important jobs, the one for which he was uniquely qualified – namely breathing life into all living things. Living things begat living things: that much was obvious. But were the tentacles that curious scientists lopped off the little creature actually living? In the normal run of things, if you lopped off a limb it died. But if these tentacles were dead, how could they come to life? It was all rather disconcerting.

Trembley's polyp helped mark a more materialistic turn in French thought in the 1740s. De Maillet's mountainous fossils and his ideas of 'transformism', and Trembley's multiplying polyp and its implications for self-(re)generating life set the tone. In 1749, the influential naturalist, Georges-Louis Leclerc, Comte de Buffon, published the first volumes of his *Histoire naturelle*, which would stretch to an impressive thirty-six volumes over the next forty years. His work drew on recent microscopic observations of seminal fluids which demonstrated, he argued, that 'living and animation, instead of being a metaphysical degree of being, is a physical property of matter.' Nature was innately active. It did not, it seemed, require God's intervention at any point.

This worried the theologians. It was less than a century since the microscope had dug them out of a 'self-generation' hole that had troubled them for years. The near-universal belief, at least at a popular level, had been that parasites like worms and flies were spontaneously generated by decaying meat. It was an uncomfortable idea, stealing a little of God's creative power from him, albeit not the most glamorous bit. The careful, magnified gaze of Antonie van Leeuwenhoek and Jan Swammerdam had proved comprehensively that this was not the case, and that maggots only appeared in rotting meat if flies had previously been there. The challenge of self-generating life was averted, at least until De Maillet and Trembley entered the fray.

La Mettrie, of course, needed no encouragement. 'As the sea perhaps originally covered the whole surface of our world,' he wrote, referencing *Telliamed*, in a book on Epicurus, 'could it not have been itself the floating cradle of all the being eternally enclosed in its breast?'[28] Maybe life began in the oceans? Polyps implied that the answer was yes. 'Look at Trembley's polyp,' he exhorted readers in *Man Machine*. 'Does it not contain inside it the causes of regeneration?' Why, then, posit a supreme being at all? Why not simply claim physical causes for life? 'Let us not limit nature's resources.' Nature was enough. Indeed, the study of nature 'can as a result only produce unbelievers.'[29] Thus did a little polyp, channelled by La Mettrie's determined materialism, attempt to overturn the entire tradition of physico-theology. The things you saw under a microscope pointed to nothing more than themselves. Life made itself. Nature was self-sufficient. Bodies were machines. Humans were animals. And the soul, on which so much divine money was bet, was lost.

'It is a monster got of a man and she-baboon': blurred boundaries

At least it was in France. The English were less sure. And the reasons why are instructive.

Descartes's philosophy, which appeared to put the human soul on such a secure, scientific footing, was commended and condemned in equal measure. Widely read and applauded in intellectual circles, it was also censured in the 1640s by the universities of Utrecht and Leiden in the Netherlands, where Descartes had been living for twenty years, and then fell foul of the authorities in France. His views on body and soul were less of an issue than his rejection of scholasticism, his alleged scepticism and, in particular, the doubt that his philosophy apparently cast on transubstantiation. That noted, his understanding of the body did worry those who thought the 'beast machine' simply cleared the way to a 'man machine'. His works were placed on the Vatican's Index in 1663, and banned from the University of Paris eight years later. When his remains were translated from Stockholm to Paris in 1666, no funeral oration was permitted.

All this, however, simply served to radicalise and politicise intellectual opposition. Louis XIV's kingdom became the model of intellectual absolutism in the later seventeenth century. The king sought to ban the teaching of Cartesian ideas in colleges and universities. Censorship was relaxed a little in 1713, after the end of the War of the Spanish Succession, and more so two years later when Louis was eventually succeeded by his nephew, Philippe, Duke of Orléans, acting as regent for Louis XV, but the political authorities still upheld intellectual orthodoxy based on Aristotelianism, with extreme force when necessary. In other European countries, the hangman burned books. In France, he disposed of their authors too.

The fact that Philippe himself was a sceptic who ruled over a court of libertines, and who had the works of Rabelais bound into his Bible so he could read them during Mass, didn't help. The choice was between a Cartesianism that at least preserved some semblance of a harmony between scientific integrity and Christian orthodoxy, but which was officially banned, and an outdated, unconvincing scholastic orthodoxy that was enforced by an autocratic and hypocritical regime. It wasn't much of a choice. As Jonathan Israel has written, France effectively destroyed 'a distinctively Christian moderate mainstream championing premises designed to reconcile faith with reason, and tradition with science'.[30] The banning of Cartesianism and the increasingly desperate attempts to shore up scholasticism did so much to undermine the Church's intellectual unity and authority. It pushed sceptics in an ever more radical direction and paved the way for a tradition of (natural) philosophy whose very purpose seemed to be to undermine the spiritual foundation of political authority. In effect, science was weaponised in an increasingly bitter struggle against Church and state.

For all that La Mettrie's work was unrivalled in its honesty, coherence and notoriety, he was really the inheritor of a half-century tradition of atheistic and materialistic natural philosophy. Most of these books were published anonymously or pseudonymously, and abroad. Many of the authors were forced to take refuge outside France, usually in the Netherlands, or in the case of Mettrie, Helvétius and a number of others, in the court of that great freethinking autocrat Frederick the Great. And most of the texts mixed science with philosophy and philosophy with politics. But that was precisely their point. Science had become a weapon in a philosophical and ultimately political battle against monarchy and priestcraft, a battle that would lead to revolution.

Not so Britain. Descartes was being read in English and Scottish universities from the 1640s and was cautiously welcomed by leading scholars, including Robert Boyle and Isaac Barrow, Newton's professorial predecessor. Descartes's thinking was as fresh as it was bold. He offered a welcome alternative to the Aristotelianism that lingered in philosophical, if not theological or scientific, circles. And he provoked none of the opposition concerning transubstantiation that he did in France. Indeed, Catholic objections to this positively recommended him to English Protestants. Admiration faded, however. His deductive philosophical method was not easily reconciled with the inductive approach, which drew general conclusions from specific observations, that was being developed by the Royal Society. Cartesian physics and cosmology compared unfavourably with Newtonian. And there was distinct scepticism about his views of body and soul. Philosophers were not persuaded by his absolute separation of the two. Anatomists were unable to understand how they might interact. And moralists, or some of them, disliked the implications.

Walter Charleton, like Willis, rejected Descartes's view of the pineal gland and criticised him for an imperfect understanding of anatomy. Theologian Henry More told Descartes that his idea of the beast machine was a 'murderous doctrine'.[31] John Ray said it was 'contrary to the common sense of mankind'. John Locke thought it was 'against all evidence of sense and reason' and taught, in his *Thoughts Concerning Education*, that those 'who delight in the Suffering and Destruction of inferior Creatures, will not be apt to be very compassionate or benign to those of their own kind'.[32] Treat a beast like a machine, and before long you will start treating man like a machine, or like a beast. All in all, in England, there was much less hanging on the pineal gland.

This criticism of Descartes's understanding of man, beast, body and soul was rooted in the stronger tradition of physico-theology in England. Endless tracts lauding God for the anatomical perfection of animals made it harder to dismiss them as mere machines. Books that drew attention to the way that animal behaviour provided a model from which humans might profit made it very difficult to separate and differentiate humans and beasts completely. 'Know'st thou not/ Thir language and thir wayes?' God asked Adam of the earth's living creatures, in Book VIII of *Paradise Lost*. 'They also know,/ And reason not contemptibly.'[33] In Eden at least, there was harmony and intercourse between man and beast.

That was before the Fall, of course, but something of the reason, language and harmony seemed to have been preserved. The English Puritan Nathaniel Holmes reasoned in 1661 that animals were able to 'express their desires with sounds or notes of voice; [and] to express their affections of love and hatred by sociableness and conflicts, of joy and sort by other notes and noises'.[34] The great diarist Samuel Pepys concurred. When, in the same year, he was invited to take a look at a 'strange creature that Captain Holmes hath brought with him from Guiny' – Pepys called it a great baboon but it was probably a chimpanzee or a gorilla – he wrote in his diary that it was 'so much like a man in most things, that though they say there is a species of them, yet I cannot believe but that it is a monster got of a man and she-baboon'. 'I do believe,' he went on, 'that it already understands much English, and I am of the mind it might be taught to speak or make signs.'[35] Pepys was writing informally, privately and unscientifically but it was nonetheless a telling rumination, given that ninety years later La Mettrie asked rhetorically in *Man Machine*, 'Would it be absolutely impossible to teach the ape a language? I do not think so.' If Pepys was anything to go by, this possibility was not quite as shocking (at least to the English) as La Mettrie imagined.

The Protestant tradition of physico-theology also preserved more of the biblical emphasis on the intrinsic value of nature. Isaiah had prophesied how, just as in Eden, eventually all nature would one day be in harmony. The law, indeed the Ten Commandments, insisted that livestock were to be given rest on the Sabbath. The book of Leviticus had decreed that even the land was to be given time to recover every seven years. St Paul himself had promised how 'the creature itself also shall be delivered from the bondage of corruption' at the resurrection. Nature could not be so easily mechanised away. The message even filtered through to sermons. John Evelyn remarked in his diary for 1677 that he heard a preacher insist that 'even the creatures should enjoy a manumission and as much felicity as their nature is capable of [because] at the last day they shall no longer groan for their servitude to sinful man'.[36] David Hartley was among a small number of eighteenth-century thinkers, which included Bishop Butler, who were prepared to take this seriously and countenance the salvation of animals. Such sentiments would, in the (very) long run, pave the way for conservationism. More immediately, they served to undermine the idea of the beast machine, to blur the boundaries between human and animal and to remove a little of the threat latent in the empirical, anatomical and, with David Hartley, proto-psychological investigation of human beings.

In addition to, and perhaps more important than this disarming of Descartes and the ongoing influence of the physico-theology tradition, was the comparative level of intellectual freedom in England and Scotland. If you need a reason why science wasn't weaponised against religion in eighteenth-century Britain in the way it was in France, it was primarily because it didn't need to be.

One should not exaggerate the level of intellectual freedom Britons enjoyed. David Hume failed to secure the chair of moral philosophy at Edinburgh in 1744 primarily because the city's ministers complained that he was an atheist. Thirty years later, he delayed and anonymised the (eventually posthumous) publication of his *Dialogues Concerning Natural Religion* because he knew what kind of reaction it would provoke. Heterodox academics still looked forward to the day when they no longer had to dissemble or jeopardise promotion for voicing unfashionable views.

Nevertheless, English and Scottish thinkers did have it easy, at least compared to French ones. The Licensing of the Press Act was allowed to lapse in England in 1695, a few years after the Act of Toleration had been passed. To publish materialistic books was to risk your reputation rather than your life, and this meant, in turn, that scientific materialists were not caught up in a larger political project of overthrowing the authorities by undermining the ideas on which they stood.

That did *not* mean that scientific materialism was not seen as a threat to Christianity and the social order. Thomas Hobbes, known today primarily as a political philosopher, had admirable scientific credentials. His life was transformed by reading Euclid, he wrote books on optics, mathematics, geometry and physics, and he met and/or corresponded with most of Europe's leading scientists, including Mersenne, Gassendi, Descartes and Galileo. Accordingly, he began his masterwork of political theory, *Leviathan*, with an uncompromising statement of human mechanisation worthy of La Mettrie: 'Seeing life is but a motion of limbs . . . what is the heart, but a spring; and the nerves, but so many strings; and the joints, but so many wheels, giving motion to the whole body, such as was intended by the Artificer?'[37]

This provoked the inevitable charge of atheism, which he vigorously denied. 'Do you think I can be an atheist and not know it?' he wrote angrily to John Wallis.[38] His critics were not persuaded, and 'Hobbism' became almost synonymous with atheism in the later seventeenth century. As with La Mettrie, this was as much to do with the apparent ethical and social implications as with the science itself. Mechanised humans were not, his critics feared, responsible for their moral actions and, absent the soul (about which Hobbes was rather unclear), they could not be bribed or bullied with promises or threats of eternal life or punishment. Licence and libertinism beckoned, and with it the decay of the whole social and political order. Scientific materialism could be as much a threat to religion in England as in France.

Hobbes himself survived, protected by his friendship with Charles II, whom he had tutored, the judicious incineration of some of his manuscripts and his lifelong association with the Cavendish family, who held the Duchy of Devonshire. But he had never needed to weaponise his scientific materialism against the establishment that was protecting him. And in any case, as his own patrons showed, scientific materialism did not need to be a weapon at all.

In 1645, William Cavendish, first Duke of Newcastle, married a formidably intelligent, independent-minded natural philosopher, Margaret Lucas, while they were both exiles from the civil war. Being a woman, Margaret's intellectual avenues were distinctly limited, but she corresponded with her older brother John, who was to become an early Fellow of the Royal Society, and through the Cavendish network would mix with Descartes, Hobbes, Mersenne, Gassendi and others. Her marriage gave her some of the status her mind deserved, so she was able to publish under her own name and was even, at one point, invited to participate in a meeting of the Royal Society.

Margaret Cavendish, as history knows her, was an uncompromising materialist. 'Nature is material, or corporeal, and so are all her Creatures, and whatsoever is not material is no part of Nature,' she wrote in one of her *Philosophical Letters*, which were published in 1664 under the subtitle 'modest Reflections upon some Opinions in Natural Philosophy'. Only the material acted on the material, and that only by immediate contact. There was no need for external, animating spiritual forces. Matter itself was eternal – '[it is] impossible . . . that something should be made or produced out of nothing' – and pregnant with potential. Minds and ideas were material things. Stuff could think.[39]

This had bracing implications for men and beasts alike. Cavendish could be withering about the 'Presumption, Pride, Vain-glory and Ambition' of mankind, and equally stirring about the capacities of animals. Rather than being Cartesian automata, beasts were possessed of a reason and language after their own kind, that was inaccessible to humans.

For what man knows, whether Fish do not Know more of the nature of Water, and ebbing and flowing, and the saltness of the Sea? or whether Birds do not know more of the nature and degrees of Air, or the cause of

Tempests? or whether Worms do not know more of the nature of Earth, and how Plants are produced? or Bees of the several sorts of juices of Flowers, [than do] Men? . . . though they have not the speech of Man, yet thence doth not follow, that they have no Intelligence at all.[40]

It was a remarkable sentiment that could have come from one of Charles Darwin's troubled notebooks, 170 years later. Cavendish went further. Why should we imagine that humans were the centre of creation, that 'all creatures were made to obey man, and not to worship God, only for man's sake, and not for God's worship, for man's use, and not God's adoration, for man's spoil and not God's blessing'?[41] Taking the Psalmist at his word, she asked rhetorically, 'why may not God be worshipped by all sorts and kinds of creatures as well as by one kind or sort?' She even ruminated on the rational capacity of non-human and even vegetative life. 'Neither can I perceive that man is a Monopoler of all Reason, or Animals of all Sense, but that Sense and Reason are in other Creatures as well as in Man and Animals.' It all made Hobbes look a bit conventional.[42]

And yet, none of this was intended to cast a shadow over the divine, as Hobbes seemed wont to do. Cavendish was remarkably close to Pascal in this respect. She pooh-poohed the nascent tradition of physico-theology, criticising the brilliant anatomist William Harvey by saying 'he doth speak so presumptuously of God's Actions, Designs, Decrees, Laws, Attributes, Power, and secret Counsels' that it sounded almost 'as if he had been Gods Counsellor and Assistant in the work of Creation'.[43] She was cruelly witty about Henry More's *Antidote against Atheism*, observing that the author's (familiar rhetorical) insistence that no one seriously denied the existence of God rendered his entire book rather pointless; why make 'so many arguments to no purpose'? Any attempt 'to prove either a God, or the Immortality of the Soul,' she wrote, 'is to make a man doubt of either'. It was only when natural philosophers tried to prove God's existence through science that people started to doubt it. In Pascalian fashion, she wrote that God was 'to be admired, adored and worshipped' rather than 'ungloriously discoursed of by vain and ambitious men'.[44]

The consequence of all this was a somewhat more chastened understanding of the divine than her age was accustomed to. God was not to be known by reason, for no created mind was capable of that, but by faith. He was to be spoken of not with arrogance and certainty but with apophatic humility: 'when we name God, we name an Unexpressible, and Incomprehensible Being.'[45] Importantly, though, however he was to be known or discussed, this God was perfectly compatible with materialism, because materialism was part of his plan to make creatures that would be capable of worshipping him.

'Antagonists think they have quenched his opinions': the tide turning

Cavendish's natural philosophy did not have the rigour of Hartley's but her work showed how the materialism towards which the new science pointed need not be the mortal threat to religion that the *philosophes* hoped and the devout feared. Her impact, sadly, was minimal. Scientific materialism remained a menace in England as it was in France. Boyle lecturers spoke on how 'Matter and Motion Cannot Think', the spectre of Hobbes was everywhere conjured to scare the spiritual horses and materialists were denounced as dangerous 'free thinkers'.

And yet, materialist ideas remained, so to speak, alive. John Locke, who was to English philosophy in the first half of the eighteenth century what Newton was to its physics, was somewhat more circumspect about materialism than Margaret Cavendish. Nevertheless, his defiantly empirical approach to human knowledge, his rejection of innate truths, his conviction that ideas were grounded solely in sensation and reflection, his belief that God could impart the power of thought to matter and his location of the self in consciousness rather than an immaterial soul all pointed in a materialist direction.

Locke provoked an angry response from many churchmen who accused him of defying the scriptures and denying the entire basis of Christian salvation. How could a temporary, psychologised self replace an immaterial, immortal soul, asked the theologians, in particular those Tory High Church theologians resident at Locke's alma mater, Oxford. How was such a self expected to survive death? And how, in that case, was God expected to reward the righteous and punish the damned? Locke was labelled a heretic.

The philosopher was, however, transparently devout and earnest – albeit, like Newton, quietly anti-Trinitarian. He grounded the entire argument of his *Letter Concerning Toleration* explicitly and repeatedly in the scriptures, did much the same for his equally influential *First Treatise on Government* and paraphrased and annotated an edition of St Paul's letters to the Galatians, Corinthians, Romans and Ephesians, which was posthumously published. It was hard to tar Locke with the Hobbist brush.

His ideas were to prove influential, in no small part due to Voltaire's enthusiasm for and relentless promotion of them. 'Mr. Locke has displayed the human soul in the same manner as an excellent anatomist explains the springs of the human body,' the great sceptic effused in his thirteenth *Letter on the English*. 'Everywhere [Locke] takes the light of physics for his guide', proceeding gradually, with intellectual humility, and reaching only tentative conclusions.[46]

By Voltaire's reckoning, Locke's 'modest philosophy . . . so far from interfering with religion, would be of use to demonstrate the truth of it', precisely because it avoided the painful dilemma on which Descartes and the churchmen sat themselves. Animals, Voltaire reasoned, have the same organs, sensations and perceptions as humans, as well as some capacity for memory and ideas. Were it 'in the power of God to animate matter, and inform it with sensation', this would all be explicable. All of creation would be animated by its creator, only in different ways and to different degrees. But those who denied the very concept of animated matter were left with either dull, inert matter in which a soul had been implanted (i.e. humans) or dull, inert matter in which it hadn't (animals). And that would mean 'either that beasts are mere machines, or that they have a spiritual soul', neither of which was an attractive option.

Not all the divines were persuaded by this argument, even as Locke's stock rose as the century wore on and many came round to his way of thinking. Their disagreement was telling. Locke, whatever one thought of him, wrote 'within the walls', so to speak. English ecclesiastical authorities, or some of them, might have deemed him a threat to faith and order but, after the Glorious Revolution, he was never going to lose his liberty, let alone his life. There were a great many authors who were considerably more subversive than Locke but few of these feared for their freedom, or even felt the need to publish anonymously or escape abroad. Indeed, it was the freethinkers, materialists and sceptics who were more likely to relocate *to* England than flee from it, and some of them, like Voltaire, who took refuge there in 1726, would spread the gospel of English intellectual toleration and flexibility Continent-wide.

So it was that when, twenty years after Voltaire left, Hartley wrote his *Observations on Man*, he found himself advocating a scientific materialism that was commonly judged consonant with religion or, at worst, was merely disputed and denounced by his co-religionists. In effect, in writing his mechanistic *Observations* he was forging a weapon that *could* be used against religion, and many feared would be. However, not only did he carefully blunt its anti-religious edge but he went as far as to insist that his science could be wielded *in support of* Christianity. And no less importantly, because of the political environment in which he wrote, there were very few British radicals sufficiently indignant to want to pick up and use his weapon against religion. Unlike in France, the kingdom of God was less ominous and therefore less in need of storming.

But there is an ironic coda to this tale of Anglo-French difference. Contemporaries took badly to La Mettrie's *L'homme Machine*. It was, in the words of one scholar, 'perhaps the most heartily condemned work in an age that saw the keenest competition for such honours'.[47] The book was read and discussed and banned and burned across the Continent; condemned by the pious, of course, but also by fellow radicals who shared La Mettrie's atheism and broadly agreed with his mechanisation of the human, but disliked his ethical conclusions, and in particular his conviction that such mechanisation legitimised, indeed demanded, a kind of hedonistic utilitarianism. This was the kind of thing that gave materialism a bad name. The godly claimed that were it not for his immortal soul, man would merely satisfy his basest desires. La Mettrie seemed to agree. Radicals despaired. Divines crowed.

And providence seemed to confirm their judgement. Too dangerous even for Leiden, La Mettrie was forced into exile again, this time with Frederick the Great in Prussia. The king welcomed him as his physician but, after three years, La Mettrie developed a gastric illness after a feast held in his honour and died. The godly gloated. La Mettrie's mechanistic ideas remained popular – his *Oeuvres philosophiques* were printed eleven times in the quarter century after his death – but the author's moral message was quietly overlooked. A radical idea was politically de-radical-ised in the name of intellectual respectability.

Something very different happened in Britain. Hartley's ideas were to prove influential largely due to the efforts of Joseph Priestley. Priestley was the greatest chemist of his age, a brilliant experimenter famed for identifying oxygen, as well as ammonia, nitrous oxide, sulphur dioxide and half a dozen other gases. He was also an important dissenting minister who studied at the nonconformist Daventry Academy and entered into ministry in Suffolk and Cheshire. In some ways, he was the model of the harmony between science and religion that characterised so much of the British Enlightenment, although as he aged he drifted ever further from Christian orthodoxy.

Priestley esteemed Hartley, republished his *Observations* in 1775 with his own accompanying essays and claimed that the book had influenced him more than any book other than the Bible. Hartley's materialist concep-tion of the brain and ideas was not only, by Priestley's reckoning, the most scientifically credible but also the best foundation for true religion. Priestley reasoned that there was nothing in the scriptures that mandated an immaterial Cartesian soul. The fact that Priestley could say all this from within the Christian fold in Britain was highly instructive. Such a position was simply not available in France, had La Mettrie's successors even wanted to occupy it.

Nevertheless, for all Britain's much-vaunted toleration, it remained a vulnerable place, and more precarious as the century drew to an end. Priestley was, in addition to his religious and scientific eminence, a prom-inent political figure. Sympathetic to the American Revolution, he spoke out against the Established Church and expressed his sympathy with the French Revolutionaries. This was not, at first, an unusual view in Britain but by 1791 the nation was increasingly anxious about the course of events in France. Priestley did not share that anxiety. In July 1791, he helped organise a dinner to celebrate the second anniversary of the storming of the Bastille. His political opponents were furious and whipped up a mob that went on a rampage, looting and then burning down Priestley's house and library. The chemist managed to escape first to London and then to America, where he died in 1804.

Hartley's reputation was to suffer alongside Priestley's, as did the cause of science and religion. When Priestley escaped to America, the future president, Thomas Jefferson, mixing his science, religion and politics, compared the chemist's plight to that of Galileo. 'Antagonists think they have quenched his opinions by sending him to America just as the Pope imagined when he shut up Galileo in prison that he had compelled the world to stand still.'[48] Two hundred years earlier, it had been the Vatican and its vindictive Pope Urban, who, it seems, had had a problem with science, imprisoning the man whom Protestants would seize on as the symbol of Catholic despotism. Now it was the pious English from whom pioneering scientists fled to intellectual freedom. At least, that was Jefferson's perspective from the New World. It was to prove a prescient one.

PART 3

EXODUS

George Combe lecturing on phrenology. Phrenology –
the idea that you could read someone's character from
their skull – was serious science in the early nineteenth
century, and Combe's *The Constitution of Man* was one
of the century's scientific bestsellers. The faithful were
not impressed with this 'science', however, because
it undermined human morality and freedom.

TEN

ABOUT TIME

'Into the abyss of time': naturalising Genesis

In March 1860, Charles Darwin learned something from Leonard Horner that surprised him. Horner was the outgoing president of the Geological Society of London and had sent Darwin a copy of his final presidential address, in which he had argued that chemistry might be able to explain geological phenomena. Grateful as he was to read this, it was neither Horner's geological expertise nor his chemical ideas that struck Darwin.

'How curious about the Bible!' he exclaimed. 'You are coming out in a new light as a Biblical critic!' Horner had remarked that public knowledge of geology was now so widespread that even the most casually acquainted could see the discrepancy between what geology claimed for the age of the earth and what the Bible did – or, more precisely, what people (like Darwin) thought the Bible did.

Horner was taking aim at the marginal notes in the standard edition of the Authorised or 'King James' translation, which was almost ubiquitous in Britain at the time. These stated that the world had begun in the year 4004 BC. Horner argued that the date be removed from future editions; it was, after all, no more than the calculation of James Ussher, a seventeenth-century bishop of Armagh. Darwin, who thirty years earlier had trained to become an Anglican clergyman, did not know this and confessed to Horner that 'I had fancied that the date was somehow in the Bible.'[1]

Archbishop Ussher assumes a small, if highly symbolic, role in the history of science and religion. His hazardously precise dating of creation – 'at the start of the evening preceding the 23rd day of October (on the Julian Calendar) 4004 B.C.' – seemingly epitomises religion's hubris, error and ultimate surrender to science. People in 1660, when Ussher was writing, believed the world was six thousand years old, on biblical grounds. Two hundred years later, they thought it was millions of years old, on scientific grounds. Case closed. Except that this particular narrative, much like the bigger story of science and religion in which it sits, obscures far more than it reveals, reducing Genesis and geology to distinct, discrete, immobile and self-evident entities that neither was, least of all in this period of scientific ferment and flux. Both Genesis and geology proved as protean and unsettled as the landscapes they purported to explain.

Christians had been calculating creation since Theophilus of Antioch in the second century, who had arrived at 5529 BC. Archbishop Ussher was unique only in as far as he was (un)lucky to have had his particular calculations entombed in a biblical authority so absolute that even Darwin couldn't distinguish text from commentary. The task of dating creation and assembling a reliable 'Universal History' had become more pressing, and more portentous since the mid-sixteenth century, when previously unknown sources from apparently ancient civilisations in Egypt, India and China began to trickle into Europe. And when Nicolas Steno convincingly argued that fossils were organic remains, a new dimension was added, and Big History was born.

In the final two decades of the seventeenth century, a new 'theory of the earth' genre emerged, which sought not so much to reconcile (natural) history with scripture – no one thought the two needed reconciling – as to use the former to amplify and explain the latter. In 1681, the writer Thomas Burnet published his *Sacred Theory of the Earth*, which put forward a complicated set of ideas about the world's formation that involved the action of the sun on the earth's crust and its subterranean waters. This was not an alternative to God's activity so much as an attempt to understand it.

In his footsteps, a number of natural philosophers ventured material explanations for the foundation and form of the world, for the existence of geological strata and fossils and for the dispersal of species. William Whiston, theologian, mathematician and disciple of Newton, wrote *A New Theory of the Earth* in 1696, which argued that the earth originated as a comet that had been captured by the sun, and that the flood was the result of vapour from a neighbouring comet's tail. John Woodward, another naturalist, who established the Woodwardian Chair of Geology in

Cambridge, published his *Essay Towards a Natural History of the Earth* the previous year, which argued that geological strata were the result of different densities of the materials that settled after the flood. (Steno had made a similar suggestion for the geology of Tuscany thirty years earlier.) Carl Linnaeus, troubled (as many were) by the variety and range of species across the planet, posited a large, vertiginous island near the equator, blessed with different climates, from which animals could disperse after the deluge. That such explanations were all wrong is irrelevant. Their importance lies in the way in which they began to naturalise the Genesis creation stories and legitimise a scientific explanation for the development of the earth.

Or rather, scientific explanations, because it would not be until the middle years of the nineteenth century (and arguably not until the twentieth) that anything like a geological consensus arose. The very word geology was not used in its modern form until the end of the eighteenth century – 'natural history' or 'cosmology' were the preferred terms – and geological explanations, when they came, tended to be somewhat heavier on theory than they were on evidence.

Unlike the other sciences, there were no accepted measurements (indeed, it wasn't even clear what one should be measuring), no objective standards, little opportunity for experimentation and none for prediction. In his great systemisation of the sciences in the 1830s, the French philosopher Auguste Comte hardly even mentioned the discipline. Geology was a comparative subject, and even though it was reasonably clear what one should be comparing – strata, fossils, etc. – interpretation was endlessly contestable.

As a result, there were radically differing 'schools' of thought. 'Catastrophists' argued that geological change occurred in occasional, rapid and intense convulsions, and pointed to volcanoes and earthquakes as evidence. In contrast, 'Uniformitarians' (another word coined by William Whewell), argued for extremely slow, cumulative effects, and pointed to sedimentation and erosion in their defence. Corresponding to this division, although not entirely coterminous, there was disagreement over which natural phenomena were responsible for change. 'Vulcanists' (or 'Plutonists') argued that heat was the primary agent of geological transformation, the centre of the earth being molten rock that occasionally extruded over the surface and created new strata. 'Neptunists' thought water the effective element, working through sedimentation, erosion and the crystallisation of minerals. There was no single geology – still less a geological orthodoxy blessed with sufficient and conclusive evidence – to topple Genesis, and indeed some geological ideas, Catastrophism and Neptunism in particular, seemed positively to support parts of the biblical narrative.

For all that evidence and argument were finely balanced and indeterminate, however, the sheer scale of time that uniformitarianism required weighed heavily against it in a religious culture that believed, like Darwin, that the earth was only a few thousand years old. The geologist Charles Lyell coined the phrase 'the present is the key to the past', and so refused to admit as explanations any processes that were not currently observable. Not everyone was quite as stringent as Lyell but, either way, such a gradualist approach to the formation of the earth was clearly impossible to reconcile with a world history that began in October 4004 BC.

Gradualism was not an entirely new idea. Robert Hooke had been fascinated by Jesuit reports of China where, he believed, sages claimed the world to be many millions of years old. Edmund Halley suggested that the (considerable) age of the earth might be estimated through measuring the (supposedly increasing) salinity of the sea. Georges-Louis Leclerc, Comte de Buffon proposed seven stages for a cooling earth (which, like other planets, had been formed from the impact of a comet on the sun), and justified his theory through a series of experiments on heated spheres. On the basis of how long each took to cool to room temperature, Buffon estimated that the earth was 75,000 years old, although privately he reckoned it might be as old as three million.

The Scottish geologist and farmer James Hutton went further than them all. In a paper submitted to the *Transactions of the Royal Society of Edinburgh* and a subsequent 1795 publication, *Theory of the Earth* – Burnet's 'Sacred' was consciously dropped from the title – Hutton argued, on the basis of his agricultural observations, that the land was reshaped and stratified by constant weathering, a process that offered 'no vestige of a beginning, – no prospect of an end'.[2] Although not going quite as far as Aristotle's eternity, Hutton's measureless time frame was worrying.

All this was clearly unacceptable on Ussher's time frame, and the orthodox said so. Buffon's estimates were condemned by the theological faculty at the Sorbonne. Hutton's theory was vigorously criticised, and he was (accurately) denounced as a deist. Such theories refused to die, however, and began to gain widespread attention and ground in Britain from the late 1820s. The timing was particularly inauspicious. Parliament passed the Catholic Emancipation Act in 1829, which removed 150 years of political and civil penalties against Roman Catholics. Protestants were furious and anxious, haunted by fears of popery and moved to defend the Bible on which their Reformed faith stood. A few years later, a movement of leading Oxford clerics – John Henry Newman and Edward Pusey most prominent among them – adopted Roman ritual, theology and eventually the Church itself, compounding such fears. Biblical Protestantism was being eroded from within as well as assailed from without.

Geology was part of this erosion and its ideas, particularly uniformitarian ones, provoked a loose school of 'scriptural' or 'Mosaic' geologists, who insisted on a literal reading of the early chapters of Genesis, and generated a minor publishing phenomenon. With greater but usually lesser levels of sophistication, clerical authors tried to fit the growing body of geological evidence into a Procrustean scriptural bed. Geology revealed the literal truth of Genesis 1–11, and if it didn't, so much the worse for geology.

And yet, the problem turned out not to lie with Genesis, which swiftly showed itself to be considerably more flexible than Procrustes' famous bed or Ussher's equally famous date. The scriptural geologists made a lot of noise, and played well to their Protestant base, but they had limited influence in ecclesiastical circles, less in academic ones and none in scientific. It soon became apparent that the Genesis time frame could be gently prised apart by the right hands. Thomas Chalmers, an endlessly energetic Scottish minister who was as famous for his pastoral work and his contribution to political economy as he was for his theology (for which he was a professor), believed in the literal days of creation but argued that there were significant gaps between God's original act of creation in Genesis 1 verses 1–2 and his earthly acts of creation in verse 3 onwards. Others dusted off their Augustine and argued that the days of Genesis 1 need not be interpreted as literal days but were better understood as epochs or ages. After all, had not the Psalmist himself written of God 'a thousand years in thy sight are but as yesterday when it is past' (Psalm 90:4). Chalmers' 'gap theory' and Augustine's 'day-age' theory would be reused and repolished by Young Earth Creationists in the twentieth century, when the scientific debate had moved on so far that the ideas now elicited only mockery and contempt. But they were less contemptible in the early nineteenth century, when evidence was piecemeal, theories contestable and geological eras speculative. Such re-reading demonstrated how creative even Protestants were prepared to be with Genesis 1 if needs be.

It was the same with the flood of Genesis 6–8, the other biblical story that geologists began to question. Virtually no one doubted its historical veracity, especially after similar flood narratives were found in other, very different cultures around the world. Nevertheless, as early as the seventeenth century, Edward Stillingfleet in *Origines sacrae* (1662) and the biblical scholar Matthew Poole in *Synopsis criticorum* were willing to suggest that Noah's flood may in fact have been localised to Mesopotamia rather than being universal. This was to remain an intensely controversial idea in Christian circles for many years. 'Diluvialism', the idea that historical, traumatic hydrological forces reshaped the earth, remained the default position well into the nineteenth century. However, by the time the Revd William Buckland recognised, in his 1836 Bridgewater Treatise on *Geology and Mineralogy*, that there were no human remains in the diluvium as one would presumably expect from a universal flood, the impetus to reinterpret Genesis 6–8 as an altogether more regional and less punitive event became overwhelming. All but the scriptural geologists were happy to oblige.

The challenge geology posed to religion was, therefore, only superficially scriptural. The faithful, or at least many of them, were willing to accommodate new readings of the relevant chapters of Genesis when the emerging science necessitated it. It was surprisingly easy to jettison Ussher and live with a gap-theory Genesis, or a day-age Genesis, or a localised flood, or even, eventually, a mythologised Genesis. Less easily accommodated was the loss of authority for the scriptures or, more precisely, for those who traditionally had the right to interpret them. This was natural philosophers vs. theologians in fourteenth-century Paris or early seventeenth-century Rome all over again.

Clerical naturalists had long prized the natural world as God's book of works, but by the early years of the nineteenth century they were confronted with the realisation that it diverged, at least in parts, from the plain meaning of his book of words. This, as noted, was not necessarily a problem; texts could always be reread. As with the question of heliocentrism among theologians at the time of Galileo, however, the questions 'How should we (re)read these texts and which interpretation should have authority?' were inextricably tied in with the wider question of 'Who was to say?'

Throughout the eighteenth century (and well into the following) the answer was obvious – the clergyman-naturalist in his rectory-laboratory. From the first decade of the nineteenth century, however, 'geology' was becoming professionalised. The Geological Society of London was founded in 1807 and although its leading members would, for a generation, be ordained ministers, its existence symbolised a subtle shift in intellectual gravity. It is telling that the scriptural geologists tended to be educated members of society – clergymen, physicians, classicists, linguists, antiquaries – for whom natural history was a passion and a pastime, rather than anyone who could claim to be anything like a 'professional' geologist. In the words of one scholar, 'Mosaic geology was . . . in part a cultural reaction to the social and cognitive exclusion of all but self-styled experts from the area of speculation that had been open to all.'[3]

Earthly questions were to be increasingly answered by chisel and hammer rather than concordance and text. Indeed, as the century wore on and German biblical criticism seeped into the seams of English intellectual life, it was even suggested that fossils and strata might constitute a *more* reliable historical archive than texts that were themselves layered upon one another or, worse, corrupted by authorial bias or even outright forgery. Science offered the intellectual certainty that religion could no longer provide.

And yet, for all that gradually professionalising geology posed a threat to the clerical naturalists who had done so much to legitimise science in eighteenth-century Britain, it was clerical naturalists themselves who endeavoured to authorise that change, supreme among them the Revd William Buckland and the Anglican layman Charles Lyell.

Buckland was to the hammer born. His father, Charles, was a vicar and enthusiastic amateur geologist in south Devon, who took the young William along and into the fossil-rich Jurassic Coast. Graduating from

Oxford in 1804, and ordained four years later, Buckland Jnr was as keen on earthly things as heavenly, and he became the university's Reader in Mineralogy in 1813. He was not to everyone's taste. Darwin judged him 'a vulgar and almost coarse man . . . incited more by a craving for notoriety . . . than by a love of science'.[4] The judgement is harsh. Buckland was undoubtedly eccentric – his Oxford rooms were crammed with bones, some taken from dinner, his dining table famed for serving mouse, rat, dog, squirrel, badger, crocodile and tiger – but he was also learned, popular and pioneering.

Educated opinion was far from convinced that studying fossils was a suitable pastime for young men who were being trained to bury the dead rather than exhume them. But Buckland did more than anyone else in the 1820s and 1830s to vindicate geology in the eyes of a sceptical Oxford. His inaugural lecture as the university's first Reader in Geology in 1819 was entitled *Vindicae Geologicae, or the Connection of Geology with Religion Explained*. By drawing attention to the anatomical parallels between modern and extinct animals (in which he was following the work of the Frenchman Georges Cuvier), writing the first full description of what would soon be called a dinosaur, recognising caves as essential to the study of the past, promoting the gap theory of creation, downplaying the significance of Noah's flood and, above all, acting as a restlessly energetic advocate and publicist for the new science, he also did more than almost anyone else to place the discipline on a scientific and socially respectable footing.

Charles Lyell was one of Buckland's students, and was to pick up the geological baton from his master, albeit critically. Darwin claimed that the discipline was more indebted to him 'than to any other man who ever lived'. He had the outward trappings of a clerical naturalist in all but ordination – classically educated, wealthy, Anglican, fascinated by the natural world – but he stepped decisively away from that tradition and fought 'to make science a profession', writing, provocatively, that 'the physical part of Geological enquiry ought to be conducted as if the Scriptures were not in existence'.[5] Lyell was responsible for professionalising the discipline. He held a Chair in Geology at the newly founded King's College in London and wrote a three-volume *Principles of Geology*, the first of which Darwin consumed on the *Beagle*, which became the foundational text for the new science. His arguments for gradual geological change helped settle the debate in favour of uniformitarianism. In doing so, he provoked the wrath of the scriptural geologists. He was unrepentant.

> We cannot sufficiently deprecate the interference of a certain class of writers on this question who ... do not scruple to promulgate theories concerning the creation and the deluge, derived from their own expositions of the sacred text, in which they endeavour to point out the accordance of the Mosaic history with phenomena which they have never studied.[6]

Buckland and Lyell were not alone, and there are others such as the French Protestant Georges Cuvier, the Revd Adam Sedgwick, the Revd William Conybeare and John Phillips – equally pious in religious faith and scientific works – who also helped establish geology as a science. But Buckland and Lyell were especially significant in the way they not only professionalised geology but also separated it from its religious scriptural twin, in the process giving both a new lease of life.

In the same year that Leonard Horner wrote to Darwin about James Ussher, a collection of essays about the reliability of the Bible, entitled *Essays and Reviews*, whose reputation for religious controversy would leave *The Origin of Species* in the shade, was published. Several chapters touched geology, and critiqued attempts to find false harmonies between it and Genesis. 'He who notices . . . that the explanations of the first chapter of Genesis have slowly changed, and, as it were, retreated before the advance of geology, will be unwilling to add another to the spurious reconcilements of science and revelation.'[7] It was good advice. 'Spurious reconcilements' had dogged the history of science and religion for ever. The two would get on better if they worked alongside one another rather than in each other's space.

And yet, even with this arrangement in place, geology still left many people unnerved. It was not that Genesis had to be re-read, or that Ussher had to be ignored, or even that clerical naturalists would have to move aside for professional geologists. Rather, it was about time. 'Time we may comprehend,' the physician Thomas Browne had written in *Religio Medici;* "tis but five days older than ourselves.'[8] No longer. The idea, implicit within

Ussher's time frame and the Universal Histories of the early modern period, was that the history of the cosmos was virtually coextensive with the history of (civilised) mankind. Time and the world began not long before our appearance, and were intimately tied up with our future. Time was about the human drama.

This was the chronological equivalent of heliocentrism, only *much* more impactful. Humans were temporally at the heart of creation. Indeed, temporally speaking, humans were pretty much the whole of creation. Revising this left many reeling, far more so than decentring the earth had done. Relocated to a stage that was 'without vestige of a beginning, without prospect of an end', human activity invariably seemed smaller and less significant. Humans were diminished as, by implication, was the God in whose image they were made. As John Playfair wrote in his biography of James Hutton, 'the mind . . . grow[s] giddy by looking so far into the abyss of time.'[9]

'A barren Golgotha': the case of the brain

'However startling the results of geological investigations may appear, the records which establish them are too authentic and precise to leave room for doubt as to their substantial truth.' So began one of Victorian Britain's scientific bestsellers.[10] Scientific truth could sometimes be unpalatable, but that didn't stop it from being true.

The book in question was not, in fact, about 'geological investigations'. Geology was simply the best place to start. It revealed an ancient, strange and changing earth, a planet on which, the 'most esteemed authorities' now agreed, 'death and reproduction formed parts of the order of nature' long before man appeared. 'Man himself' was 'to a certain extent an animal in his structure, powers, feelings, and desires', an animal that was 'adapted to a world in which death reigns, and generation succeeds generation', a world that 'seems not to have been changed at his introduction'.[11] He was, though, an animal that could at least grasp the truth of his condition through a new scientific discipline.

That discipline was not evolution and the book was not by Charles Darwin. Around the time that Hutton and Buckland were advancing the cause of geology, another powerful and popular science was emerging in

Europe. It was determinedly empirical. It firmly rejected metaphysical speculation in favour of careful observation, measurement and, where possible, anatomical investigation. Ideas of 'mind', let alone 'soul', it insisted, were vague, unscientific and ultimately untenable.

Like all sciences, it had its premises and hypotheses. In this instance, the vital ones were that 'the brain is the organ of the mind', as this best-seller put it, and that the brain was modular, 'different parts of it manifest[ing] distinct faculties'.[12] The science went on to postulate that the use or neglect of these faculties could cause the respective regions of the brain to enlarge or atrophy, much in the way that the right hippocampus of London taxi drivers would be shown, many years later when neuro-imaging technology enabled it, to be enlarged with use. Although there was debate over how many 'modules' or distinct regions were present in the brain, the belief was that they separated into different categories or classes, some being responsible for more 'primitive' faculties like sexual impulse and self-preservation, others for more 'advanced' ones, like benevolence, wisdom and religious sentiment. What really marked this new scientific discipline out, however, and enabled it to pour 'a great flood of light' on 'the past barrenness of mental science' was the realisation that organs differed between people and that these cognitive and perceptual differences had cranial correlates. Or, less obscurely, that you could under-stand the organs and faculties of people's brains by tracing and measuring the size and shape of the skull. The new science was phrenology.

That phrenology is today universally dismissed as quackery blinds us to the seriousness with which it was treated in the early nineteenth century. It was a science that was widely judged to have brought the hopeful but vague materialisation of the human mind in the eighteenth century into a more empirical realm in the nineteenth. Its origins lie in the work of Franz Joseph Gall – yet another in the eighteenth-century line of would-be priests turned medics – and his disciple J. G. Spurzheim. Gall had been struck, in his training, by the anatomical similarity between man and ape, and began to wonder about the way in which the anatomical structure of different species reflected their cognitive capacities. If thought and feeling were brain activities, and the brain was a material object, it stood to reason that thought and feeling should take some material form and leave some material mark. Discovery of this would obviate the need for philosophical (and theological) speculation about human character. The physical brain, which Gall meticulously anatomised, would be sufficient.

Gall collected animal and human skulls, began to lecture and developed a set of theories about the correlation between the brain and its case. He was first ignored and then banned in Vienna, his works placed on the Index on the grounds of their materialism and irreligiosity. He persevered, travelling Europe on an increasingly successful lecture tour of scientific societies. His ideas began to seep into the scientific mainstream.

There remained dissenters, and when his theories first reached British shores, via an 1815 article in the respected *Edinburgh Review*, the response was critical. Much of the objection was rooted in the discipline's apparent determinism, but when Spurzheim visited Edinburgh the following year, illustrating his master's ideas with a careful dissection of the brain, many sceptics were won over. Moral fears should not be enough to deny scientific truths. However upsetting the results of phrenology might be, the scientific evidence was too authentic and precise to leave room for serious doubt.

Edinburgh became a launch pad for the new science. The city started a Phrenological Society in 1820, and a journal three years later. Darwin attended discussions on the topic when he was studying medicine there in the late 1820s, and gave it little credence at first, until the Scottish politician James Macintosh persuaded him otherwise. By the time Darwin left on the *Beagle* in 1831, the science was extremely popular. There were twenty-nine phrenological societies across the country. Employers commonly asked to see a character reference from a local phrenologist as testimony to a potential employee's virtues. Darwin himself almost lost his place on the *Beagle* when Robert FitzRoy, the ship's captain and a keen phrenologist, was initially put off by the shape of the naturalist's nose.

At the heart of the discipline's success lay a book entitled *The Constitution of Man and Its Relation to External Objects*, one of the two scientific bestsellers of the century (we shall come to the other in the next chapter). It was by an Edinburgh lawyer-turned-scientist, George Combe. Combe had read the critical 1815 review of Gall's work but had been among those won over by Spurzheim's visit the following year. He helped launch the Edinburgh society, edited the journal, and wrote and lectured on phrenology in a way that made it sound like a forerunner of

twentieth-century behaviourism, the theory that behaviour was determined by external conditioning rather than internal will or thought, both
of which sounded a bit vague and metaphysical. Will and moral responsibility, Combe argued in his early *On Human Responsibility as Affected by
Phrenology*, were merely terms that described human behaviour, rather
than having any substance in themselves.[13]

Combe was keen to place the new science within a wider framework
and in 1828 published *The Constitution of Man*. After a slow start, it proved
phenomenally successful, selling over 125,000 copies in the UK and over
350,000 worldwide, being translated into Swedish, French, German,
Polish, Bengali and an edition for the blind. It remained continuously in
print to the end of the century. 'No book published within the memory of
man, in the English or any other language, has effected so great a revolution in the previously received opinions of society,' enthused the *Illustrated
London News* in 1858. Popular, well argued, readable, self-assured, scientifically literate and critical of religion, it was in many ways *The God
Delusion* of its time.

The Constitution put phrenology in the bigger scientific context of
fixed, universal, harmonious natural laws. By Combe's reckoning, there
were three kinds of laws, and humans were subject to them all. The physical laws regulated the entire universe, the organic laws governed life and
the moral and intellectual laws dictated human nature. It was on the last
of these that phrenology could shed new light, clearing away the longstanding confusion of 'philosophers and divines' concerning 'the number
and functions of the human faculties' by identifying the true material
structure and function of the brain.

Combe identified numerous 'orders'. Some were feelings that man
shared with 'lower animals', such as 'amativeness' (sexual urges) and
'philoprogenitiveness' (the desire for offspring). Others were sentiments,
such as self-esteem or benevolence, that were similarly common. Still
others, such as veneration, conscientiousness, hope and wonder, were
unique to man. Phrenology could help identify which was which, where
they resided in the brain and, in the light of all this, help determine an
accurate and empirically robust understanding of human nature. Having
done that, Combe was in a position to outline a scientifically based
approach to personal and social flourishing. Just as material things were

damaged when they ignored the physical law of gravity, and living things were harmed when they ignored organic laws (such as the need for sufficient food, light, air and exercise), so human things were injured when they ignored moral and intellectual laws. A great deal of misery could be avoided if only people adopted a scientific approach to human nature.

Take corrupt bankers, for example. 'A mercantile house of great reputation, in London, was [recently] ruined and became bankrupt, by a clerk having embezzled a large amount of funds,' Combe wrote. Or vexatious servants. 'How many little annoyances arise from the misconduct of servants and dependants in various departments of life?' Or corrupt postmasters. 'It is said that depredations are constantly committed in the post-offices of the United Kingdom, although every effort is made to select persons of the best character.' Such social ills were simply the result of flouting the moral and intellectual laws of life by placing the wrong people – 'decidedly deficient in moral or intellectual qualities' – in the wrong situations. Mercifully, science could help. 'It is obvious that the evils here enumerated may, to some extent, be obviated by the application of Phrenology.'[14]

Phrenology could also help shed light on wider social and global issues. Just as people differed in their characters, brains and skulls, so did ethnic groups. Indeed, the mental differences identified by phrenology 'however faint or obscure [they] may appear in individual cases, [become] absolutely undeniable in nations'.[15] The Phrenological Society's collection of 'Hindoo, Carib, Esquimuax, Peruvian, and Swiss' skulls apparently proved, beyond doubt, that differences of national character were visible in 'national brains' and that the European brain 'possess[ed] a favourable development of the moral and intellectual organs [compared with] Hindoos and native Americans, whose brains are inferior'. This wasn't an absolute rule. The Society had the skull of a 'Negro, named Eustache', which 'displayed a degree of shrewdness and disinterested benevolence very rare even in Europe'. Such examples underlined how wicked the slave trade was, Combe reasoned. But this was a moral conclusion reached by science, not by philosophical, let alone theological, deduction.

The Constitution's attitude to religion was not entirely contemptuous. Indeed, superficially it was respectful, particularly in later editions when controversy more or less compelled Combe to emollience. God is 'intelligent, benevolent, and powerful', he wrote towards the end of the book, and 'we may reasonably conclude that creation is one harmonious system, in which the physical is adapted to the moral, the moral to the physical.'[16] This was straight from the physico-theological playbook.

Yet it was also clear that that playbook had now lost any semblance of biblical Christianity, which was, if anything, an encumbrance to human progress. Religion was as liable to excite 'morbid excitement of the organs of Wonder and Veneration' as it was to improve the human lot.[17] Death was 'obviously a part of the organic law', an ultimately benevolent removal of 'the old and decayed', rather than the punishment for sin, as theologians believed.[18] Ethics should heed the utility of pain and pleasure, and the various laws of nature, rather than alleged divine revelation. Religion's proper role was to hear and affirm the moral lessons that science discerned from nature, not replace them. Religions needed a 'second Reformation', a 'new Christian faith' in which people recognised and integrated the true *scientific* picture of human character, intelligence and morality.[19]

The religious response to this was mixed. There were many Christians, schooled in the principles of physico-theology, for whom Combe's identification of the organs of Veneration and Wonder in the brain served as further confirmation of God's wisdom and benevolence. The tradition, now a century and a half old, of finding God's fingerprints all over the material world now crossed the threshold of the skull.

There were others who were attracted by the reformist direction in which, Combe insisted, phrenology pointed. The laws of creation, he argued, worked both ways. The neglect of physical and organic laws could affect moral wellbeing, just as much as personal immorality could materially ruin a person. This meant that certain material circumstances – 'insufficient food and clothing, unwholesome workshops, dwelling-places and diet, and severe and long-protracted labour' – could damage the 'higher feelings and faculties of the mind', including 'religious emotions'.[20] Spiritual betterment therefore demanded social reform. Phrenologists campaigned for education, exercise, temperance, vegetarianism, prison reform and better working conditions, and they often found liberal Protestants in their ranks.

For all this endorsement, however, the principal religious response was hostile. From the mid-1830s, the point at which *The Constitution's* popularity took off, British Christians, and in particular evangelicals, preached sermons, wrote articles, pamphlets and entire books and founded societies denouncing phrenology. In America, where phrenology was to be more popular and for longer, the Christian reaction was even more critical. Phrenology had turned 'the beautiful region of mental philosophy into a barren Golgotha, or place of skulls', fumed the *Christian Examiner*. It was 'a carnal philosophy', weighed down with 'weary dogmatism' that sought 'to supplant . . . the sublime philosophy of the Bible, and to sit in judgement on the Infinite and the Eternal'.[21]

As the *Examiner*'s colourful denunciation illustrates, it was phrenology's materialism that lay at the heart of religious antipathy. Its reduction of mind to brain and of ethics to utility denied the existence of the soul (obviously) but also human agency and moral responsibility (it was never clear how the science's apparent determinism was to be squared with its energetic social reformism). There were dimensions other than materialism at play too, however. Socially, the very fact that so many freethinkers and anti-clericals championed the new science made it suspect to believers. Evangelistically, this was the age when the British mission societies – Baptist Missionary Society (founded 1792), London Missionary Society (1795), Church Missionary Society (1799), British and Foreign Bible Society (1804) – were beginning to pour significant money and resources into work that was predicated on the idea that 'primitives' and 'savages' were capable of reading and receiving the gospel just as much as Europeans. Any science that implied the missionary societies were wasting their time was unlikely to win Christian adherents.

Scientifically, some of the more literate religious critics questioned phrenology's credentials, in which task they were not alone. Phrenology was not admitted to the British Association for the Advancement of Science (BAAS) when it was founded by the Revd William Vernon Harcourt in 1831, so it launched its own – the Phrenological Association – later in the decade. Philosophically, the idea that it was even possible to derive a moral system from nature vexed theologians. C. J. Kennedy, a Scottish minister, wrote an entire book against *The Constitution*, a chapter of which was dedicated to countering 'the alleged possibility of deducing a system of morality merely from the natural laws'.[22] For all these reasons – social, religious, scientific, philosophical – phrenology was to prove a far greater flashpoint in the relationship between science and religion in the nineteenth century than geology ever was. We have simply forgotten this because we have forgotten phrenology.

Behind all these different dimensions, however, there was another, familiar, underlying source of friction. In spite of their hostility to phrenology (or indeed to geology), evangelicals like C. J. Kennedy were generally positive about science that they understood, in the ancient formulation approvingly repeated by Kennedy, as 'the handmaid of religion'.[23] By Combe's reckoning, however, the handmaid had clearly now become the mistress.

The earliest editions of *The Constitution* were largely free from religious discussion (the lengthy chapters in later editions were due to the religious controversy the book provoked). This, in itself, is telling. As far as Combe, phrenology and the realm of science for which he spoke were concerned, religion – in the form of revelation, Bible, theology and Church, as opposed to the naturalistic deism he favoured – was largely irrelevant because it was largely redundant. 'Science is an exposition of the order of Nature,' he wrote in his chapter on science and religion in a later edition, 'and the order of nature is just another form of expression for the course of God's providence in the affairs of this world.' Rarely had the claim that scientists were the true priests of nature, so popular in the later seventeenth century, been made with greater force or clarity.

There was more. 'Until science shall discover her own character and vocation,' he wrote in the same chapter, 'that she is the messenger of God, speaking directly to these sentiments . . . she will never wield her proper influence over society for the promotion of their moral, religious, and physical welfare.' Religion was not so much challenged by science but replaced by it.

'Gradual and slow improvement of human nature': a new future

'Phrenology is of German origin [and] Vienna was its birthplace . . . but it was in France that it acquired its European *éclat*.' So wrote George Lewes, philosopher, critic and soulmate to novelist George Eliot, in *Blackwood's Magazine*, in 1857. He was right. The new discipline was more popular in

Britain and America, but it was more promising in France, where science was needed to shore up the intellectual and moral foundations and hopes of a nation shattered by a generation of revolution and war.

French science recovered remarkably quickly after the chaos of the early 1790s. The Jardin du Roi, where Georges-Louis Leclerc had served as director for nearly fifty years, was rapidly reconstituted as the Jardin des Plantes, part of the National Museum of Natural History. The École Polytechnique, which became the model for technical universities across Europe, was founded in 1794. The Académie des sciences, suspended by the National Convention, was revived as part of the National Institute of Sciences and Arts in 1795. When Napoleon invaded Egypt three years later, he took with him 167 historians, engineers, botanists, draughtsmen and artists to capture the scientific and historical fruit of the campaign. French politics would take a decidedly conservative turn in the wake of the Napoleonic Wars, but that did not entail a turn against science as it might have done in the eighteenth century.

French science could be religiously antagonistic. When asked by Napoleon about the role of God in his *Celestial Mechanics*, Pierre-Simon Laplace (1749–1827) famously replied, 'I have no need of that hypothesis.' The phrase has become iconic, passing into science and religion lore as a milestone, or perhaps a tombstone. In reality, it is far from certain what Laplace actually believed about God, what he meant by the statement, or even whether he uttered those exact words. Moreover, however significant the so-called 'French Newton' was, other leading scientists such as André-Marie Ampère, Georges Cuvier and Nicolas Léonard Sadi Carnot – the 'fathers' respectively of electromagnetism, palaeontology and thermodynamics – were somewhat more placatory and even sincere in their attitude to religion. French science of the nineteenth century was not as weaponised as that of the eighteenth.

What early nineteenth-century French science lacked in an anti-Christian edge, however, it made up for in secular confidence. A century later, the economist F. A. Hayek would criticise the École Polytechnique as the source of European scientific hubris.[24] He was writing from a position of instinctive hostility to any 'sciences of society' that facilitated social engineering, but the accusation retains some historical power. The brilliance of scientists, the multiplication of disciplines, the success of observations and experiments and above all the progress they all promised strengthened existing convictions, firstly, that scientific knowledge was uniquely reliable, and secondly, that the future, once thought to rest only in God's hands, now lay malleable in scientific ones.

Long inchoate, this understanding was first systematised in post-Napoleonic France by a man who had served as a soldier and then as a financial speculator before turning to philosophy and science. Henri de Saint-Simon inherited the French *philosophes'* adulation of science but criticised them for what he judged to be a primarily destructive agenda. Where the Encyclopedists brought disorganisation, Saint-Simon would bring order.

Saint-Simon believed that a complete edifice – of philosophy, morality, politics and ultimately religion – could be raised on the scientific foundations that had been laid by mathematics, physics and astronomy. Like George Combe, he sought to extend the scientific method to include the study of 'man' and society. In a memoir on 'the science of man', written in Napoleon's last years but published only much later, he proposed a new Encyclopedia, and new academies of moral and social science that would parallel the Academy of (natural) Science. Only in this way could society, finally, be established on a sound scientific basis.

Like many who had lived through a generation of political upheaval, Saint-Simon also worried about the direction of history, which seemed so much more dynamic and unpredictable than it had in his youth. History, he concluded, was characterised by the accumulation of knowledge and, accordingly, of moral progress. He posited a three-stage process – religious, metaphysical and finally scientific – through which societies passed as knowledge became more assured. Humanity was heading for an age in which we would understand mind, morality, society and economy with as much confidence and precision as we did the orbit of the earth or the course of projectiles.

For all his dedication to science, however, Saint-Simon recognised that experiment and reason alone were not enough. People needed devotion and passion and emotion in their lives. Religion had its uses and so, rather than abolish it, Saint-Simon sought to replace it with a more credible, scientific institution, one that was better suited to his utopian socialist vision for the future. His final book was called – bluntly – *Nouveau*

Christianisme and it replaced deity with humanity, clergy with scientists, faith with scientific knowledge and advocated a wholesale social reorganisation centred on a cult of Newton.

By this stage in his life, Saint-Simon was a pitiful figure. Having been born into the aristocracy – his valet used to wake him each morning with the words, 'Remember, Monsieur le Comte, that you have great things to do' – and then made vast sums buying vacant property during the Revolution, he had lost almost everything, spent time in an institution and attempted suicide. He died in penury, before he had managed to finish *Nouveau Christianisme* or achieve any of the 'great things' his valet had once urged on him. He had, however, befriended a young man who first worked as his secretary, then lived as his disciple and finally coined a word for the kind of scientific study of society that Saint-Simon had been pursuing for twenty-five years.

Auguste Comte was a brilliant thinker, albeit one of questionable mental stability. Brought up in a Catholic family, he was educated at the École Polytechnique before he fell under Saint-Simon's spell. The younger man prized and adopted his master's devotion to science and its capacity to remake the world. One of his first books, written when he was only twenty-four and which he considered seminal, was entitled *Plan for the scientific studies necessary for the reorganisation of society*.

Comte was not an easy man to get on with. He fell out acrimoniously with his master when the latter took his quasi-religious turn (somewhat ironically, given where Comte's own science would lead him), and subsequently refused to acknowledge any intellectual debt. Having suffered a mental breakdown in 1826, Comte withdrew for a time from public life and began work on what would become a six-volume masterwork entitled *Course of Positive Philosophy*. In the 1830s, just as geology was being professionalised and Combe's work on phrenology popularised, Comte systematised the physical sciences and established a new scientific discipline on which genuine social reform could be founded.

He began, following his master, by explaining that the study of human thought through the ages revealed how knowledge passed successively through three stages or 'methods of philosophizing'. These were the Theological ('or fictitious'), the Metaphysical ('or abstract') and the Scientific ('or positive'). The first of these could be divided still further

– Comte was a great subdivider – into fetishist, polytheist and monotheist phases of differing sophistication, but all were characterised by the most 'primitive' form of philosophy in which the human mind 'supposes all phenomena to be produced by immediate action of supernatural beings'. In the second stage, supernatural agents were replaced by 'abstract forces', 'personified abstractions' that were judged capable of producing all the phenomena that had previously been explained by direct divine intervention. By the time mankind had reached the third stage, the mind had given up on 'the vain search after Absolute notions', and instead committed itself solely to reasoning and observation within a framework of natural laws.[25]

The cause of positive knowledge had been advancing, slowly but with unrelenting determination, for centuries. The first three volumes of Comte's masterwork were dedicated to surveying and structuring the extent of true scientific knowledge – mathematics, astronomy, physics, chemistry and biology – each of which was dutifully divided into subcategories, like geometry and mechanics (for mathematics), or optics and electrology (for physics). In this way, Comte not only mapped out the ever-spreading terrains of scientific knowledge but showed how they were all linked together.

His real objective, however, was to understand 'the most complicated' level of reality, that of 'social phenomena'. Such phenomena had heretofore been the provenance of metaphysical or, worse, theological 'knowledge'. Comte was on a crusade to change this. 'Now that the human mind has grasped celestial and terrestrial physics . . . there remains one science, to fill up the series of sciences of observation, — Social physics.'[26] Accordingly, the final volumes of his *Positive Philosophy* were dedicated to this new science of 'sociology', outlining the methods – of observation, experiment, comparison, and what he calls the 'Historical Method' – by means of which this 'social science' would be elevated to the status of a genuine positive philosophy.

All of this was self-consciously scientific, although far from disinterested. The way in which Comte grounded human knowledge in a tripartite history (a division which came ultimately from a theological division of history into three eras of Father, Son and Holy Spirit) underlined how his final, positivist era was not simply an observation of the facts, or even an exhortation to surer knowledge, but was the mark of the approaching culmination of human history. In religious language, Comte was engaged in a kind of 'eschatology' – the part of theology concerned with the direction and destination of humans and history – only his was a resolutely secular, scientific kind of eschatology.

Comte was, at least at first, cautious about saying too much about this. In Volume 6 of his *Course of Positive Philosophy*, he poured cold water on contemporary debates about human perfectibility. Arguments about 'the absolute happiness of Man at different stages of civilization' constituted precisely the kind of 'metaphysical controversy' that positive philosophy was supposed to rescue us from. And yet, he had no doubt that bringing the study of man and society out of its theological and metaphysical darkness and into the glorious light of science would lead not only to an amelioration of the 'conditions of human existence' but also – crucially – to the 'gradual and slow improvement of human nature'.[27] Science promised social and personal progress, redemption and perhaps even salvation.

All this was implicit in Comte's philosophy, but it didn't stay implicit for long. Like Saint-Simon, and indeed many thinkers in the period of the Bourbon Restoration (to 1830) and the July Monarchy (to 1848), Comte was worried about cultural coherence. Without the nation's historic Catholicism, the centre could not hold and anarchy threatened to engulf the nation once more. And yet, Catholicism itself was, obviously, no longer a serious option on which to centre the nation's intellectual gravity. A modern people, emerging into the light of social science and positive philosophy, needed a suitably enlightened religion to give it shape and coherence.

Shortly after finishing the final volume of his *Course*, Comte broke with his first wife, who had nursed him back into mental health after his breakdown. Two years later he met the married, but at the time abandoned, Clotilde de Vaux. They formed an intense relationship that was never consummated and which ended, in 1846, with Clotilde's death. Devastated, and still pursuing a scientific religion that would help unify society, Comte turned his Positive Philosophy into a fully-fledged Religion of Humanity, which made Saint-Simon's *Nouveau Christianisme* seem like a restrained Protestantism by comparison.

Comte had long esteemed phrenology, which he, like Combe, believed had finally rescued the mind from the metaphysicians. 'It was not till our own time that modern science, with the illustrious [Franz Joseph] Gall for its organ, drove the old philosophy from this last portion of its domain,' he crowed.[28] The discipline, he claimed, dispelled confused ideas about human nature, such as 'the imaginary conflict between nature and grace [mentioned by] St. Paul', and replaced them with 'real opposition[s]', such as, in this instance, 'between the posterior part of the brain, the seat of our personal instincts, and its anterior region, the seat both of our sympathetic impulses and our intellectual faculties.'[29] Phrenology was not above criticism, of course, and in his later writings in particular, Comte would recognise the science's failure to locate accurately the functions of the brain. Nevertheless, he honoured the discipline, venerated its founder and drew on his ideas for both his science of sociology and his Religion of Humanity.

Comte prescribed this religion in excruciating detail. An adherent should pray three times a day, once to each of his household goddesses: mother, wife and daughter. He was to cross himself by tapping his head with his finger three times in the place where, according to phrenology, the impulses of benevolence, order and progress were to be found. The Religion of Humanity had nine sacraments, beginning with presentation (a form of baptism), and going through initiation, admission, destination, marriage (at a specified age), maturity, retirement, transformation, and then, seven years after death, incorporation. Comte set out a new calendar, with each of its thirteen months named after great men, from Aristotle and Archimedes to Caesar and St Paul, and festivals that were the scientific-secular equivalent of Saints' Days. (Gall had the 28th day of Bichat, after the anatomist Xavier Bichat, dedicated to him.) He specified the duties of various ranked, positivist clergy, their stipends rising in neat mathematical progression. He commissioned new hymns, celebrating holy Humanity. He designed new clothing, most famously waistcoats that buttoned only at the back and could thus only be put on and removed with others' help (thereby inculcating altruism, another word he coined). As the Grand Pontiff of this new church, Comte regulated all this piety, elevating Clotilde as a kind of Virgin Mary, and Humanity in place of God.

It didn't catch on. Indeed, Comte managed to achieve the near-impossible by uniting the religious and irreligious against him. Allies fell away. John Stuart Mill, a keen admirer of Comte's *Positive Philosophy*, broke with him on the religion. Darwin's 'bulldog', Thomas Henry Huxley, spoke for many professional scientists when he mocked Comte's efforts as 'Catholicism minus Christianity'. (Adherents replied it was really 'Catholicism plus science'.) Comte was unbowed. He dedicated his final years to this scientific religion, dying in 1857, disappointed but far from despairing, having recently reached out to the Jesuits as improbable allies in his endeavour. The Religion of Humanity limped on, planting its deepest roots in South America, but Comte slipped from history, more ridiculed than remembered.

And yet, his contribution to the story of science and religion is far from ridiculous, in spite of the eccentric fripperies of his scientific Catholicism. His forensic systematisation of science, and his elevation of 'social science' on to the level of its physical counterpart, was not only impressive in itself – Mill admired him for a reason – but historic for the way in which it finally brought the scientific method into the field of human affairs, a move long-anticipated but never yet satisfactorily delivered.

His core idea that humanity rather than divinity should be worshipped remained a strangely persistent sideshow in the story of science and religion. Britain, hardly receptive to Roman Catholicism at the best of times, let alone in this new scientific guise, did not welcome the Religion of Humanity, and the movement only gained a meagre toehold among a handful of intellectuals. Richard Congreve, a history fellow at Wadham College, Oxford, assumed its leadership and founded the London Positivist Society in 1867 but the British movement soon endured splits. In the final years of the nineteenth century, one strand morphed into the so-called Ethical Movement, which would, in turn, become modern secular humanism. There is a nice irony in Comte's scientific Religion of Humanity turning by degrees into a contemporary 'Humanism' – now presided over by leading scientists like Professors Richard Dawkins and Alice Roberts – animated by its commitment to humanity, and providing a range of quasi-religious rituals, with accredited celebrants officiating at naming, marriage and funeral ceremonies, and humanist chaplains providing non-religious 'spiritual' support in schools and hospitals. Comte would have approved.

Perhaps most significantly, for all that his Positivist Calendar was hubristic and is ridiculous in retrospect, it did signify another important moment in the science–religion history: the changing of time. Geology may have been a Christian discipline in its origins, but it nonetheless shook Christian audiences by disturbing their idea of history, stretching time past to heretofore unimaginable lengths, shrinking the human presence on the world's temporal stage and leaving awkward questions hanging over humanity's significance.

Phrenology was not a Christian discipline, for all that its founders tried to assuage their religious critics. Its challenge to believers lay not in the past but the present, and in particular in how people were now supposed to conduct themselves. Its apparent determinism reduced the human capacity for moral autonomy and responsibility and, in George Combe's hands, it threatened to replace religion as a guide for society. Phrenology boasted a science of living, of morality, of character, more reliable than anything the clerics or philosophers could produce.

Comte's sociology was explicitly anti-Christian and intentionally posed a direct challenge to the very future of religion. Even without the extravagancies of the humanist religion and its new calendar, positive philosophy promised a new future, in which society could be constructed and ordered on scientific grounds, and progress guaranteed. Knowledge would advance and with it would come scientifically organised societies and human moral progress. Social science promised a new future, a new heaven and a new earth or, more precisely, a new heaven on the old earth.

Bishop 'Soapy Sam' Wilberforce and Thomas 'Darwin's Bulldog' Huxley. Wilberforce was not called 'Soapy Sam' for the reasons people think, and Huxley was not called 'Darwin's Bulldog' at all in his lifetime. Only now, with the discovery of an unknown transcript of the event, can we know what they said during their famous clash in Oxford in 1860, and why they said it.

ELEVEN

THE BALANCE

'Better than a dog anyhow': Marry. Not Marry.

In the years after he returned to England, having circumnavigated the world on the *Beagle* and gathered evidence that caused him to doubt the fixity of species, Charles Darwin found himself wrestling with a painful and momentous decision. Should he marry?

Marriage was not inevitable. Nor was it necessarily desirable. Science, to which Darwin was clearly now affianced, could no longer be treated as a clerical naturalist's biddable handmaid. She demanded commitment. Darwin weighed the pros and cons in one of the notebooks he was filling up post-*Beagle*. Across the top of two adjoining pages, he scribbled the words 'This is the Question', and then drew up two columns, one headed 'Marry' and the other 'Not Marry':

Marry	Not Marry
Children – (if it Please God) – Constant companion, (& friend in old age) who will feel interested in one, – object to be beloved & played with. – better than a dog anyhow. – Home, & someone to take care of house – Charms of music & female chit-chat. – These things good for one's health. – ~~Forced to visit & receive relations~~ *but terrible loss of time.* – ~~W~~ My God, it is intolerable to think of spending ones whole life, like a neuter bee, working, working, & nothing after all. – No, no won't do. – Imagine living all one's day solitarily in smoky dirty London House. – Only picture to yourself a nice soft wife on a sofa with good fire, & books & music perhaps – Compare this vision with the dingy reality of Grt. Marlbro' St. [where he was living]	No children, (no second life), no one to care for one in old age. – What is the use of working 'in' without sympathy from near & dear friends – who are near & dear friends to the old, except relatives Freedom to go where one liked – choice of Society & little of it. – *Conversation of clever men at clubs –* *Not forced to visit relatives, & to bend in every trifle.* *– to have the expense & anxiety of children – perhaps* *quarelling – Loss of time. – cannot read in the* *Evenings – fatness & idleness – Anxiety & responsi-* *bility – less money for books &c – if many children* *forced to gain one's bread. – (But then it is very bad* *for ones health to work too much)* *Perhaps my wife wont like London; then the* *sentence is banishment & degradation into indolent,* *idle fool –*

There were arguments on both sides but it was clear which option won. 'Marry – Marry – Marry Q.E.D.' he wrote triumphantly at the bottom of the page.

Darwin's marital deliberations have been conjecturally dated to July 1838, by which time he had been charmed by his cousin Emma Wedgwood. Emma was intelligent, striking and devout. Darwin was beginning to doubt his (admittedly only ever lukewarm) faith. Ever candid – 'he is the most open, transparent man I ever saw', Emma would tell her aunt – Darwin broke the news to her.[1] Without underestimating the seriousness of his confession, Emma dealt with it well. 'My reason,' she wrote to him in a pre-marital note, 'tells me that honest & conscientious doubts cannot be a sin.'

The same notebooks that bore his matrimonial debate bear witness to those honest and conscientious doubts. Darwin was not particularly given to metaphysical speculation: 'I feel most deeply that the whole subject is too profound for the human intellect,' he told his friend Asa Gray many years later. 'A dog might as well speculate on the mind of Newton.' And yet, in the notebooks he began in 1837, in which we can trace the evolution of his theory, he touched on metaphysics, morals and even, on occasion, theology alongside his discussions of geology and anatomy. He was, at the time, reading John Locke, Thomas Browne, David Hume and (at least a review of) Auguste Comte's *Positive Philosophy*, and sometimes weighed their ideas alongside his scientific intuitions, and his liberal Anglican Christianity. As with marriage, there were arguments on both sides.

Darwin was by now convinced that the world was ancient and Genesis not a historical account of creation. He was, after all, a geologist before a biologist and, in his own words, 'a zealous disciple of [Charles] Lyell's [uniformitarian] views.'[2] He was also convinced that species were not created in their current form by God, and were subject to some kind of 'transmutation', an idea that, as Lyell later put it, had long been hanging tensely in the air.

It was not necessarily a ruinous idea. Indeed, it could be a liberating one. The heretofore dominant account of 'special creation' – the idea that the species were created (by God) in their current form – was, scientifically speaking, an empty vessel. 'The explanation of types of structure in classes – as resulting from the will of the deity, to create animals on certain plans,– is no explanation,' Darwin noted in response to John MacCulloch's book *Proofs and Illustrations of the Attributes of God*. 'It has not the character of a physical law & is therefore utterly useless. – it foretells nothing because we know nothing of the will of the Deity.' A workable theory of species transmutation would liberate the creation account in Genesis from explanatory responsibilities for which it was ill-fitted.

Ill-fitted . . . and theologically demeaning. A 'scientific' reading of Genesis was not only unpersuasive but undignified. 'If we were to presume that God created plant[s] to arrest earth, (like a Dutchman plants them to stop the moving sand),' Darwin went on, 'we lower the creator to the standard of one of his weak creations.' 'Has the Creator since the Cambrian formation gone on creating animals with [the] same general structure?' he asked rhetorically in Notebook B. What a 'miserable limited view'.

A theory of transmutating species offered to do to the God of the earth what Newton (and his predecessors) had done to the God of the heavens. Once upon a time, people had believed 'that God ordered each planet to move in its particular destiny'. Newton demonstrated that they, and every other heavenly body, were in fact subject to the same, limited number of elegant laws. Why not life on earth too? 'How much more simple & sublime powers let attraction act according to certain law[s] such are inevitable consequen[ces]. Let animals be created, then by the fixed laws of generation,' he wrote in Notebook B.

Notebook D saw him ruminate in a similar vein, sketching the closest he could to a complete system. This stretched from 'astronomical causes' through 'geography & changes of climate' to 'changes of form in the organic world'. How much 'grander' was this 'than [the] idea from cramped imagination that . . . since the time of the Silurian [God] has made a long succession of vile molluscous animals'? Such a vision of God was hardly very divine. 'How beneath the dignity of him, who is supposed to have said let there be light & there was light.'

This approach was better too, at explaining humanity's religious sensibilities. Theologians and philosophers might want to claim that 'the innate knowledge of creator has been implanted in us . . . by a separate act of God', but how much more elegant would it be if that 'knowledge' were to have evolved in humanity like everything else? Why was it not 'a necessary integrant part of his most magnificent laws'?

There was, of course, a question of how innate those religious sensibilities actually were. Referring to his brother-in-law, Darwin remarked in Notebook C, 'Hensleigh says the love of the deity & thought of him or eternity [is the] only difference between the mind of man & animals.' 'Yet how faint in a Fuegian or Australian!' he went on, referring to some of the indigenous peoples he had seen on his travels, whose state and behaviour had shocked him.

Even if the faintness of these religious sensibilities was a problem, however, it was not insurmountable. 'Why not gradation?' he asked. Surely it was 'no greater difficulty for [the] Deity to choose when [primates became] perfect enough for future state', than to choose 'when [humans were] good enough for Heaven or bad enough for Hell'. Could not mankind's spiritual awareness have evolved? Indeed, he added in parentheses, he may even have witnessed such evolution on his journey, citing how the Fuegians whom the *Beagle* had formerly brought back to Britain (the ship doubled as a missionary enterprise) showed 'glimpses bursting on mind & giving rise to the wildest imagination & superstition'.

Such ideas filled 'Column A' in Darwin's mind. Evolution by natural selection was a daunting prospect, but it by no means finished God off. There was, however, a Column B. In March 1838, Darwin went to the Zoological Society in London where he saw 'the Ourang-outang in great perfection'. It was a fascinating encounter.

> The keeper showed her an apple, but would not give it her, whereupon she threw herself on her back, kicked & cried, precisely like a naughty child. – She then looked very sulky & after two or three fits of passion, the keeper said, 'Jenny if you will stop bawling & be a good girl, I will give you the apple.' – She certainly understood every word of this, &, though like a child, she had great work to stop whining, she at last succeeded, & then got the apple, with which she jumped into an arm chair & began eating it, with the most contented countenance imaginable.

The encounter stayed with him, provoking further speculations in Notebook B.

Let man visit Ourang-outang in domestication, hear expressive whine, see its intelligence when spoken [to]; as if it understands every word said – see its affection – to those it knew – see its passion & rage, sulkiness & very actions of despair; let him look at savage, roasting his parent, naked, artless, not improving yet improvable & let him dare to boast of his proud pre-eminence.

Evolution might elevate God from the position of Dutch gardener fiddling around in the Silurian mud, but it didn't do much for human dignity. 'Man – wonderful man . . . with divine face turned towards heaven . . . he is not a deity, his end under present form will come . . . he is no exception', he wrote in Notebook C, sounding a bit like an Old Testament prophet.

Humans could not even comfort themselves with the claim that their particular nature was somehow better. 'It is absurd to talk of one animal being higher than another', Darwin wrote in Notebook B. 'We consider those [with] . . . the intellectual faculties . . . most developed, as highest.' But such a view was inexcusably anthropocentric. 'A bee doubtless would [consider differently].' In this light, 'better than a dog' might have been more complimentary than it sounded.

Such speculations bruised human pride, but there was more. Evolution had implications beyond merely denting mankind's sense of superiority. 'He who understands [a] baboon would do more toward metaphysics than Locke', Darwin suggested in Notebook M. Plato's *Phaedo* claims that 'our "imaginary ideas" arise from the pre-existence of the soul, [and] are not derivable from experience', he wrote elsewhere, before adding decisively: 'read monkeys for pre-existence'.

As with the human dignity, and the soul, so with thought itself. 'Why is thought being a secretion of brain, more wonderful than gravity a property of matter?' he mused in Notebook C. The brain being material, thinking lost its distinctive qualities. 'Thought, however unintelligible it may be, seems as much [a] function of organ, as bile of liver', he speculated. Ultimately, he thought, if 'the mind is [a] function of body', and thinking were merely an 'instinct . . . [that] result[s] from [the] organization of [the] brain', how could we trust our thoughts? The notebooks do not offer much of an answer.

And then there was morality. Perhaps that was merely an instinct too? 'Our descent, then, is the origin of our evil passions!!' Darwin wrote. 'The

Devil under form of Baboon is our grandfather!' Liberal Anglican that he still was, Darwin did not make the connection between instinctive 'evil passions' and Original Sin, as a Catholic or evangelical might have done. He was too worried about the idea that all morality was relative, rooted not in some transcendent idea of good and evil but in the accidents of natural selection.

Human dignity, the soul, thought, morality: such were the ideas troubling him as he courted Emma. 'Love of the deity [is merely the] effect of organization,' he berated himself. 'Oh you materialist!' However much Emma knew, it did not deter her. Shortly after they were married on 29 January 1839, unable to 'say exactly what I wish to say', she wrote him her fullest note on the subject. Probing and perceptive, she recognised that 'while you are acting conscientiously & sincerely wishing & trying to learn the truth, you cannot be wrong.' However, she admonished gently, there was a danger that he was demanding proof for a subject that did not lend itself to proof. 'May not the habit in scientific pursuits of believing nothing till it is proved, influence your mind too much in other things which cannot be proved in the same way, & which if true are likely to be above our comprehension.' Darwin read her note and sometime afterwards he scrawled in ink underneath: 'When I am dead, know that many times, I have kissed & cryed over this.'

'Like confessing a murder': between Erasmus and Paley

Nearly five years after Emma's note, Charles wrote to his friend Joseph Dalton Hooker. Hooker was an increasingly eminent botanist who would end up as director of the Royal Botanic Gardens at Kew. He had just agreed to describe Darwin's *Beagle* specimens in his *Flora Antarctica* and Darwin spent much of his letter on this matter. Towards the end, however, he risked opening up.

Since returning home, he confessed, he had been engaged on a task that was simultaneously 'presumptuous' and 'foolish'. Struck by the distribution of organisms on Galapagos and mammalian fossils in South America, he had determined 'to collect blindly every sort of fact, which could bear any way on what are species'. Having supplemented his collection with 'heaps of agricultural & horticultural books', he had come to a startling conclusion. 'I am almost convinced (quite contrary to opinion I started with) that species are not (it is like confessing a murder) immutable.'

Some more excitable commentators have posited that the murder to which Darwin was privately confessing was that of God, the divinity lying slain by the sword of evolution. It was not. The real meaning of Darwin's conspiratorial parentheses is hinted at by what he went on to say. Hardly drawing breath, Darwin reassured Hooker, 'Heaven forfend me from Lamarck's nonsense of a "tendency to progression" ' – the idea that animals somehow passed on to their offspring the physical characteristics acquired in their lifetime. The real reason for the confession was that he thought he had found '(here's presumption!) the simple way by which species become exquisitely adapted to various ends' – the Holy Grail of evolution: how it actually worked. He acknowledged the response that this admission was liable to provoke – 'You will now groan, & think to yourself "on what a man have I been wasting my time in writing to" ' – so he apologised, again, and then signed off, without offering Hooker any more details.

Evolution, in 1844, was a dangerous place to linger, a cross between a scientific graveyard and a political minefield. Darwin knew this better than most. His paternal grandfather was Erasmus Darwin, a disciple of the notorious French *philosophes*, a prophet of progress and a poet of liberty, an evolutionist, a free thinker and a rationalist, disposed towards materialism and atheism if not quite advocating either. For Erasmus, science happily replaced religion. His poems worshipped at the altar of nature and suggested (vaguely) that life did not need a creator but arose 'without parent by spontaneous birth', thereafter advancing of its own volition, governed only by the laws of nature.[3] Darwin admired Erasmus's biological work *Zoonomia* but wrote in his autobiography that he 'was much disappointed, the proportion of speculation being so large to the facts given'.[4] Erasmus's evolution, unlike his grandson's, was not science. Charles knew that and needed to show that he knew it.

If not science, though, it was certainly politics. Erasmus's earlier writings, even as they departed from Christian orthodoxy, received comparatively little condemnation. From the mid-1790s, however, as Britain entered war with a French nation that had seemingly lost its mind and moral compass, public opinion hardened. Scientific threats became indistinguishable from political or religious ones. Mind and soul, Church and society, government and God hung together, or they would hang separately. The *British Critic* slammed Erasmus's *Zoonomia* for depriving man of his soul, and by 1803 the critics were near-unanimous in savaging his posthumous, evolutionary poem *Temple of Nature* for its 'unrestrained and constant tendency to subvert the first principles and most important precepts of revelation'.[5]

This was the reaction to which Thomas Jefferson had been referring when he had compared Priestley's plight to that of Galileo. By the time of Erasmus's death in 1802, science had become inextricably caught up in the cause of national defence. The right kind of science served that defence well. The wrong kind – self-generating life, transmutating species, a plastic and protean natural order – led, at least in the British mind, to disorder, revolution and violence. Things had a right order, a right place, ordained by God and guaranteed by the ordained. Darwin knew this too. 'Once grant that species . . . may pass into each other,' he mused in Notebook C, '& the whole fabric totters & falls.'

There was salvation, however. When Darwin arrived in Cambridge in 1828 to train for the Anglican ministry, no rooms were available in his college, so he took lodgings nearby. By an appealing quirk of fate, the rooms in which he lived had once been occupied by William Paley. Paley was the most influential British theologian for the first quarter of the nineteenth century. His *Principles of Moral and Political Philosophy* (1785) and *Evidences of Christianity* (1794) were compulsory texts for ordinands but it was for his *Natural Theology* (1802) that he was to remain famous.

Subtitled *Evidences of the Existence and Attributes of the Deity, Collected from their Appearances in Nature*, Paley's book was in many ways the apotheosis of the physico-theology tradition. In twenty-seven confident chapters Paley drew together and updated a century of close observation to show that nature contains 'every indication of contrivance [and] every manifestation of design'.[6] Although he touched on astronomy and the elements, Paley's passion was for the exquisite anatomy and instincts of plants, insects, birds, fish, mammals and, of course, humans. He was a meticulous observer and could be forensically detailed. His chapter on the human frame examined the neck, forearm, spine, chest, kneecap, shoulder blade, ball and socket joints, hinge joints, ankle, shoulder, blood vessels, gristle, cartilages, mucilage, bones and skull. Everything had its place and its purpose.

Deploying an argument that would prove remarkably long-lived, he was at pains to stress that such seemingly miraculous entities could not have come about either by chance or gradually. Take the perfectly designed epiglottis, for example. 'There is no room for pretending that the action of the parts may have gradually formed the epiglottis,' he reasoned. 'The animal could not live, nor consequently the parts act, either without it or with it in a half-formed state.'[7] Whoever heard of a half-epiglottis?

All this was deployed in the cause of the deity. 'Design must have had a designer,' he said with a flourish at the end of chapter 23. 'That designer must have been a person. That person is God.' And not just any disinterested deistic God. Design could establish a great deal about the characteristics of the deity, such as his eternity, necessity, self-existence, omnipotence, omniscience, omnipresence, goodness, benevolence and spirituality. Paley was nothing if not ambitious.

The book was successful, although not immediately. By 1816, when it passed out of copyright, there were a respectable 20,000 copies in print. Thereafter, in the post-Napoleonic reaction, it was repeatedly reprinted, often in popular, cheap editions, so that an increasingly literate but restive population might access its comprehensive, unarguable and reassuring message.

Darwin was impressed by Paley. He was required to study his *Principles of Philosophy* and *Evidences of Christianity* as part of his degree and later remarked that 'I could have written out the whole of the *Evidences* with perfect correctness, but not of course in the clear language of Paley.'[8] He also read by choice, and delighted in, Paley's *Natural Theology*, commenting in his autobiography that it 'gave me as much delight as did Euclid.'[9] High praise indeed.

By the time Darwin was reading him, however, Paley's star was beginning to fall. In 1829, Francis Egerton, the 8th Earl of Bridgewater, left a bequest that enabled the publication of a series of eight works, to which a ninth was later appended by the mathematician and engineer Charles Babbage. Their objective was to demonstrate 'the Power, Wisdom, and Goodness of God, as manifested in the Creation'. These so-called Bridgewater Treatises, published between 1833 and 1840, saw, among others, Buckland write on geology, William Whewell on astronomy, William Prout on chemistry and the ever-industrious Thomas Chalmers on *The Adaptation of External Nature to the Moral and Intellectual Condition of Man*. Some built explicitly on Paley's work, others implicitly; sometimes the tone was as self-assured as that of *Natural Theology* but generally it was a little less insistent. Several of the authors sought to introduce the still-controversial awareness of deep time. None, of course, denied that nature did anything other than declare the glory of God – that was, after all, their commission. But few quite captured *Natural Theology*'s hubris.

Paley was also losing ground from other directions. Christian ethicists pointed out that his moral and political philosophy owed more to Jeremy Bentham's utilitarianism than to the Sermon on the Mount. Theologians drew attention to how his remarkably comprehensive understanding of God's character seemed to be able to dispense with revelation altogether. His vision of creation also began to ring false. Paley's was 'a happy world', recognisably that of an eighteenth-century Anglican rector, whose parish positively teemed with 'delighted existence'. 'In a spring noon, or a summer evening,' he remarked, 'on whichever side I turn my eyes, myriads of happy beings crowd upon my view.'[10] This was not quite how St Paul saw it when he wrote how 'the whole creation groaneth and travaileth in pain' (Romans 8:22), let alone the creatures who were forced to eke out their brief lives in England's newly industrialising cities.

If the tradition of Paleyian natural theology was scientifically and theologically on the wane, however, it was still *socially* useful. The alternative, in which somehow – no one had a credible idea of how – life generated itself and species slipped across permeable boundaries, remained politically explosive. Transmutation was still the doctrine of the French, who had just overthrown another monarch. It was the doctrine of radicals, who circulated penny papers attacking Paley, advocated the overthrow of the Church and overturned episcopal carriages and stoned the bishops' palaces when they voted against the Great Reform Act in 1832. It was the doctrine of Chartists who raised the spectre of a British revolution with their incendiary People's Charter. Evolution was revolution.

The other – sometimes overlooked but incalculably important – change that was occurring in this period was the popularisation of science. At the start of the century, 'natural philosophy' was still primarily a matter for rectories and gentlemen's clubs. There were exceptions. Both Erasmus Darwin and Joseph Priestley were committed 'lunatics', members of the Birmingham-based 'Lunar Society', whose interest in scientific (and social and political) affairs extended, if not quite to the working classes, at least to those industrialists, dissenters and experimenters excluded from Anglican Oxbridge. Such exceptions notwithstanding, in 1800 natural philosophy remained an elite affair. Thanks in part (and ironically) to the foundation of church schools and the slowly rising levels of literacy across Britain, by the time Darwin was scribbling in his notebooks this was changing and there was growing popular interest in science.

The material culture met the need. Books became cheaper. Periodicals multiplied. 'Shilling monthlies' penetrated the growing middle-class market. Religious Tract Societies and Radical lending libraries proliferated. These were not all, by any means, scientific texts; indeed, the lion's share was to remain religious or literary. And, of those that were scientific, most were not, by any means, 'materialist' or 'transmutationist'. The greater proportion of science books published in this period, and for several decades to come, presented the sciences within a broadly Christian framework, albeit not necessarily one as assured as William Paley's. The point was not that science was becoming 'dominant' over religion in terms of reading matter, or that it was becoming 'secular'. It was that science was becoming *public*, and to publish on a prominent scientific idea, let alone one as contentious as evolution, was to take a stand, thereby inviting not only professional criticism but public opprobrium.

Darwin did not have the stomach for this kind of controversy – literally, it seems, as he was plagued for decades by painful, mysterious and often debilitating stomach pains – and a few months after he wrote to Hooker, he was to be reminded why. Robert Chambers was a Scottish autodidact, bookseller and publisher, who maintained a lively personal interest in science, in particular medicine, phrenology and geology. In the early 1840s, he wrote a book whose title deliberately echoed Scottish geologist James Hutton's already famous remark about geology having 'no vestige of a beginning'. *Vestiges of the Natural History of Creation* was published in October 1844, anonymously. It became one of the few books to rival George Combe's *The Constitution of Man* in sales, scope and notoriety.

The book was popular science at its most inspiring, painting a picture that stretched from the formation of planets to the 'mental constitution' of animals, and bringing the law-governed self-generation of life and the transmutation of species kicking and screaming into a hundred thousand living rooms. It was brilliantly written, scientifically patchy and very popular.

The scientific professionals were not impressed. His 'geology strikes me as bad, and zoology far worse', Darwin lamented to Hooker in 1845. The religious professionals were not impressed, though their criticism was couched primarily in scientific terms. Darwin's early mentor, the Revd Adam Sedgwick, was particularly savage. 'If the book be true, the labours of sober induction are in vain,' he wrote in the *Edinburgh Review*. 'Religion is a lie; human law is a mass of folly, and a base injustice; morality is moon-shine; our labours for the black people of Africa were works of madmen; and man and woman are only better beasts!'[11]

But the public *was* impressed. Hooker labelled *Vestiges* a 'nine days wonder', but it turned out to be a 'nine days wonder' in much the same way as the Genesis creation story was a six-day wonder. Each day seemed to last an inordinately long time, as one corrected, revised edition followed another, each a slight improvement on the last, in a neatly evolutionary way.

Chambers benefited financially but never acknowledged authorship. When asked why by his future son-in-law he pointed to his house, in which his many children lived and said 'I have eleven reasons'.[12] Darwin was to have ten children and was considerably more financially secure than Chambers, but what he wouldn't lose financially, he could lose socially and scientifically. He chose silence.

'An incalculable waste': grandeur and grief

It must have been a hard decision, because by the time *Vestiges* was published, Darwin had turned his fevered notes into (two) extended essays. These were to form a template for *The Origin of Species* and like Darwin's iconic book they avoided metaphysical conjecture and religious controversy.

Or rather they *nearly* did, because in an extended concluding paragraph in the 1842 essay, Darwin allowed himself a brief flight of theological speculation. In doing so, he drew on many of his notebook ideas. By comparison to special creation, evolution was an elegant and ingenious mechanism, and better suited to any God worthy of the name. 'It accords with what we know of the law impressed on matter by the Creator, that the creation and extinction of forms, like the birth and death of individuals should be the effect of secondary [laws] means.'[13] It was positively 'derogatory that the Creator of countless systems of worlds should have created each of the myriads of creeping parasites and worms which have swarmed each day of life on land and water on [this] one globe'.[14] Far more fitting was the simple, lawful elegance of evolution by natural selection.

It may at first 'transcend our humble powers, to conceive laws capable of creating individual organisms, each characterised by the most exquisite workmanship and widely-extended adaptations', but the existence of such laws was not only defensible but 'should exalt our notion of the power of the omniscient Creator'.[15] 'There is a simple grandeur in the view of life,' he continued, with words that passed largely unaltered into *The Origin of Species*, 'with its powers of growth, assimilation and reproduction, being originally breathed into matter under one or a few forms.' It astonished him to think that 'from so simple an origin, through the process of gradual selection of infinitesimal changes, endless forms most beautiful and most wonderful have been evolved.'[16]

And yet, such elegance and grandeur and beauty had a grave, if hidden, cost. Darwin had been steered decisively towards natural selection when, in September 1838, he read the Revd Thomas Robert Malthus's *Essay on the Principle of Population*. Malthus's world was not Paley's, despite the fact that both came from the same social, intellectual and established religious milieu. Malthus argued that population growth was 'indefinitely greater than the power in the earth to produce subsistence'.[17] Whereas population increased at a geometric rate, doubling every twenty-five years or so among humans, the resources on which it depended increased only arithmetically. The result was too many mouths to feed. Among irrational animals, the superabundance of organisms was controlled 'by want of room and nourishment'. In humans, reason and morality (i.e. sexual self-restraint) could solve the problem. But if they did not (and they tended not to), nature would, in the form of 'the whole train of common diseases and epidemics, wars, plague, and famine'.

The mechanism was central to natural selection but it haunted Darwin's imagination, casting the darkest of shadows over Paley's happy spring noon and summer evening. Cruelty was as much a part of creation as beauty. Some animals were, it seems, 'directly created to lay their eggs in bowels and flesh of other'. Others were deceived to their doom, 'led away by false instincts'. Every year there was 'an incalculable waste of eggs and pollen'.[18] This was not the world that natural theology had conjured. The tradition carefully laid out by John Ray, William Derham, William Paley and their ilk stumbled and fell among the briars of natural selection. For all there was grandeur in this view of life, there was grief too.

The question was, was it worth it? There was undoubtedly waste and pain, unavoidable and on an enormous scale, but there was also pleasure and joy. 'Some writers . . . are so much impressed with the amount of suffering in the world,' he wrote years later in his *Autobiography*, 'that they doubt, if we look to all sentient beings, whether there is more of misery or of happiness; – whether the world as a whole is a good or a bad one.' The world was poised, suspended precariously between grandeur and grief. But as far as Darwin was concerned, grandeur won through. 'According to my judgment,' he went on, 'happiness decidedly prevails.'[19] This was not sentimentalism. There were, he thought, good scientific arguments for it. 'The sum of such pleasures as these, which are habitual or frequently recurrent, give, as I can hardly doubt, to most sentient beings an excess of happiness over misery, although many occasionally suffer much.'[20]

The problem of pain still nagged, however, and invited another question. Could what resulted from that balance of life ever justify or excuse the pain inherent in it? Darwin's answer to this in 1842 appeared to be a qualified 'yes'. 'From death, famine, rapine, and the concealed war of nature we can see that the highest good, which we can conceive, the creation of the higher animals has directly come.'[21] This was 'the Question'. If 'higher animals' – with all their splendour and sophistication, their grace and their grandeur, and ultimately their minds, metaphysics and morality – if they were indeed 'the highest good, which we can conceive', then was that enough for the price tag that came with evolution by natural selection? Everything hung on how the scales were balanced between life's evident grandeur and its potential for grief.

And there it stayed hanging for Darwin through the 1840s. He worked patiently away through the decade, making a forensic study of barnacles, and reading the occasional book of theology, including, in 1848, Andrews Norton's two-volume *Evidences of the Genuineness of the Gospels*, which he marked as 'good'. He and Emma filled the house at Downe with children, and by the time Darwin was reading Norton, they were expecting their seventh.

Darwin's own health had deteriorated in the 1840s, and when conventional medical advice proved unhelpful, he tried 'hydropathy' or water treatment, taking the whole family to Dr James Gully's hydropathic establishment in Malvern. It seemed to work – or at least didn't make his stomach ailments worse – so when his eldest daughter Annie fell ill in 1851 he decided to take her there.

Annie was an affectionate and sensitive ten-year-old, 'cordial, frank, open . . . without any shade of reserve'. She had never quite recovered from scarlet fever in 1849, so when she sickened again, in March 1851, Darwin's mind turned to Dr Gully. Emma was eight months pregnant, so Charles took Annie and left her there with her nurse, governess and sister Henrietta. A fortnight later he got an urgent message. Annie had taken a turn for the worse and was vomiting badly. Darwin rushed back, arriving in Malvern on Maundy Thursday, 17 April. Annie looked wretched, although she did at least recognise her father when he arrived. Dr Gully had eighty-seven other patients and could not spare the Darwins much time. Still, he assured Charles, Annie was 'several degrees better' than she had been.

Her condition worsened in the night. Her pulse became irregular and she slipped into semi-consciousness. It was 'from hour to hour a struggle between life & death'.[22] Gully believed she was dying and stayed with Charles all night. Somehow, she survived. She suffered throughout the following day, Good Friday. 'Her one good point is her pulse, now regular & not very weak,' Darwin recorded at midday. 'Excepting for this there would be no hope.' 'She keeps the same,' he wrote three hours later. 'We must hope against hope.'[23]

Friday night was better. Annie slept 'tranquilly . . . throughout the whole night'. At one point she said 'Papa' 'quite distinctly', although by now he hardly knew her. 'You would not in the least recognize her,' he told

Emma, 'with her poor hard, sharp pinched features; I could only bear to look at her by forgetting our former dear Annie. There is nothing in common between the two.'[24]

Despite all this, things began to look up. Dr Gully came to see her and, noticing how peacefully she slept, told Darwin 'she is turning the corner'. She sat up and took 'two spoonfuls of tea with evident relish, and no sickness'. Darwin was almost delirious with joy. 'I then dared picture to myself my own former Annie with her dear affectionate radiant face.'[25]

The weekend ground on. Annie vomited again, badly, on Easter Sunday morning. Her bladder was now paralysed and a catheter had to be inserted. In spite of this, the signs were still deemed positive. Mr Coates, the medical officer, took her pulse and immediately said, 'I declare I almost think she will recover.' Recent fevers such as the one Annie was experiencing, he explained, often appeared very bad but had tended not to be fatal. 'Oh my dear was not this joyous to hear,' Darwin wrote at 10 a.m. Her condition picked up throughout the morning. Her senses recovered and she called for him again. Darwin's spirits rose higher but he was by now completely exhausted. 'These alternations of no hope & hope sicken one's soul,' he told Emma.[26]

The emotional roller coaster continued into Monday. Annie emptied her own bladder and bowels for the first time in over a week, which Darwin judged 'very good'. Desperate for encouragement, he thought again of the little girl he loved so dearly. 'An hour ago I was foolish with delight & pictured her to myself making custards (whirling round) as, I think, she called them. I told her I thought she would be better & she so meekly said "thank you". Her gentleness is inexpressibly touching.'[27] He seized on any sign of recovery. 'She asked for orange this morning, the first time she has asked for anything except water . . . Fanny gave her a spoonful of tea a little while ago, & asked her whether it was good & she cried out quite audibly "it is beautifully good".'[28]

In spite of such signs, her condition worsened on Tuesday. Diarrhoea set in. Darwin could hardly take any more and his own stomach troubles erupted. Annie was sinking, slipping into unconsciousness. 'Twice amid her wanderings she made pathetic attempts to sing.'[29] By Wednesday morning her breathing was shallow, her body wasted. Eventually, she died at midday. 'Our poor dear dear child . . . went to her final sleep most tranquilly, most sweetly,' he wrote to Emma later in the day. 'She expired without a sigh.' The memory of her was almost too painful to recall. 'How desolate it makes one to think of her frank cordial manners . . . I cannot remember ever seeing the dear child naughty. God bless her.'[30] Gully's death certificate recorded 'bilious fever with typhoid character'.

Annie's death devastated Darwin. 'She was my favourite child,' he told his cousin William Fox frankly. 'Her cordiality, openness, buoyant joyousness & strong affection made her most loveable. Poor dear little soul.'[31] A week after her death, he recorded his memories of her, hoping to capture her 'strong affection' and 'buoyant joyousness' before they faded from his mind. He owned a daguerreotype which was 'very like her', but it was already two years out of date and failed entirely to capture the character of the daughter he had lost.[32] He remembered how she pirouetted before him as they walked round the garden at home; how 'she would spend hours in comparing the colours of any objects with a book of mine'; how 'she used sometimes to come running down stairs with a stolen pinch of snuff for me'; how 'she would at almost anytime spend half-an-hour in arranging my hair, "making it" as she called it "beautiful"'.

And he remembered her final, debilitating illness. 'She never once complained; never became fretful; was ever considerate of others; & was thankful in the most gentle, pathetic manner for everything done for her.' She was to be the solace of their old age. 'I always thought, that come what might, we should have had in our old age, at least one loving soul, which nothing could have changed.' It was not to be. 'She must have known how we loved her;' he concluded. 'Oh that she could now know how deeply, how tenderly we do still & shall ever love her dear joyous face.'

Annie was not the first child Darwin lost, nor the last. But nothing was the same after she died. Nearly a decade later, in September 1860, he wrote to his friend Thomas Huxley, whose four-year-old son had just died of scarlet fever. 'I know well how intolerable is the bitterness of such grief. Yet believe me, that time, & time alone, acts wonderfully. To this day, though so many years have passed away, I cannot think of one child without tears rising in my eyes; but the grief is become tenderer & I can even call up the smile of our lost darling, with something like pleasure.'[33]

There it was, the balance of pleasure and grief, happiness and pain. But whereas, twenty years earlier, Darwin tentatively argued that such a balance might indeed justify 'the creation of the higher animals', even a decade after Annie's death the answer was different.

'I am sharpening up my claws': and so to Oxford

Darwin went back to his barnacles, dissecting and writing about them in an effort that would earn him the Royal Society's prestigious Royal Medal in 1853. He then returned to his big idea, corresponding with naturalists across the world and building a case he hoped would prove irrefutable.

One of his correspondents was Alfred Russel Wallace, then travelling through the Malay Archipelago. Darwin had written to him in December 1855 and then again in May 1857. A year later, innocent of Darwin's thinking, Wallace sent him the manuscript of a paper entitled 'On the tendency of varieties to depart indefinitely from the original type'. Darwin was devastated. 'I never saw a more striking coincidence. If Wallace had my M.S. sketch written out in 1842 he could not have made a better short abstract!' he wrote to Charles Lyell.[34]

Darwin was in a conundrum – not least as Wallace had asked him to pass his paper on to Lyell – and saw the prospect of 'all my originality . . . smashed'.[35] Lyell and Hooker, two of the very few who knew of Darwin's ever-expanding manuscript, counselled a compromise in which Wallace's paper and an abstract of Darwin's would be presented jointly at a meeting of the Linnean Society on 1 July 1858. The reaction to the world's first public airing of the theory of evolution by natural selection was underwhelming, and it was not until November the following year, when a shortened version of Darwin's big book sold out on its first day, that the tide of public interest began to rise.

Early reviews were mixed. Darwin had advised Wallace to 'avoid [the] whole subject [of man]' because, although 'it is the highest & most interesting problem for the naturalist', it was also 'surrounded with prejudices'.[36] He was as good as his advice, and humans were largely absent from *The Origin of Species*. Not that you would have known that from the reviews.

The earliest, which appeared in the *Athenaeum* five days before publication day, set the tone. Referencing Lady Constance Rawleigh from Benjamin Disraeli's novel *Tancred*, who 'inclines to a belief that man descends from the monkeys', the reviewer commented that although this 'pleasant idea' was only hinted at in Chambers' *Vestiges*, Darwin had now 'wrought [it] into something like a creed'. On his account, man 'was born yesterday [and] he will perish tomorrow. In place of being immortal, we are only temporary, and, as it were, incidental.' Readers might have mistakenly thought they were reading a book of metaphysical speculation. Having worked so hard to extract such distracting controversy from his book, Darwin was indignant. 'The manner in which [the reviewer] drags in immortality, & sets the Priests at me & leaves me to their mercies, is base,' he wrote to Hooker a week later. 'He would on no account burn me; but he will get the wood ready & tell the black beasts how to catch me.'[37]

Several other reviews were equally critical. The *Saturday Review* – popularly known as the 'Saturday Slasher' – lacerated him for 'manifest errors' in his geological calculations. The *Daily News* accused him of stealing from Robert Chambers' *Vestiges of Creation*, a barb that went deep. 'One cannot expect fairness in a Reviewer,' he sighed to Hooker.[38] Not all were so negative, and some such as those in the *Gardener's Chronicle*, the *National Review* and the *Edinburgh Journal*, were positively encouraging, albeit written by friends and admirers (Hooker, Huxley and Robert Chambers respectively). Perhaps most surprisingly, *The Times*, the voice of the Victorian establishment, carried a glowing appraisal on Boxing Day, ignoring any religious and human dimension and stressing the book's scientific credentials. 'Mr. Darwin abhors mere speculation as nature abhors a vacuum . . . and all the principles he lays down are capable of being brought to the test of observation and experiment.' This was a book of science and should be judged as such.

The new decade opened and reviews proliferated, initial short notices giving way to the longer and more penetrating analyses in the monthly and quarterly press. Reviewers began to acknowledge that the book was becoming a publishing phenomenon, albeit not on the scale of Chambers or Combe. 'Mr. Darwin's long-standing and well-earned scientific eminence probably renders him indifferent to that social notoriety which passes by the name of success', began a 15,000-word piece in the *Westminster Review*. Nevertheless, the reviewer continued, 'he must be well satisfied with the results of his venture in publishing the "*Origin of Species*". The review went on not only to laud the book but to deploy its ideas against the Church and its tradition of clerical naturalism, with exuberant aggression.

> Who shall number the patient and earnest seekers after truth, from the days of Galileo until now, whose lives have been embittered and their good name blasted by the mistaken zeal of Bibliolaters? . . . Extinguished theologians lie about the cradle of every science as the strangled snakes beside that of Hercules; and history records that whenever science and orthodoxy have been fairly opposed, the latter has been forced to retire from the lists, bleeding and crushed if not annihilated; scotched, if not slain.

Barely six months after he had broken cover, Darwin was being enlisted to fight alongside Galileo in the ranks. There was certainly some truth in the reviewer's claims. 'If you were to read a little pamphlet which I received a couple of days ago by a clergyman,' Darwin would later write to the now former vicar of Downe, Brodie Innes, 'you would laugh & admit that I had some excuse for bitterness; after abusing me for 2 or 3 pages in language sufficiently plain & emphatic to have satisfied any reasonable man, he sums up by saying that he has vainly searched the English language to find terms to express his contempt of me & all Darwinians.'[39]

And yet, in spite of what the *Westminster Review* claimed, this clerical reaction was not entirely typical. One of the earliest letters Darwin received was from the author and clergyman Charles Kingsley, shortly to become Professor of Modern History at Cambridge, who gushed 'all I have seen of it awes me', and part of whose text Darwin integrated into later editions. The explicitly religious press were as divided as the mainstream. *The Rambler*, a liberal Catholic journal, was severely critical, the *English Churchman* rather more supportive.[40] By the time Darwin was writing to Innes, with whom he had long enjoyed a warm friendship, learned Christian opinion was largely comfortable with the idea of evolution. In the words of the British historian James Moore, author of the definitive book tracing the reception of Darwinism in Britain and America in the nineteenth century, 'with but few exceptions the leading Christian thinkers in Great Britain and America came to terms quite readily with Darwinism and evolution.'[41] It was almost as if the *Westminster* reviewer were inciting a war rather than describing one.

As with the first newspaper reviews, the later monthly and quarterly ones were equally mixed. For every *Westminster* there was a *Quarterly Review*, whose piece on the book was even longer and rather more critical, if far from unintelligent. It did not condemn the idea outright, nor seek cheap reconciliation. 'Few things have more deeply injured the cause of religion,' the reviewer sagely observed, 'than the busy fussy energy with which men, narrow and feeble alike in faith and in science, have bustled forth to reconcile all new discoveries in physics with the word of inspiration.' It did, however, volunteer some penetrating criticism of the new theory. Darwin gave credit where it was due. 'It is uncommonly clever', he admitted to Hooker, and 'picks out with skill all the most conjectural parts, & brings forwards well all difficulties.'[42]

By this time, it was clear that if *The Origin* was not quite the book of the moment – that award went to *Essays and Reviews*, the volume of biblical revisionism that was scandalising Victorian society – then it was certainly a book that demanded attention. And so the British Association for the Advancement of Science, meeting that summer in Oxford, gave it some.

Darwin didn't have the stomach to appear in the debate on his book that was scheduled for Saturday 30 June. His friend and advocate Thomas Henry Huxley took his place. Huxley had much in common with Darwin: a brilliant mind (a Fellow of the Royal Society by twenty-five, a Medal winner at twenty-seven), a passionate naturalist and a fully paid-up evolutionist. But unlike him, he was poor (he came from a middle-class family that had slipped down the social scale, and was forced to leave school at ten), an autodidact, agnostic (he coined the word), repelled by Oxbridge privilege and clerical superiority and hostile to amateurism. Above all, he was an aspiring *professional* scientist.

Coming from where he did, this was a demanding aspiration, even with his genius. At one point, when trying to secure a job, he had asked Richard Owen, Hunterian Professor of comparative anatomy and physiology at the Royal College of Surgeons, superintendent of the natural history departments in the British Museum and probably the country's leading anatomist, for a reference. It failed to appear. Huxley persisted and happened to spy Owen in the street. 'I was going to walk past,' he later recalled, 'but he stopped me, and in the blandest and most gracious manner said, "I have received your note. I shall grant it."' The condescension maddened Huxley who remarked that 'if I [had] stopped a moment longer I must knock him into the gutter.'

Huxley had authored the piece in the *Westminster Review*, and indeed the one for *The Times*, whose normal reviewer Samuel Lucas had been unable to make head or tail of the book. He was ready for a fight. 'I trust you will not allow yourself to be in any way disgusted or annoyed by the considerable abuse & misrepresentation which . . . is in store for you,' he reassured his friend. 'Some of your friends . . . are endowed with an amount of combativeness . . . I am sharpening up my claws & beak in readiness.'

His opponent, at least on the surface, was to be Bishop Samuel Wilberforce, who had preached against Chambers' *Vestiges* the last time the BAAS was in Oxford, in 1847. Wilberforce was, like Huxley, a highly intelligent and educated man but he was different in two crucial ways. First, he was establishment through and through. Third son of abolitionist William Wilberforce, Oxford-educated (by Buckland among others), canon of Winchester, chaplain to Prince Albert, dean of Westminster, and now bishop of Oxford, he was known popularly as 'Soapy Sam' – ostensibly for his prolonged vacillation over an appointment to the see of Hereford in 1847, but really for the way he rubbed his hands together when speaking, *and* because he was, in Disraeli's colourful phrase, 'unctuous, oleaginous, saponaceous'.

Second, for all that he was a more than competent natural philosopher, a vice president of the BAAS and a Fellow of the Royal Society, Wilberforce was no scientist. He had written the 'uncommonly clever' review of *The Origin* in the *Quarterly* but had needed help in doing so. That help had come from Richard Owen, whose lectures Wilberforce had assiduously attended in the 1840s. Two days earlier, Owen and Huxley had both attended a lecture on 'the final causes of the sexuality in plants', and had managed to get into another argument, somewhat bizarrely over the similarity between human and primate brains.

Owen himself had been down to chair the Saturday debate but had tactically withdrawn at the last minute in favour of Darwin's longstanding friend the Reverend Henslow. There were other ways of skinning this cat, and Darwin's circle easily detected Owen's hand behind Wilberforce's performance. 'Sam Oxon got up & spouted for half an hour with inimitable spirit uglyness & emptyness & unfairness,' Hooker later related to Darwin. 'I saw he was coached up by Owen & knew nothing & he said not a syllable but what was in the Reviews.' Between the three of them – Huxley, Wilberforce and Owen – there were many scores to settle.

The Huxley–Wilberforce clash has passed into legend as the British equivalent of Galileo's trial. Unfortunately, in this case, source documents are fragmentary and our knowledge of the debate has had to be pieced together from a handful of newspaper accounts and a number of gossipy letters. What is clear is that public interest was intense. Upwards of seven hundred people were crammed into the library of the new University Museum, on a hot day, the experience made no more pleasant by the long and dull lecture – on 'The Intellectual Development of Europe with reference to the views of Mr Darwin' – to which the protagonists were supposedly responding.

When they came to speak, Wilberforce tried to introduce a bit of ill-judged levity into what was becoming a rather tedious afternoon. According to the myth, he turned to Huxley and facetiously asked him whether he would prefer to be descended from an ape on his grandfather's or his grandmother's side. According to the same myth, Huxley whispered to a neighbour, 'The Lord hath delivered him into mine hands', and then replied that he would rather have an ape for an ancestor than a bishop. Another myth had him say that he would rather be an ape than a bishop, though Huxley later denied this. Whatever the precise nature of his retort, as Isabel Sidgwick later recalled in *Macmillan's Magazine* 'the effect was tremendous'. The crowd erupted. An audience member, Jane Purnell, Lady Brewster, fainted. And Sidgwick 'for one, jumped out of my seat'. Wilberforce, the myth goes, was undone and forced to retire from the lists scotched, if not slain.

The odd thing about the whole affair was that the supposedly momentous event largely disappeared from public awareness for a generation and only reappeared when the 'warfare' narrative of science and religion was gaining traction in the 1890s. Once the immediate dust had settled, none of the protagonists dwelt on the affair and even at the time the fullest account we have for the debate concluded that the two protagonists 'each found foemen worthy of their steel'. Much in the way that the trial of Galileo would be used by Protestants to make their case against Catholic dogmatism, so it seems the battle of Oxford was later worked up by anti-religious writers who wished to do the same to Christianity.

And yet, for all that historical scholarship has defused the Oxford debate and shown it to be no more indicative of the alleged warfare between science and religion than was the trial of Galileo, it *was* a water-shed of sorts. A hundred and fifty or so years after the event, as Victorian periodicals and newspaper archives were being digitised, an American scholar, Richard England, noticed a previously unknown and remarkably full account of the debate published in the *Oxford Chronicle and Berks and Bucks Gazette*. This was clearly the source text for the *Athenaeum*'s account as there was considerable overlap, but the *Gazette* offered significantly more detail, not to mention a rather charming running commentary on the crowd's response.

The new account shows Wilberforce insisting that 'everywhere sterility attended hybridism' and that 'the permanence of specific forms was a fact confirmed by all observation', although it also shows that he was willing to acknowledge exceptions among certain species of plants. It shows Huxley arguing that natural selection was more than a mere hypothesis and complaining that Mr Darwin's opponents never 'attempt[ed] to bring forward any important fact against his theory'. But what really emerges from the *Gazette*'s account was how preoccupied the debate was by the two matters that had haunted the histories of science and religion over the centuries.

The first was the question of the dignity of humanity. For all that Wilberforce was willing to countenance evolution in certain limited circumstances, that did not – *could not* – extend to humans. His almost hysterically emotive language showed how much this was *the* issue for him. 'It [is] a most degrading assumption – (hear, hear) – that man, who, in many respects, partook of the highest attributes of God – (hear, hear) – was a mere development of the lowest forms of creation. (Applause.) He could scarcely trust himself to speak upon the subject, so indignant did he feel at the idea.'

Huxley countered with the perfectly reasonable point that the vague-ness of the division between humans and other species was no different to that between humans and their own embryonic formation. 'It must be remembered man himself was once a monad – a mere atom of matter – and who could say at what moment of his development he became consciously intelligent. (Hear, hear.)' Wilberforce was not persuaded. For him, this was the fight for human dignity.

For Huxley, by contrast, the debate was about authority. Huxley had already remarked, earlier in the week, that he 'did not think that a general audience, in which sentiment would unduly interfere with intellect, was a fit place for such a discussion', and he began in a similar way on the Saturday, 'alleging the undesirability of contesting a scientific subject involving nice shades of idea [*sic*] before a general audience'. This was not, one would have thought, a foolproof tactic for winning over said audience. Science should be decided by scientists, and not crowds, let alone bishops. Huxley was indignant that an amateur clerical naturalist like Wilberforce should consider himself capable of adjudicating on such matters and protested 'against this subject being dealt with by amateurs in science. (Applause.)'

This was where *his* indignation lay, and he had good reason for it. Revealing what happened at the crux of the famous debate, the *Gazette* records how Wilberforce, referring to Huxley's discussion with Owen two days earlier, wondered 'if he had any particular predilection for a monkey ancestry, and, if so, on which side – whether he would prefer an ape for his grandfather, and a woman for his grandmother, or a man for his grandfather, and an ape for his grandmother'. This dignitary of the Established Church had asked Huxley, in front of nearly a thousand people, which of his grandparents would he prefer to have had sex with a monkey. Wisely, once his opponent had gone so low, Huxley went high. 'If the alternative were given him of being descended from a man conspicuous for his talents and eloquence, but who misused his gifts to ridicule the laborious investigators of science and obscure the light of scientific truth, or from the humble origin alluded to, he would far rather choose the latter than the former. (Oh, oh, and laughter and cheering.)'

The *Gazette*'s diligent report of reactions shows how the audience swung between protagonists. The *Athenaeum* was right in judging the encounter a score draw. And yet Wilberforce's cheap shot – and his subsequently feigned innocence: 'he regretted that Professor Huxley had taken umbrage at what he had said' – merely confirmed everything Huxley was fighting for. Unburdened by scientific authority, Wilberforce chose to relinquish his moral authority too.

Wilberforce ridiculed Huxley's 'appeal to authority'. He hinted that there were authorities (such as 'Professor Owen, and other eminent men') who were against the theory. And he observed that, in any case, proper science should precisely avoid such appeals to authority. But Huxley clarified it wasn't authority per se that was the problem. 'What he . . . deprecated was authority like the Bishop's, authority derived from a reputation acquired in another sphere. (Hear, hear, and laughter.)'

Bishops, no matter how admired and erudite, were not and should not be in the position to adjudicate on issues of science. It was something that scientists, Catholic, Protestant and sceptical, had been saying since the time of Galileo. But it was only by the time Huxley took on Wilberforce in Oxford, in 1860, that it was becoming a reality. In effect, the Oxford debate of 1860 was the moment at which newly professionalised science knocked clerical naturalism 'into the gutter'. And it seemed to work. In the thirty or so years before the Oxford debate, nearly forty Anglican clergy had presided over the various sections of the British Association for the Advancement of Science. In the thirty or so years after, three did. Science, having been born of theological parents, had finally left home.

Four young African men posing with spears at the 1904
World Fair. Ota Benga, on the right, would subsequently
be displayed in the monkey house at the New York
Zoological Park as the missing Darwinian link. 'Our
race . . . is depressed enough, without exhibiting one of
us with the apes,' protested the Revd James H. Gordon
of the Colored Baptist Ministers' Conference.

TWELVE

GLOBALISATION

'Illuminate the dark paths': missionary science

Charles Darwin's very first piece of published writing was a defence of Christian missionary activity.[1] Towards the end of the *Beagle*'s journey, the ship docked in southern Africa, and Darwin and Captain FitzRoy co-authored an article for the *South African Christian Recorder*. The early 1830s were a tumultuous time in southern Africa, with economic disruption following slave emancipation in 1833, the sixth war between the Xhosa kingdom and European settlers in 1834 and the 'Great Trek' of Dutch colonists east from the Cape in 1835. Much of the blame for this turbulence was directed at the mission societies.

Darwin had been impressed by the missionary activity he had seen in the South Pacific and New Zealand. 'Dishonesty, intemperance, and licentiousness have been greatly reduced by the introduction of Christianity,' he wrote in his *Voyage of the Beagle*. Those who attacked missionary work were either confusing it with the activities of English settlers – 'men of the most worthless character . . . addicted to drunkenness and all kinds of vice' – or comparing it unfairly with 'the high standard of Gospel perfection'. So it was that, when the opportunity presented itself on the last leg of their journey, the increasingly pious captain and his increasingly sceptical scientific companion penned 'A letter, containing remarks on the moral state of Tahiti, New Zealand, &c', in which they vigorously defended missionary activity.

This strange marriage of science, religion and mission was not, in fact, that strange at all. Indeed, by the time FitzRoy and Darwin were writing, it was already a quarter of a millennium old. From the late 1580s, the

Jesuits' mission in China had drawn on much mathematical, geometrical, astronomical and mechanical learning. Under the leadership of the Italian Matteo Ricci, Jesuits had earned the trust and then the respect of the Imperial Court, as much for their science as their faith. The Jesuits introduced clocks, telescopes, astrolabes and modern astronomical tables. Ricci himself helped translate Euclid's *Elements*. The German Johann Schreck published the first book in Chinese detailing European mechanical knowledge in 1627 and the Polish Jan Mikołaj Smogulecki taught mathematics and introduced logarithms in the early 1650s.[2] The mission was erudite, culturally sensitive, international and scientific.

The accuracy of the Jesuits' science was impressive, despite the fact that, with only a few exceptions, the missionaries propounded Tycho Brahe's geo-heliocentric model, the Copernican one being, of course, banned. It soon became clear to them that the Chinese calendar needed correction. The court was, at first, resistant to outside intervention in this, so vital to the performance of Chinese rites and the authority of the emperor was the calendar. In 1629, however, two years after becoming emperor, Zhu Youjian, fearful of the implications of miscalculating another solar eclipse, invited Schreck and others to submit new calculations. When it became obvious that these were more accurate than traditional Chinese ones, the emperor introduced reform and invited the German Jesuit Johann Adam Schall von Bell to become a director of the Imperial Bureau of Astronomy. The fall of the initially welcoming Ming dynasty fifteen years later threatened the Jesuit position, but the succeeding Qing dynasty, and especially the long-reigning Kangxi emperor (r. 1662–1722), consciously appropriated Jesuit science, which he himself studied as a way of establishing and legitimising his authority.[3] Jesuits were to head up the Imperial Bureau for well over a century.

The mission's scientific traffic was not all one way. For over a century, it was Jesuit missionaries who introduced Chinese thought to Europe, translating Confucian texts into Latin, writing the first Chinese–Latin dictionary and offering detailed accounts of Chinese territory, history and ideas. European intellectuals, such as Robert Hooke, were fascinated by this alien thought world, and disturbed. The Jesuits left no one in doubt that Chinese culture was ancient, moral, reasoned, highly sophisticated, technologically literate and easily comparable to European Christendom.

But was it theistic? Early sceptics thought not, and as Sinophilia swept through European coffee houses in the early decades of the eighteenth century, some, like Pierre Bayle, delighted in portraying Confucians as 'Virtuous Atheists', a contradiction in terms (according to European Christians) that threatened to pull the keystone from their entire edifice.

The Jesuits argued otherwise. There were deep theistic sources within Confucian thought, they claimed, albeit made unfamiliar by strange terminology and centuries of cultural accretion. Confucians were certainly virtuous but not atheistic. The issue was to prove divisive, initially within the Jesuits but then exacerbated by the long-running dispute between them and other Catholic orders. Franciscans and Dominicans, scenting blood, pursued the issue until Pope Clement XI eventually passed judgement. The result made ongoing Jesuit activity in China much more difficult. A few continued the work of the early mission. The Frenchman Jean Marie Amiot, a correspondent of the Académie des sciences, wrote a fifteen-volume *Memoire concerning the history, science and art of China* that was published as late as the 1780s. But the Chinese mission was a shadow of its former self when the entire Society of Jesus fell afoul of Church politics and was suppressed in 1773.

By this time, Protestant missionaries were beginning to pick up – or wrest – the missionary baton from their Catholic brethren. The Anglican Society for the Propagation of the Gospel in Foreign Parts had been in existence since 1701, but its energies were directed primarily at the newly independent American colonies. Over the intervening years, the evangelical movement had reinvigorated British Christianity and its commitment to mission, and by the last decade of the century, aided by a rapprochement between denominations, and a waning of strict Calvinist theology, the gospel was ready to be taken to the heathen. Science went with it.

The publication of Captain Cook's *Voyages* in 1777 opened up a new world to British readers, just as they were losing control over their first New World. Evangelicals were inspired. William Carey, who founded the Baptist Mission Society, later admitted that 'reading Cook's voyages was the first thing that engaged my mind to think of missions.'[4] Thomas Haweis, an Anglican evangelical, was similarly inspired, and referred to the *Voyages* in his inaugural sermon before the London Missionary Society (originally The Missionary Society), whose activities he steered towards the South Pacific.

Cook's voyages sought, among other things, to advance British science, by observing the transit of Venus across the sun, mapping the geography of the southern oceans and collecting flora and fauna for examination back home. Joseph Banks was a naturalist on Cook's first voyage, after which he went on to become president of the Royal Society for forty-one years. He had a distinctly lukewarm, establishment Christianity and was ambivalent at best about Christian activism.[5] He disliked the evangelicals' enthusiasm, their moral crusading, their political campaigns and in particular their abolitionism. The natives he encountered on Cook's first voyage were essentially specimens to him. 'I do not know why I may not keep him as a curiosity, as well as my neighbours do lions and tigers,' he wrote of Tupai'a, a Pacific islander he brought back to London.[6]

He was, however, impressed by missionary attempts to 'illuminate' foreign lands. 'You & I certainly differ in opinion respecting the things we each deem necessary for salvation,' he told Haweis, when the two discussed the doctrine of justification by faith. However, he went on, 'I [do want to] express my satisfaction at the attempt of your Society to illuminate the dark paths of our pagan Brethren,' and he offered to facilitate the London Missionary Society activities in the South Pacific as best he could.[7]

In this way, science was at the heart of global Protestant missionary activity from its earliest days. Whereas historians once considered evangelicalism to be a reaction against the Enlightenment's emphasis on reason, it is now widely judged to be part of it. For all they prized scriptures and the personal experience of the Holy Spirit, the evangelicals who launched missions across the world from the 1790s were also self-consciously 'reasonable' men and women. Though few were made in the image of William Paley, they nonetheless honoured God's book of works, saw in it proof of his might and wisdom and were committed to its study. 'The more the mind of the Christian is enlarged and strengthened by scientific pursuits,' reasoned one missionary in 1813, 'the better he is fitted to understand, believe, and defend the truths of the Bible.'[8] The idea of a division, let alone a war between science and missionary religion, was anathema. As the historian Sujit Sivasundaram put it in his study of missionary science in the South Pacific, 'the distinction between men of science and missionaries should be shed, because scientific speculation was seen to be the rightful preserve of the theologically minded before the professional vocation of the scientist was formed.'[9]

This missionary amalgam of science and religion took numerous forms. Some missionaries were collectors. William Carey, a keen botanist from childhood, founded the Agri Horticultural Society of India and edited Dr William Roxburgh's *Hortus Bengalensis* and his posthumous *Flora Indica; or Descriptions of Indian Plants*. When an American missionary supporter threatened to withhold funds unless the Baptists focused solely on saving souls, Carey indignantly responded that he had never heard anything 'more illiberal'. 'Pray can youth be trained up for the Christian ministry without science?'[10]

Many made small but helpful contributions to emerging fields of knowledge. Darwin's early work on coral reefs, published in 1842, drew repeatedly on the work of the LMS missionary, John Williams, who had been killed and eaten on Eromanga Island three years earlier. Some were respectable scientists in their own right. The LMS missionary Samuel James Whitmee published numerous notes and articles in *Nature* and the *Proceedings of the Zoological Society* on the natural history, botany and anthropology of the South Pacific. Tellingly, such contributions continued to the end of the century and even beyond, by which time science had become thoroughly professionalised. British South Pacific missionaries alone contributed over 200 articles to scientific journals between 1869 and 1900.

Others, particularly in later missions, were medics. When the missionary and physician Peter Parker arrived in China from America in 1834, he set up the Canton Ophthalmic Hospital and, three years later, the Medical Missionary Society of China. The demand for his work so overwhelmed his capacity that he began to despair of achieving any conversions. Parker was a pioneer. While there were no more than fifty medical missionaries working abroad in 1849, by 1925 Protestant missions alone could boast over two thousand doctors and nurses. Perhaps the most famous missionary of the century, David Livingstone, decided to become a Christian missionary precisely because it enabled him to combine his religious and scientific vocations. 'My view of what is missionary duty is not so contracted,' he told the secretary of the LMS, who was worried he was going too far off-piste, 'as those whose ideal is a dumpy sort of man with a Bible under his arm.' He served Christ, he went on with gloriously Victorian confidence, as much 'when shooting a buffalo for my men or taking an astronomical observation' as when preaching a sermon. How shooting buffalo served Christ was left unclear.

As with the earlier Jesuit mission to China, the traffic was two-way, although unlike the earlier enterprise, there was often a pronounced cultural condescension at work in this encounter. Science was frequently deployed as a badge of European intellectual and moral superiority. Agronomy, a science close to Joseph Banks's heart, was often used by early missionaries as a means of 'improving' native agriculture, and several LMS missionaries were appointed with the specific title of 'agriculturist'. Their success was mixed. What worked in southern England did not always work in the South Pacific. On other occasions, scientific disciplines could be employed as a way of undermining indigenous religious beliefs. Astronomy, botany and geology were all used to show natives that planets, plants and mountains were not in themselves divine but merely creations of the divine. When confronted by islanders who believed that a nearby coral reef was 'a rib of one of their gods', the LMS missionaries George Bennet and Daniel Tyerman patiently explained that 'these marvellous structures are formed by multitudes of the feeblest things that have life, through ages working together, and in succession, one mighty onward purpose of the eternal God.'[11]

Success was not always guaranteed, not least when missionaries found themselves among religious cultures as ancient and confident as their own. In the late 1820s, Daniel Poor, an American Presbyterian missionary, found himself in dialogue with Vicuvanata Aiyar, an elderly Brahmin and astrologer in Sri Lanka. Poor had detected miscalculations in Vicuvanata's almanac, which predicted a lunar eclipse on 20 March 1829, but Vicuvanata refused to accept his corrections. When the appointed night came, the two stood with a crowd of witnesses and the predicted eclipse duly failed to appear. Poor then did his best to win Vicuvanata over, only to realise that the latter was more interested in adopting his science than his religion.

There were similar stories from China. When the American Presbyterian missionary, Calvin Mateer, arrived in China in 1864, he was convinced that the scientific knowledge he brought with him would 'dissipate faith in idols and lead from nature up to nature's God'.[12] Like Poor, he soon discovered that Confucian scholars were happy to keep the discussion focused on nature. As a result, his next forty-four years were divided between Bible translation and setting up Tengchow College, which taught algebra, geometry, navigation, mathematical physics and astronomy. As the Jesuits had learned two hundred years earlier, having better mathematics did not necessarily persuade others you had better religion.

Just as missionaries carried European science to India, Africa and beyond, they carried information about the rest of the world back to the West. Indeed, perhaps the most common and effective scientific role performed by missionaries was as informants for the growing band of professional scientists back home. It was a global enterprise, as oblivious to national borders as the Jesuits had been two centuries earlier. Darwin cited scientific insights from the Revd Robert Everest (a keen geologist and author of papers on the geology and climate of India, who lived in Calcutta) about dogs in India; the aforementioned Tyerman and Bennet (about pigs in the Pacific); the Revd Jakob Erhardt (a German missionary who unsuccessfully requested funds for an expedition through Zanzibar from the Royal Geographical Society) about domesticated poultry and animals in East Africa; the Revd Leonard Jenyns (the original choice of naturalist for the *Beagle*) about an 'almost globular' breed of goldfish imported from China; James Digues La Touche (an Irish cleric, geologist and palaeontologist, with interests in entomology, botany, meteorology and astronomy) about apples from Canada; and Thomas Bridges (cate-chist to the Anglican Mission in Tierra del Fuego) about dogs' ability to catch crustaceans – and that was only in his 1868 book *The variation of animals and plants under domestication*.

Perhaps most notoriously, missionaries transmitted information and ideas back to curious European scientists about the people whose souls they were there to save. This role has been subject to much scrutiny and criticism over recent years, often with good reason, although as with most elements in the histories of science and religion what one finds is far from straightforward.

There can be little doubt that missionaries held the natives among whom they lived to be morally, religiously and usually intellectually inferior. In the earliest days, missionaries had expected to find South Pacific islanders who conformed to the European enlightened ideal of the noble savage: peaceful and peaceable, innocent and uncorrupted by sinful civilisation. The view did not long survive contact. Idolatry, theft, violence and cannibalism shocked missionaries, at least those who had read Rousseau. Whether violent or peaceful, however, such natives were usually judged 'primitive' and 'savage', unlettered and uneducated. The LMS missionary William Ellis spent nearly a decade in the South Pacific after the Napoleonic Wars, learning languages, observing land and culture in detail and publishing his multi-volume account *Polynesian Researches* on his return. He knew the region, its inhabitants and its missionaries as well as anyone at the time. He spoke for many when he observed that as the missionaries became better acquainted with the natives, they beheld them with 'their deepest commiseration . . . not only wholly given to idolatry, and mad after their idols, but sunk into the lowest state of moral degrada-tion and consequent wretchedness'.[13] Such a state should catalyse mission-aries' acquisition of native languages so that 'they might speedily instruct them in the principles of Christianity, and thereby elevate their moral character'.

At the same time, however, missionaries were uncompromising mono-genists, committed to a single, common origin for all peoples. When Isaac La Peyrère had, 150 years earlier, published his book *Men before Adam*, arguing, *contra* Genesis, that different races had different ancestors, he was denounced and his book publicly burned by the hangman in Paris. More recently, British Protestant missionary societies had grown up alongside the campaign to abolish the slave trade, and then slavery itself, campaigns for which monogenesis was absolutely central. 'Oh fool!' exclaimed the freed slave and abolitionist, Olaudah Equiano. 'See the 17th chapter of the Acts, verse 26: "God hath made of one blood all nations of men, for to dwell on all the face of the earth".[14]

The natives among whom the missionaries mixed were created by God, like the missionaries themselves. They were sinners, like the missionaries. They had eternal souls, like missionaries. They needed saving, like missionaries. The same Lord Jesus Christ had died for their sins. Natives were as capable of repentance and belief as those who brought them the gospel. They were capable of moral improvement. They were capable of reading the scriptures. And their native tongue was every bit as capable at accommodating the word of God as English. However much missionaries might have patronised natives or used science to undermine their religious beliefs, they still adamantly affirmed their fundamental equality, dignity and humanity. Their work was characterised, in the neat formulation of historian Jane Samson, by a dual process of 'othering' and 'brothering'. As the century went on and science and religion slowly separated, with the former claiming ever-greater intellectual authority over the question of human nature, there was to be rather more 'othering' than 'brothering'.

'The vileness and degeneracy of Europe': Western science
For all that they hoped to bring the world to the knowledge of the one true living God, missionaries inadvertently helped create a lot of religions.

As we saw in Part 1, 'religio' meant something rather different in the Middle Ages, when it was understood as a virtue, an internal disposition to devotion. At the Reformation, different confessional blocks – Lutheran, Catholic, Reformed – spread across Europe. Each had different approaches to devotion and each became a religion in its own right, identifiable by its particular confession, practices, hierarchy and territories. When European traders and missionaries subsequently went south and east, they encountered traditions that they interpreted through this lens. The world became populated by different religions. The word Buddhism was coined in 1801, Hinduism in 1829, Taoism in 1838 and Confucianism in 1862. These were new categories, discrete entities, characterised by identifiable practices, personal beliefs and settled hierarchies. It's an understanding we still struggle with today.

Such a conceptualisation of religion lent itself to conflict with science. In as far as it was possible to discern their beliefs at all, these other religions were apparently unscientific and structured around primitive myths, ineffective practices and erroneous views of the universe. Moreover, if science was in harmony with European – and now, following the Second Great Awakening in America, *Western* – missionary activity, it meant, almost by definition, that it was at odds with 'Other' religions, those from which missionaries sought conversions. In this way, the story of science and religion, when it went global in the nineteenth century, was, in effect, a sub-plot of the bigger story of the West and the Rest. Science was Western, so the way in which these newly identified religions reacted to it could not but be shaped by the way they reacted to the Westerners who brought it.

The manner in which colonial science came into contact with world religions certainly invited 'conflict' but it did not necessitate it. In the first instance, some religious cultures were harder to dismiss as unscientific than others. The Society for the Diffusion of Useful Knowledge in China was founded by a missionary in 1834, who sought to publish 'such books as may enlighten the minds of the Chinese, and communicate to them the arts and sciences of the West'. In actual fact, European science in China at the time was a thoroughly collaborative affair, relying on Confucian learning as much as Western.[15]

It was a similar story in India. Hindu mathematics was exceptionally advanced. 'Every scrap of Hindu science is interesting,' wrote Edward Strachey of the East India Company's Bengal Civil Service in the Preface to his 1813 translation of the *Bijaganita*, a twelfth-century treatise on algebra.[16] 'We are not cleverer than the Hindu,' wrote the historian John Seeley at the other end of the century.

> Our minds are not richer or larger than his. We cannot astonish him, as we astonish the barbarian, by putting before him ideas that he never dreamed of. He can match from his poetry our sublimest thoughts; even our science perhaps has few conceptions that are altogether novel to him.[17]

These might have been atypically generous views but they nonetheless underline how there was no inevitability to conflict between the (Christian) science of the West and the non-Christian religious traditions of the Rest.

In the second instance, science had yet to be professionalised in the early decades of missionary activity, and at least until the mid-nineteenth century it was quite possible for the scientific encounter with the other religions to be as harmonious as it had been with Christianity in the eighteenth century. The terminology illustrates this. In Chinese, the term *gezhixue*, meaning 'investigating things and extending knowledge', was initially used to refer both to traditional Chinese and to Western sciences.

The term was borrowed from the Jesuit translation for 'natural philosophy' and captured the harmonious integration of that term within a Chinese context. It was only later on that there emerged a sharper terminological distinction between 'Western science' and 'Chinese learning', and by the twentieth century there were wholly different terms used for the two: *kexue* (literally, 'classified learning based on technical training') for science, and *gewu* ('investigation of things') for 'natural philosophy'. For much of the nineteenth century, there was as much conceptual harmony here as there had been in the heyday of European physico-theology.[18] So it was, then, that for all the *potential* for tension there was in the relationship between science and 'other' religions in the nineteenth century, the reality oscillated between harmony and unrest, much as it did in Europe. There is no single history here either.

The complexity is seen particularly clearly in the case of Islam, if only because the Islamic and Christian worlds had so long threatened each other. Since the Treaty of Karlowitz in 1699, European Christendom and the Ottoman empire had lived in uneasy harmony. From 1798, Napoleon's campaign in Egypt and Syria opened up new channels of communication. Fifteen years later, the East India Company had its charter tied to educational and missionary permissions. Islamic scholars visited the West, missionaries moved east and civilisations intermingled through the Levant and South Asia. Protestant colleges were founded and Europeans transformed the Muslim landscape of learning, founding mission colleges, polytechnics, medical schools and naval and military academies. 'Whereas communities of knowledge previously served religious (or scholastic) functions first and bureaucratic functions second, these new institutions were directed mainly to the needs of the state.'[19] The Christian West acquired a new identity in Islamic eyes, and vice versa, at precisely the moment that science was emerging as a distinct, and for the first time demonstrably useful, discipline. Science was engineering and communication, railways and electricity, steam, guns and all that made colonialism possible.

This did not necessitate a hostile view for either party. As we saw in Part 1, Europeans adopted some very different approaches to the relationship between Islam and science in the nineteenth century. One of the earliest academic monographs on the topic, by the accomplished German philologist Franz August Schmölders, was contemptuous of the very idea of Islamic science. Muslims were characterised by intellectual torpor. What was admirable in the achievements of their 'Golden Age' – a seemingly sympathetic phrase that somehow also insinuated centuries of intellectual decay – was derived almost entirely from the Greeks. It would become a familiar refrain.

Others were less hostile. In a forty-page review of Schmölder's thesis, the philosopher Heinrich Ritter demolished his dismissive approach. There were even European and American historians later in the century who enthusiastically claimed that Islam was superior to Christianity in its incubation of the sciences, though they often did so as a means of attacking the Catholic Church.

There was a similar spectrum in evidence among Muslims themselves, although the cultural associations of science loomed far larger in Islamic eyes than they did in Western; your view of science could not but be taken to indicate your view of the West. This was an association that grew in its intensity over the nineteenth century. In the earlier decades of missionary and colonial contact, science was still a part of natural philosophy. Arabic then underwent the same disambiguation as Chinese. At first, the word *ilm* could still be used to encompass both scientific and religious knowledge. However, when the Egyptian intellectual Rifa'a Rafi' al-Tahtawi visited Paris as part of an official delegation in the 1820s and described modern Europe to his peers, he explained how 'the French divide human knowledge into two sections, the sciences and the arts. The sciences consist of verified facts by established proofs, whereas art is the knowledge of techniques . . . when they say *'alim* [scholar] in France, they do not mean that he is well versed in religion but that he is well versed in another subject.'[20] The fact that al-Tahtawi published his account in 1834, the year in which Whewell coined the English word 'scientist', is coincidental but instructive. The same forces that were pulling science and religion apart in Paris or London were at work, at a distance, in Cairo.

For some schools of Islamic thought, this was simply pushing at an open door. For nearly a century before al-Tahtawi visited Paris, an austere reform movement had dominated the Arabian peninsula, as the theology of Muhammad ibn Abd al-Wahhab merged with the political power of Muhammad bin Saud. Denouncing what it saw as the idolatry and moral laxity of contemporary Islam, it was already ideologically alienated from Ottoman Muslims, who sought to learn from the kind of non-religious *ilm* that al-Tahtawi had seen in Paris.

This was not a typical reaction across the rather variegated Islamic world, and the incursion of European science often precipitated both a measure of soul-searching and a reappropriation of Muslim thought from earlier centuries. Al-Tahtawi, charged by the Ottoman governor to harvest French ideas, science and technology as a means of modernising Egypt, reasoned that, however advanced the French were, they were essentially carrying a scientific torch that had been lit by Muslims centuries earlier. Others returned to the heyday of Islamic Spain. A generation after al-Tahtawi, the Ottoman writer Ziya Pasha published sections from Louis Viardot's history of Moorish Spain under the title *The History of Andalusia* which, drawing on additional sources, placed singular emphasis on the Muslim contribution to science. Around the same time, another Ottoman intellectual, Namik Kemal, made the same point, forcibly, to a sceptical European audience. 'You still declare our religion an obstacle to progress,' he berated them. 'Some wise men among you state "The Arabs of Andalusia were the teachers of knowledge to Europe." Weren't they Muslim? Wasn't it Islam that advanced and revived rational knowledge?'[21]

Many Islamic scholars found specific scientific precedents within this intellectual history, and indeed in the Qur'an. Some drew on the rationalising theology and the atomistic views of the Mu'tazilites and the Mutakallimun as a legitimisation of the Western scientific approach, which they claimed had been anticipated a millennium earlier. When Isma'il Mazhar translated Darwin's work into Arabic he compared the ideas of Ernst Haeckel, Darwin's German disciple, with those of the tenth-century philosophical sect the Ikhawan al-Safa' or Brethren of Purity, and the ideas of Jean-Baptiste Lamarck and Herbert Spencer – both evolutionists, though very different to Darwin in argument and style – with the fourteenth-century philosopher Ibn Khaldun.[22]

Some Islamic scholars pointed out that Muslim philosophers had long referred to the idea that species or 'kinds' (from the Arabic term *anwa'*) could change over time. Others claimed that the Qur'an had more flexibility than the Genesis creation story. Q7.54 could be translated to mean that

Allah created the heavens and the earth over 'six periods of time', rather than days, and in any case Q32.5 spoke of 'a day whose span is a thousand years in your reckoning' (echoing Psalm 90:4 and 2 Peter 3:8). This seemed positively to encourage the day-age theory, over which literal-minded Christians were then fretting. Still others, more pragmatically minded, were content to point out how modern science, in the form of medicine, now proved the health benefits of Islamic practices, the physician Mehmed Fahri describing how fasting and praying had measurable health benefits to the pious.[23] Such responses demonstrated an acceptance of 'Western' science, albeit a critical one.

None of this was in principle any different from what was simultaneously happening in other ancient religious cultures that found themselves confronted by European power, science and Christianity. Indian visitors to England marvelled at what they saw. Two Parsis, Jahangir Nowrojee and Hirjeebhoy Merwanjee, visited in the late 1830s to study the science of ship construction, and wrote a *Journal of a Residence of Two Years and a Half in Great Britain*, which dedicated an entire chapter to 'scientific institutions'. It covered steam engines, daguerreotypes, polytechnic institutions, mechanical institutes and the 'blessings' of coal and iron. What they found resonated with them, both culturally and religiously:

> There can be nothing conceived more interesting to persons like ourselves, who having from an early age been taught to believe that next to our duty of thankfulness and praise to our God and Creator, that it is the duty of every man to do all that he can to make all mankind happy; we were early instructed that the man who devoted his energies to the works of science and of art deserved well of his fellow men.[24]

Science was what enabled Britain to colonise India. But it also encouraged Indians to reappropriate their own scientific heritage, often locating Western scientific ideas in ancient Hindu texts. Late nineteenth-century Bengali intellectuals, for example, argued that Darwin's theories supported longstanding Hindu cosmological beliefs, while others claimed that the positivism that characterised Western scientific thought was to be found in Hindu theories of creation.[25]

There was a similar process at work in China. Chinese intellectuals compared Confucian ideas about the perfectibility of the cosmic order with modern evolutionary principles. The scholar and translator Yen Fu presented Thomas Huxley's *Evolution and Ethics* as merely a modern version of much older Daoist ethical debates.[26] The science of the Christian West could undermine, or challenge, or stimulate, or reinvigorate the non-Christian religious traditions with which it came into contact.

And yet, precisely because the science of the Christian West was a core part of its political, economic and military power, the relationship between (Western) science and (other) religions was also a messy and fractious one. All too often, science was treated as synonymous with those alleged Western civilisational characteristics – materialism, immorality, colonialism – that Islamic authorities feared and derided. There was a live political element in this antagonism. The Ottoman empire, which was self-evidently poorer, weaker and more vulnerable than its European neighbours, and which finally lost control over Egypt in 1882, was worried about the loyalty of its young men. When they expressed admiration for European science, the state feared they were also surreptitiously declaring political loyalty. Such fears acquired a distinctly religious framework, despite the fact that it wasn't primarily, or at least exclusively, a religious confrontation. Jamal al-Din al-Afghani, an influential Muslim reformer who travelled through India, the Middle East and Europe agitating against colonialism and for Islamic political independence, became one of the first major Islamic scholars to reject Darwinism. He was far from 'anti-science'. 'There was, is, and will be no ruler in the world but science,' he once wrote. Moreover, he was insistent that 'there is no incompatibility between science and knowledge and the foundation of the Islamic faith', and indeed that 'the Islamic religion is the closest . . . to science and knowledge'.[27]

And yet, he still wrote a scathing attack on evolution that was to prove lastingly popular. Originally published in Persian as *The Truth about the Naturalists*, it was translated into Arabic under the informative if unwieldy title *A Treatise on the Corruption of the Materialists' School and Proof That Religion Is the Origin of Civilisation and Unbelief the Source of Civilizational Decline*, which was subsequently shortened to *The Refutation of the Materialists*. It is far from clear that al-Afghani had read *The Origin of Species* or that he understood what natural selection entailed. His description of it was highly idiosyncratic. There was no doubting, however, that it was naturalistic – *nayshiriya*, a derogatory neologism at the time. It denied any spiritual dimension to life, blurred the boundary between the organic and the inorganic, diminished agency and ignored divine revelation. Its adherents were *dahriyin*, meaning impious and materialistic, who denied God's creation, his providence and his punishment.

Framed in his way, al-Afghani's attack pitted Darwin's science firmly against Islamic religion. His ideas were to prove immensely popular, an Arabic translation of his book being republished four times before 1903 in Cairo alone, and then in every decade of the twentieth century. Nonetheless, as its full title made clear, his was primarily a *civilisational* polemic, targeted at the perceived threat of Western materialism.

For all that al-Afghani also wanted to modernise Islamic society and to broaden its education, he was clear that modernity should not come at the expense of diluting Islamic principles or making peace with colonial authorities. Darwinism was part of a bigger picture. It epitomised the foreign, irreligious element against which his pan-Islamic movement struggled. The result was a conflict between science and religion, though one that was entirely shaped, and misshaped, by the circumstances of the time.

'Our race is depressed enough': human science

Somewhat surprisingly, given the tenor of his attack in *The Refutation*, al-Afghani changed his mind about evolution. Always open to finding contemporary scientific ideas in classical Islamic sources, he came to believe that the transmutation of species had been posited centuries before. He had red lines, however. First, God had created life itself, a religious conviction he found confirmed by the science, and in particular Darwin's words, in the final paragraph of *The Origin of Species*, where he claimed that 'life . . . [was] originally *breathed* into a few forms or into one' – a phrase that the author came to regret using.

Second, evolution could not fully explain mankind, which transcended the merely material, another conclusion that al-Afghani thought was supported by Darwin's almost-complete refusal to talk about 'man' in *The Origin*. In this way, the issue that so concerned Samuel Wilberforce – the status of humanity under an evolutionary lens – also vexed al-Afghani and his peers. The difference, however, was that, as a member of a culture over which European science was casting its imperial and imperious gaze, al-Afghani had a little more reason for concern.

Scientific racism had a long prehistory. The greatest minds of the Enlightenment were confident in their white supremacy. Voltaire categorised the 'Caffres, the Hottentots, the Topinambous' as 'children' in his *Philosophical Dictionary*. David Hume had confessed that he was 'apt to suspect the negroes and in general all other species of men . . . to be naturally [as opposed to culturally] inferior to the whites'.[28] As far as Kant was concerned, different human 'races' could be ranked in order from 'white' through 'yellow' to 'Negroes' who, while 'strong, fleshy, and agile', were also 'lazy, indolent, and dawdling'.

This was philosophical rather than scientific racism, although such thinkers aspired to scientific reasoning. 'The profusion of iron particles, which are otherwise found in the blood of every human being, and, in this case, are precipitated in the net-shaped substance through the evaporation of the phosphoric acid . . . is the cause of the blackness that shines through the epidermis,' Kant wrote in 1777, adding in parentheses, 'which explains why all Negroes stink.'[29] It was only in the final years of the eighteenth century, however, that science began to throw its growing authority behind such views. Phrenology paved the way, enabling the 'scientific' measurement of cranial, and therefore cognitive, differences. But phrenology simply begged bigger questions. These apparently huge and self-evident differences cried out for a deeper biological explanation. Climate and culture seemed inadequate and, more importantly, unscientific.

The whole endeavour was frustrated by the Christian doctrine of monogenism, doggedly defended by the physician and pioneer ethnologist James Cowles Prichard. Prichard studied medicine at Edinburgh, where he wrote his thesis on the 'varieties of mankind'. The work became, in due course, *Researches into the Physical History of Man*, and was first published in 1813, but it expanded in both influence and size until, by the 1830s, it comprised five volumes.

Prichard was born into a Quaker family but subsequently converted to evangelicalism. His monogenism was theological in its foundation but justified by scientific arguments. Physiology, zoology, ethnography, linguistics and even, tentatively, geology were all drawn on to defend monogenism and explain 'racial' difference through environment and transmutation. Prichard went further, however, by claiming (at least in the first edition) that evidence suggested that all people were descended from African progenitors. 'The process of Nature in the human species is the transmutation of the characters of the Negro into those of the European . . . [which] leads us to the inference that the primitive stock of men were Negroes.' This, he acknowledged in one of the scientific understatements of the century, was 'a conclusion which may be questioned.'[30]

Prichard had authority – biblical and ecclesiastical – on his side. John Bird Sumner, bishop of Chester, future archbishop of Canterbury, cousin to the Wilberforce clan, drew on Prichard for his *Treatise on the Records of Creation and the Moral Attributes of the Creator*, which used common descent to argue that 'the European is not more unlike the Caffre [Black African], than the Caffre differs from the Bojesman [bushman], or the Hottentot, from whom they are separated only by a range of hills.'[31] Humans were different, but one.

Others were not persuaded, and some tried to work polygenism into Genesis as a means of explaining what they judged otherwise inexplicable. There were two distinct creation narratives in Genesis, argued Edward King, a Fellow of the Royal Society, and this surely implied there were (at least) two separate moments of human creation. Increasingly, however, with the literal reading of Genesis undermined by the science of geology, such negotiation with the scriptures was becoming unnecessary. Josiah Nott, a respected American physician, argued forcefully for polygenism both on the 'scientific' grounds of anatomy and archaeology and on the (anti-)religious grounds that the Bible was unscientific, outdated and irrelevant to such matters.[32]

As with pretty much everything in the history of science and religion this wasn't simply a black and white issue, so to speak. There were Christian phrenologists – like George Murray Paterson, who founded the Calcutta Phrenological Society in 1825, to collect and study 'hindoo skulls' – for whom the difference between 'races' ran deep into nature.[33] Conversely, there were scientists like the Heidelberg physiologist Friedrich Tiedemann, whose article for *Philosophical Transactions* in 1836, entitled 'On the Brain of the Negro, compared with that of the European and the Orang-Outang', argued defiantly that 'there are no well-marked and essential differences between the brain of the Negro and the European'.[34] Darwin, for all that he was wedded to a scientific understanding of mankind, was adamantly opposed to Nott and was a passionate monogenist his entire life, although for reasons that had more to do with his abolitionist heritage than any scientific rationale.[35]

Wherever intellectuals came down on the question of the unity of mankind, however, the growing cultural authority of science shifted the conversation about humans from a theological to a scientific register. The question of who was 'man' was now finally becoming a scientific rather than a religious one, science assuming the prestige of being the sole source of *reliable* knowledge about the world. If people were serious about understanding human nature, they needed to do so scientifically.

In reality, this meant that the question itself changed subtly. *Who* was 'man'? was the kind of thing that theologians asked because 'who' already assumed a great deal of the object under examination. 'Who' implied a person, with agency, moral awareness, a first-person perspective; a someone rather than a something; in the Christian tradition, a someone that bore the image of God. This was not the language of science. Indeed, it sounded hopelessly mired in metaphysics and theology. When science asked the question, it became *what* was 'man'? What could be understood of humans objectively, definitively, universally and by means of detached observation and experiment, rather than, say, through attachment and communication. There was no necessary tension between these two approaches. It was (and is) quite possible to understand the human as a 'who' and a 'what'. But if one approach becomes hegemonic and delegitimises the other, the effect is disastrous. As historian Constance Clark has written, 'human societies that seemed foreign could [now] be categorised just as naturalists classified animals and plants', and they were.[36]

This shift can be traced through Victorian societies. After the abolition of slavery in British colonies, a number of Quakers and evangelicals came together to found the Aborigines' Protection Society (APS) in 1837. The organisation set out to expose colonial atrocities (particularly in Australia) and to Christianise imperial policy. Its motto was *Ab Uno Sanguine* ('Of one blood') and it enthusiastically proclaimed the Adamic unity of all peoples on earth. James Prichard was an early member.

A few years later, following a disastrous expedition up the river Niger on behalf of the African Civilisation Society, a cloud gathered over such humanitarian enterprises. The African Civilisation Society was wound up and the Aborigines' Protection Society shifted its focus towards the scientific study of aborigines, in theory as a way of ensuring their protection. It created the Ethnological Society of London, with the new organisation splitting the APS's scientific and linguistic interests from its religious and humanitarian ones. Prichard was again a founder member.

In its early years, the Ethnological Society shared much with its parent, including a number of members, a vaguely philanthropic ethos and a firm commitment to monogenism. However, following Prichard's death in 1848, its commitments and interests began to diverge and in 1863 a number of dissidents split from it to form the Anthropological Society of London (ASL).

The split was genealogical, moral and political. The new anthropologists disliked the quasi-biblical monogenism, rejected the sentimental idea that 'the Negro is a man and a brother', and threw their weight behind the Confederacy in the American Civil War. Josiah Nott, whom the society called 'the greatest living Anthropologist of America', was given an Honorary Fellowship.[37] But the split was also professional and scientific. ESL membership had comprised too many civil servants and clergymen. The new society was for scientists, people who wanted to establish the biological facts about race, unencumbered by humanitarian sensitivities. Science was about the facts; feelings were an irrelevance; theology worse.

Darwin's own monogenetic work was a stumbling block to this new 'scientific' understanding of race, in spite of his own well-evidenced theory of natural selection. There would be no progress in the application of the Darwinian principles to anthropology 'until we can free the subject from the unity hypothesis', wrote the speech therapist and leading anthropological light James Hunt in 1866.[38] It was not that the anthropologists denied evolution or even descent from apes. It was the fact that this theory simply couldn't explain the gulf that apparently stretched between whites and other races.

Salvation seemed to come in the form of a book, *Lectures on Man*, written by the highly respected German-Swiss scientist Carl Vogt. Vogt was a zoologist, a geologist and a physiologist, as well as an active politician and an energetic atheist and scientific materialist. Contemptuous about religious belief for its soft-headedness – he called certain simian-looking heads 'Apostle skulls' – he was a member of a small group of German scientists who scandalised religious opinion by arguing that everything, including everything that was considered quintessentially human or religious, could be reduced to matter. 'Thoughts come out of the brain as gall from the liver, or urine from the kidneys,' Vogt wrote.[39] The universe was blind, governed by unalterable law and necessity, and humans were no exception. This was precisely the kind of hard-headed view that the anthropological debate needed. Vogt's *Lectures on Man: his Place in Creation and in the History of the Earth* was published by the Anthropological Society in 1864 and argued that, while humans had evolved from apes, different races of human had evolved from different species of ape. In a flourish, materialism and evolution could be combined with a kind of polygenesis so as to provide a satisfactory scientific explanation for human racial difference.

Darwin referenced Vogt's *Lectures* on numerous occasions in his *Descent of Man*, which was published in 1871, the year in which the Ethnological and the Anthropological societies finally overcame their differences and merged to form the Anthropological Institute. While recognising his erudition and appreciating his early willingness to apply natural selection to human beings, Darwin was distinctly lukewarm about Vogt's strange hybrid, poly-monogenist proposal. It is possible, he wrote, that 'the early progenitors of man might at first have diverged much in character, until they became more unlike each other than are any existing races' – though, he added, it was 'far from probable'.[40]

By this time, however, Darwin had developed a theory that appeared to do what natural selection failed to: namely, explain racial differences without resorting to polygenism. 'I suspect that a sort of sexual selection has been the most powerful means of changing the races of man', he had written to his fellow evolutionary pioneer Alfred Russel Wallace in May 1864, though 'I do not expect I shall convince anyone else', he later confessed to Hooker, having failed to convince Wallace. His 1871 book

– its full title *The Descent of Man and Selection in Relation to Sex* – was his best attempt to do so.

The book upset many religious believers, especially those who had clung doggedly to Paley's natural theology after *The Origin* had been published. James Grant, a fishing tackle maker from Grantown in Scotland, spoke for many when he wrote to Darwin having read his book, and John Tyndall's *Fragments of Science*, asking 'if you would, in two or three words, simply tell me if your doctrine of the descent of man destroys the evidence of the existence of a God looked at through nature's phenomena.'[41] But it delighted those who, like Darwin, sought a reliable (if not necessarily test-able) hypothesis for racial difference, without embracing polygenism, with its tenacious and uncompromising racism.

And yet, for all that Darwin was passionately committed to monogen-esis, to common descent and to the brotherhood of man, and for all that his theory dismantled ideas of racial fixity or 'essentialism', the book left no doubt that some groups were superior to others and that biology, for all its variability, was still destiny. Darwin quoted the essayist William Rathbone Greg on how the 'the careless, squalid, unaspiring Irishman multiplies like rabbits'. He was subsequently taken to task for this by a polite, anonymous Irish correspondent who wrote to him in 1877. '[*Descent*] is a great Scientific work', he said, and it is 'destined to go to all Time and into all languages, [but] the passage to which I refer you, is . . . quite unworthy of such a book, and of you.'[42] How Darwin responded is not known but he did not excise the remarks from future editions.

More troublingly, he wrote, entirely in his own voice, that 'at some future period, not very distant as measured by centuries, the civilised races of man will almost certainly exterminate and replace throughout the world the savage races.' Had he known how such ideas would be used as a justification by generations of racists and eugenicists, the abolitionist, humanitarian Darwin would have been nauseated. But it was nonetheless an unfortunate and revealing statement. Immersed as he was in the perva-sive Christian culture of Victorian Britain, Darwin could not have envis-aged how authoritative his own work would become and how, in a strange way, it would replace the Bible in (some) people's minds. This was not simply because it put forward a rather more accurate explanation of human origins than that provided by Genesis. Rather, it was because in a culture where science was now intellectually authoritative, his ideas offered an account of human origins and nature whose accuracy was confused with its sufficiency. In knowing *what* 'man' was – how mankind had descended, how its intellectual and moral faculties had developed, to which other species it was related, and so forth – the conviction was that we also now knew *who* 'mankind' was. Or, more precisely, that the *who* question was now essentially redundant, the provenance of metaphysicians, theologians and other dusty remnants of pre-scientific history.

The consequence of this was visible – literally – in New York a few years later. In September 1906, William Temple Hornaday, the first new director of the New York Zoological Park, put Ota Benga, a 23-year-old Congolese pygmy, on public display in the monkey house. He was in the same cage as an orang-utan called Dohung, and next to one that contained a gorilla called Dinah.[43] Benga was a survivor of a massacre in the Congo Free State in which he had lost his wife and children. Sold into slavery he was (willingly) brought to America by Samuel Verner, another missionary turned anthropologist, and housed in the American Museum of Natural History, before he was moved to the zoo. At first, he helped keepers look after animals but after a while Hornaday had the bright idea of moving him in with them. A sign outside his cage accurately described his age, height, weight and provenance.

The exhibit drew an enormous crowd. Benga, many speculated, might be the famously elusive missing link between humans and other primates. He certainly seemed to have much in common with his cell mate, the newspapers claimed. 'The pygmy was not much taller than the orang-utan, their heads are much alike, and both grin in the same way when pleased,' observed the *New York Times*.[44]

Ota Benga's own impressions were scarcely recorded but hardly a secret. He went out of his way to be as disruptive as possible to Hornaday, at one point brandishing a knife at his keepers. When the prodding, poking crowd asked him what he thought of America, he was heard to reply 'Me no like America, me like St Louis.' (where he had been living before being transported to New York).[45] His voice was ignored, even as his feelings were evident.

Visitors were fascinated and enthused but not everyone was impressed. The display drew a furious response from the Colored Baptist Ministers'

Conference, the Revd James H. Gordon complaining that 'our race . . . is depressed enough, without exhibiting one of us with the apes.' He went on, 'we think we are worthy of being considered human beings, with souls.' Other, non-black ministers agreed. 'The person responsible . . . degrades himself as much as he does the African,' said the pastor of the Calvary Baptist Church. 'Instead of making a beast of this little fellow, we should be putting him in school for the development of such powers as God gave him . . . We send our missionaries to Africa to Christianize the people and then we bring one here to brutalize him.'[46]

Hornaday was unrepentant. 'I am a believer in the Darwinian theory', he replied, and demonstrate such specimens 'purely as an ethnological exhibit'. The *New York Times* supported him. 'The revered colored brother should be told that evolution . . . is now taught in the text books in all schools, and that it is no more debateable than the multiplication table.' Its readers, or at least some of them, concurred. The exhibit should be praised for its 'scientific character', one correspondent wrote. It was to be hoped that it would 'help our clergymen to familiarise themselves with the scientific point of view so absolutely foreign to many of them'.[47]

The lines were not perfectly drawn. There were many anthropologists who insisted that Benga's display was scientifically nonsensical, as well as morally disgusting. And there were Christian ministers who objected to the Benga display not so much because it was grotesquely racist but because it endorsed Darwinian evolution.

Moreover, one did not need Darwinism to treat other humans as interesting specimens: Banks had wanted to do just the same to 'his' Pacific Islander Tupai'a, nearly 150 years earlier. Nevertheless, the fate of Ota Benga brought together the two themes that ran through the varied historical landscapes of science and religion – the question over who or what was 'man', and the tussle over who had the authority to adjudicate. In effect, what was the nature – or what were the natures – of the beast, and who got to say?

The two sides in Ota Benga's dispute were in effect speaking at cross purposes, almost in different languages. Hornaday and his defenders talked science – Darwinism, evolution, ethnology, the 'scientific point of view'. Gordon and his fellow ministers talked religion, justifying their indignation in terms – 'souls', God-given gifts – that could never have had any purchase on Hornaday's scientific worldview. They were two views of seeing the world or, in this case, a Congolese pygmy.

Benga himself was released after a few weeks and went to stay with Gordon at the Howard Colored Orphan Asylum. Thereafter, he moved to Lynchburg, where he started elementary school at the Baptist seminary. His plan was to return to what was by then the Belgian Congo but the start of World War One made this impossible. He grew depressed and on 20 March 1916 he borrowed a gun and shot himself through the heart.

John William Draper, chemist, would-be intellectual historian,
speaker at the infamous Oxford debate between Huxley and
Wilberforce – and responsible, more than any other figure,
for developing the myth of the history of the conflict between
religion and science – he literally wrote the book on it.

THIRTEEN

PEACE AND WAR

'The science of dead matter': peace

A year before he died, Darwin wrote to the philosopher William Graham. He had recently read Graham's book *The Creed of Science* and was, unusually for him, writing to the author cold to thank and praise him. 'It is a very long time,' he confessed, 'since any other book has interested me so much.'

The Creed explored how 'the chief conclusions reached by Modem Science' – such as evolution by natural selection and the conservation of energy – related to 'the central questions of religion, morals, and society'. Popular and widely admired, it was an example of a genre within late nineteenth-century publishing that neither rejected new scientific ideas nor collapsed into despondent unbelief, but engaged science and religion in sympathetic conversation. It is a genre now almost completely forgotten.

The book covered many of the issues with which Darwin had been wrestling for years – indeed it mentioned the great naturalist more than forty times – but it was not so much this that impressed Darwin as Graham's account of the physical universe. 'You have expressed my inward conviction, though far more vividly and clearly than I could have done, that the Universe is not the result of chance.'[1]

Given the views of Laplace, whom Graham discussed early in his book, that God was a hypothesis for which (his model of) the universe had no need, Darwin's verdict may seem odd, not least coming from a now-convinced agnostic. But paradoxically, Laplace's view – or at least the sentiments it conveyed – made the interaction between religious belief and the physical sciences, such as astronomy, chemistry, physics and, underlying them all, mathematics, *more* rather than less constructive in the nineteenth century.

There were (at least) three reasons for this. First, there was less riding on the physical sciences. Biology had become the battleground in part because William Paley and his contemporaries had planted their flag there. Natural theology had invested a great deal in earthly bodies. Twenty of the relevant twenty-two chapters in Paley's book were dedicated to the organisation and structure of vegetables, animals and insects. Celestial bodies, by comparison, got a single chapter. The truth or otherwise of religion rested on the science of life rather than the science of matter – at least that is what the natural theologians implied – and the result was that there was less reason to fight God battles over cosmology or physics.

Second, it was already patently clear by the early years of the century that mathematics and the physical sciences were now the territory of trained professionals. There was little chance that amateur natural philosophers sitting in country rectories could have described the refraction of light as did Leonhard Euler or developed the law of the conservation of energy as did James Joule. This, in turn, meant that there was no comparable tussle over professionalism and authority such as afflicted the life sciences through the nineteenth century. It didn't mean that believers could not be authoritative physical scientists. It simply meant that there was no question about where the foundations of their authority lay.

Third, the physical sciences were less amenable to the kind of anti-religious reductionism and positivism that affected biology and anthropology in the nineteenth century. Next to no physical scientists envisaged direct divine intervention in the way that Newton had done, but they didn't have to. Fundamental concepts, like the universal lawfulness of the cosmos, and fundamental tools, like mathematics, were not easily reducible to 'gall from the liver, or urine from the kidneys' as Vogt had put it. The lawfulness of creation and the reality of mathematical concepts did not demand belief in God, but they were easy to reconcile with it, and this was arguably easier than with belief in a random, materialistic universe. At times, mathematics and physics could even feel like distant cousins to theology and metaphysics.

The consequence of all this was a kind of constructive concord. Religion and the physical scientists neither fought over the same toys nor ignored one another like estranged relatives, but got on with their lives in a measure of broadly edifying and harmonious peace. How edifying and how harmonious could vary, however. Peace came in different forms.

Michael Faraday was, by anyone's reckoning, one of the most important and most consequential scientists of the nineteenth century. A brilliant experimentalist and popular lecturer at the Royal Institution in London, he was an accomplished chemist who discerned the laws of electrolysis and synthesised compounds of chlorine and carbon. However, it was for his work on magnetism and electricity, which was to become the basis of the electric motor and the dynamo, that he was and is best known.

Faraday's brilliance was underlined by the fact that he had next to no formal education. His father was a blacksmith, the family was poor and Michael was taught only basic literacy and arithmetic before he had to go out to work. He read through the rudiments of science during a seven-year apprenticeship as a bookbinder, before being taken on by the chemist Humphry Davy as his assistant at the Royal Institution. It meant that, unusually, his scientific work didn't begin until his thirties or flourish until his forties.

Through all of this, Faraday was extremely devout, belonging to 'a small and despised sect of Christians known, if at all, as Sandemanians', as he later described them.[2] The group originated as a splinter from the Church of Scotland during the First Great Awakening, the evangelical revival beginning in the 1740s, and pursued a simple evangelical piety with considerable moral energy. Faraday served as an elder to the City of London congregation for many years, to whom he also preached (though not, by some accounts, very well). He acquired a reputation for near saintliness, and was renowned for being generous, charitable and possessed of considerable integrity. He refused to advise the government on chemical weapons during the Crimean War, twice refused the presidency of the Royal Society, declined the invitation to be buried in Westminster Abbey and turned down a knighthood, all because such honours violated his evangelical aversion to worldly titles.

For all that his Christianity suffused his personal life, however, it did not shape his public one. Faraday's peace was the peace of silence. His friend and biographer John Tyndall, who admired Faraday's piety while certainly not sharing it, wrote that 'never once during an intimacy of fifteen years did he mention religion to me, save when I drew him on to the subject [when] he spoke to me without hesitation or reluctance.'[3] Faraday seems to have compartmentalised his religion and his science. As an evangelical, indeed coming from the fringes of evangelicalism, he had not been marinated in the tradition of natural theology that had once so influenced Anglicans. Paley's would-be scientific arguments for the character of God were alien to him. 'There is no [natural] philosophy in my religion . . . I do not think it at all necessary to tie the study of natural science and religion together,' he once wrote.[4]

There was conviction as well as circumstance in this view. Faraday was of Pascal's school in believing that human reason could only carry you so far, and could certainly not establish, let alone prove, God's existence or character. As he said in one of his few sermons that found its way into print, the knowledge of 'God's truth [is] far beyond anything we can have in this world'.[5] And yet, even with Faraday there were moments when science and religion conversed in more substantial fashion. An unpublished paper on the nature of matter, written in 1844 and held in the Library of the Institution of Electrical Engineers until its discovery over a century later, is witness to Faraday's inner thoughts on the matter. 'As a natural philosopher', he began, I have 'purposely limit[ed] my object to the investigation of the phenomena presented by the material creation'. Nevertheless, I 'feel constrained to form some idea of matter', even if that meant transgressing on to marshier, metaphysical ground. The paper then proceeded to speculate on how force and matter acted upon one another, on the basis that God 'governs his material works by definite laws resulting from the forces impressed on matter'. Faraday was acutely conscious of the conjectural nature of his musings, which is one of the reasons why the paper remained unknown, but it seems that ultimately even a scientist as private about his faith as Faraday could not, as Tyndall put it, keep his 'religious feeling and his philosophy . . . apart'.[6]

More loquacious than Faraday on the relationship between science and religion – although not much more – was Scottish mathematical physicist James Clerk Maxwell. If Faraday qualifies as one of the most important scientists of the nineteenth century, Maxwell is one of the most important of all time. Einstein had pictures of both Faraday and Maxwell on his wall but it was on Maxwell's shoulders that he claimed he stood.

Maxwell's formative years were as pious as Faraday's. He grew up on the remote estate of Glenlair in Kirkcudbrightshire, and his mother, whom he later described as 'guided by religious thought [but] very independent of the exhortations of acquaintances, clerical or lay',[7] immersed him in the Bible, which he was able to quote extensively from a young age. Unlike Faraday, there was a natural theological dimension to Maxwell's religion. His mother encouraged him to 'look up through Nature to Nature's God'[8] and by the time he was studying mathematics at Cambridge in the early 1850s he was writing poems which proclaimed:

> Through the creatures Thou hast made
> Show the brightness of Thy glory,
> Be eternal Truth displayed
> In their substance transitory,
> Till green Earth and Ocean hoary,
> Massy rock and tender blade
> Tell the same unending story –
> 'We are Truth in Form arrayed.'[9]

For all this might have sounded like Paley rendered by a cathedral choir, it was not the basis of Maxwell's faith, which developed an intense and much-examined personal dimension. In the year he wrote the poem he fell ill, and while convalescing at the home of the Revd Charles Tayler, a friend's uncle, he experienced an evangelical conversion. 'I have the capacity of being more wicked than any example that man could set me,' he wrote to his friend, 'and that if I escape, it is only by God's grace helping me to get rid of myself.'[10] Thereafter, he vowed 'to let nothing be wilfully left unexamined', as he wrote to another friend who was to become his first biographer.[11] His subsequent correspondence with his wife, both before and after marriage, was intensely theological, often reading like a commentary on the New Testament. When they lived in London, the couple attended a Baptist church on account of its preacher, despite the fact that 'we believe ourselves baptized already.'[12] When back in Glenlair, he led devotions with the estate's workers and families, with whom he was very popular. Maxwell's religion saturated his life.

As did science. Publishing academic papers from the age of eighteen, a Fellow of Trinity College, Cambridge at twenty-four and Professor of Natural Philosophy at Aberdeen the following year, Maxwell's brilliance was evident to all. At the age of twenty-nine, he took the Chair of Natural Philosophy at King's College, London, and began studying electricity and magnetism in earnest, integrating the two forces in a theory of electromagnetic radiation that came to be recognised as the second great 'unification' in physics, after Newton's work on gravity. So much of what we take for granted today – from radio through microwave ovens to X-ray machines – is based ultimately on Maxwell's work.

Maxwell corresponded with the aged Faraday as he developed his ideas on electromagnetism. Where the elder man was an exemplary experimenter, Maxwell was primarily a theoretician, an outstanding mathematician whose equations probed depths of reality that were only confirmed empirically decades later (many were considered *merely* theoretical at the time). Nonetheless, the two men had deep respect for one another, without ever being very close, and when Maxwell wrote the essay on Faraday for the *Encyclopaedia Britannica* a decade or so after the latter died, he paid tribute not only to his science but, obliquely, to his religion. 'When, on rare occasions, he was forced out of the region of science into that of controversy, he stated the facts, and let them make their own way.'[13]

For his part, Maxwell was a little less reticent about the relationship between science and Christianity, and a bit more willing to direct the facts to their speculative ends. One of the Glenlair household prayers that were found among his papers after he died recorded him asking God to 'teach us to study the works of Thy hands that we may subdue the earth to our use, and strengthen our reason for Thy service.'[14] The words might have come from Sprat's history of the Royal Society two hundred years earlier.

Maxwell's religious ruminations could be more precise, however. His inaugural lecture in Aberdeen in 1856 explained that the laws of nature were 'not mere arbitrary and unconnected decisions of Omnipotence' but 'essential parts of one universal system in which infinite Power serves only to reveal unsearchable Wisdom and external Truth.'[15] A decade later, when he turned his attention to the second law of thermodynamics – the irreversible dissipation of energy and the linear history of the universe that it implied – he homed in on its wider metaphysical implications. The physics seemed to suggest that there was a moment of creation, Maxwell positing something like a Big Bang. 'If, therefore, there was ever an instant at which the whole energy of the universe was available energy, that instant must have been the very first instant at which the universe began to exist.'[16] Moreover, it also suggested an irreversible direction to time and history. 'Our experience of irreversible Processes . . . leads to the doctrine of a beginning & an end instead of cyclical progression for ever.'[17] A few years after this, when speaking at the Royal Institution about his electromagnetic theory of light in a lecture entitled 'Action at a Distance', he concluded by placing this new understanding of causal mechanisms in the universe within a longstanding religious debate. 'The vast interplanetary and interstellar regions will no longer be regarded as waste places in the universe, which the Creator has not seen fit to fill with the symbols of the manifold order of His kingdom.'[18]

Most substantially and controversially, in an 1873 BAAS lecture 'On Molecules' he argued, making oblique reference to Thomas Chalmers' Bridgewater Treatise from forty years earlier, that the nature and permanence of molecules throughout the history of the universe could not be explained simply by invoking 'nature', and he pointed, instead, in the direction of the kind of supernatural (though not necessarily interventionist) architect whose work structured creation.

> They continue this day as they were created – perfect in number and measure and weight, and from the ineffaceable characters impressed on them we may learn that those aspirations after accuracy in measurement, truth in statement, and justice in action, which we reckon among our noblest attributes as men, are ours because they are essential constituents of the image of Him Who in the beginning created, not only the heaven and the earth, but the materials of which heaven and earth consist.[19]

This was technically a deistic rather than a Christian point but the passing reference to the apocryphal Old Testament book *Wisdom of Solomon* ('thou hast ordered all things in measure and number and weight' (11:20)) left no doubt about what Maxwell was inferring.

Not everyone was persuaded or impressed. Joseph Hooker wrote to Darwin complaining that Maxwell's lecture was 'dull dry & singularly unintelligible' and 'without a point or significance to the outside world'.[20] Having read it, Darwin was more charitable, replying to Hooker that 'I think if you *read* Clerk Maxwell's lecture [Hooker had only been in the audience] you will alter your opinion.'[21] More substantially, the most important scientific materialist of the age, Faraday's biographer John Tyndall, critiqued the lecture in his controversial *Belfast Address* as the president of the British Association in August 1874, and commented how 'Professor Maxwell finds the basis of an induction which enables him to scale philosophic heights considered inaccessible by Kant, and to take the logical step from the atoms to their Maker' – a remark that was left out of the official printed version of the lecture.

Whether or not Maxwell had overstated his case, he did recognise that there were limits to the science–religion dialogue. A few years after his Molecule lecture, he received a letter from the Anglican bishop C. J. Ellicott, who was turning some lectures into a book that was to be published under the title *Modern Unbelief: its principles and characteristics*. Impressed by Maxwell's work, he wrote asking him about his understanding of 'the creation of light' in Genesis 1.

Maxwell was hesitant. If forced to make a comparison, he said that it would be very tempting to say that the light of the first day means an 'all-embracing æther, the vehicle of radiation, and not actual light, whether from the sun or from any other source'. However, he stressed that this was in accordance with the science of 1876 'which may not agree with that of 1896', and went on say that he would be very sorry 'if an interpretation founded on a most conjectural scientific hypothesis were to get fastened to the text in Genesis'. The 'rate of change of scientific hypothesis' was 'much more rapid than that of Biblical interpretations', so that 'if an interpretation is founded on such an hypothesis, it may help to keep the hypothesis above ground long after it ought to be buried and forgotten'[22] – or conversely drag a religious hypothesis down when it fell. Not for Maxwell the kind of artificial and usually damaging 'concordism' that had dogged the science–religion interaction since the time of John Philoponus in the sixth century.

Maxwell's reticence was also visible in his unwillingness to join the Victoria Institute, an organisation set up in 1865 to reconcile science and religion. It wasn't that Maxwell was unsympathetic to its aims. 'I think men of science as well as other men need to learn from Christ, and I think Christians whose minds are scientific are bound to study science [so] that their view of the glory of God may be as extensive as [possible],' he wrote to the secretary of the Institute.[23] However, he continued, 'I think that the results which each man arrives at in his attempts to harmonise his science with his Christianity ought not to be regarded as having any significance except to the man himself, and to him only for a time, and should not receive the stamp of a society.' The reason was the same as that which had caused him to dampen Bishop Ellicott's apologetic ardour. Science and religion flourished in dialogue but as soon as that dialogue hardened into a formal, institutional concordat, disharmony beckoned. Maxwell's peace wasn't as private as Faraday's but its public profile remained low, and its nature reserved and flexible.

Others were less nervous. George Gabriel Stokes, a friend of Maxwell's, used the Victoria Institute as a platform to emphasise the harmony of science and religion. Stokes is largely unknown today but he was a major scientific figure in late Victorian Britain. Born in County Sligo in Ireland, he studied mathematics at Cambridge before being appointed as Lucasian Professor, the chair held by Isaac Newton, at the age of thirty. Stokes was impressively if not uniquely young (Nicholas Saunderson had been twenty-nine, Newton twenty-six and Edward Waring twenty-four) but he did go on to hold the position for an astonishing fifty-four years. Over such an extended period he made original contributions to fluid dynamics and spectroscopy, while also serving as secretary of the Royal Society for thirty-one years, president for five and Member of Parliament for Cambridge between 1887 and 1892, before being made a baronet. He was, in short, a living embodiment of late Victorian establishment science.

And religion, for over his many years at Cambridge he also worked for numerous evangelical organisations, including as vice president of the British and Foreign Bible Society. Stokes was rather more of the natural theological school than either Faraday or Maxwell, and as late as 1880 he explained to an audience that he understood a designed universe 'much in the same way that was mentioned long ago by Paley in his *Natural Theology*'.[24] As such, he was willing to use the opportunity that the Victoria Institute offered him to parade the harmony of science and religion.

The Institute had originated in the 1860s as a reaction against Darwinism or – as the Institute described it with reference to 1 Timothy 6:20 – the challenge to 'revealed truth from "the oppositions of science, falsely so called" '. This did not augur well for fruitful dialogue – the satirical magazine *Punch* lampooned it as the 'Anti-Geological Society' – and for its first few years the Institute was dominated by precisely the kind of amateur scientists that were losing their cultural authority. One distinguished professor refused to attend a meeting on the grounds that the society's honorary secretary, James Reddie, a civil servant rather than a practising scientist, did not even believe in the Newtonian theory of gravity.[25]

Reddie died in 1871 and the Institute subsequently if slowly achieved wider credibility. This was largely due to Stokes. Reluctant at first, he agreed to present a paper in March 1880 and another ('On the Absence of Real Opposition Between Science and Revelation') three years later. He accepted the Institute's presidency in 1885, just as his star was at its brightest, having also just been elected president of the Royal Society. His unimpeachable authority rubbed off on the organisation, which attracted some of the greatest figures in late Victorian science, former and future Royal Society presidents among them, to his presidential addresses.

Stokes was more forthright than Maxwell on the ways in which contemporary science supported religious belief. Like Maxwell and Faraday before him, he understood the universe as not so much designed and controlled directly by God as operating according to fixed laws of nature that were ultimately God's doing. Creation was still, however, ultimately teleological: even if the eye, that classic natural theological example, could be said to have developed from earlier, less complicated forms, it remained testimony to a remarkably intricate, complex yet dependable lawful creation, one that was set up to enable sight; one that, so to speak, saw the eye coming.

Moreover, just as the eye was built into the purposeful laws of the universe, so were the humans who used it. 'To me it seems to be simplest to suppose that man's mental powers, as well as his bodily frame, were designed to be what they are.' Precisely how that design was carried out, he brushed over a bit too easily: 'We have no means of knowing, and it does not concern us to inquire.' Moreover, as far as Stokes was concerned, there is 'no difficulty in supposing that man's innate sense of right and wrong was as much impressed upon him . . . as his bodily frame'.[26]

This was heady stuff, a world away from the early years of the Victoria Institute (let alone Paley) in the way it was prepared to understand that seemingly quintessential human characteristic – morality – *within* an evolutionary framework. Yet it was still a risky enterprise, in danger of drifting away from fruitful conversation between science and religion and towards basing the latter on the former.

This is arguably what Stokes's friend, William Thomson, Lord Kelvin, our final great late Victorian mathematical physicist, did. Although not quite in Maxwell's league, Thomson did much to unify physics, formulating the laws of thermodynamics, determining the lowest possible level of temperature (now measured in Kelvin) and working on the transatlantic telegraph project. He succeeded Stokes as president of the Royal Society, won its Copley Medal and was the first British scientist to be elevated to the House of Lords. He was every bit as establishment as Stokes.

He was also more forthright than his friend, let alone Maxwell and Faraday, and rather more bullish in tone. 'Do not be afraid of being free-thinkers,' he wrote in response to a popular lecture by the botanist and Christian apologist George Henslow, son of Darwin's friend and mentor John Stevens Henslow. 'If you think strongly enough you will be forced by science to the belief in God.'[27]

Thomson, like Maxwell, saw in the second law of thermodynamics confirmation of a beginning, a direction and an end to creation, all of which offered a welcome alternative to the unbiblical timelessness of uniformitarian geologists. Thomson was triumphant, writing as if the physical sciences – 'the science of dead matter, which has been the principal subject of my thoughts during my life' – could now finally confirm biblical truth. 'St Peter speaks of scoffers who said that "all things continue as they were from the beginning of the creation"; but the apostle affirms himself that "all these things shall be dissolved". It seems to me that even physical science absolutely demonstrates the scientific truth of these words.'[28] This was a dangerous move, however, as it effectively placed religious truth on scientific foundations while completely ignoring Maxwell's sage advice that those foundations shifted and the science of one generation was not necessarily the same as that of the next.

This came to pass in Thomson's attitude to Darwinism, which was ambivalent, to put it mildly. His ambivalence was, on the surface at least, for good scientific reasons. Evolution by natural selection needed time, a lot of it. Thomson was convinced that the earth didn't offer enough. If the planet began as molten rock, he calculated that, on the basis of the known thermal conductivity of rocks near Edinburgh, the earth could be no more than 400 million years old. Thomson entered the fray against Huxley but ended up reducing rather than increasing his estimate, initially to 100 million years and then to around 20 million. Whatever the precise figure, it wasn't enough. 'The limitations of geological periods, imposed by physical science, cannot, of course, disprove the hypothesis of transmutation of species,' he said in a lecture in 1869, 'but it does seem sufficient to disprove that transmutation has taken place through "descent with modification by natural selection".'[29]

This was not nonsense; anything but, in fact. Thomson knew whereof he spoke and his calculations were as sound as any, given the state of the physical sciences in the 1860s. They worried Darwin enough for him to revise down and then remove early estimates for the time natural selection needed, and to introduce elements of Lamarckian evolution to speed up the process.

Nevertheless, they were wrong. The discovery of radioactive decay in the early 1900s, together with the new source of energy it introduced, began to lengthen the earth's history enough for natural selection to perform its wonders. Thomson became cognisant of the new science in his final years but it was not until after he died that hard numbers were put on the calculation, and many years later that his other objection – that the sun would have burned itself out in the time needed for natural selection – was finally satisfactorily answered by the discovery of nuclear fusion. Ultimately and once again, it was Darwin – who had privately written to Hooker saying 'I feel a conviction that the world will be proved rather older than Thomson makes it' – who was proven right and Thomson wrong.[30]

Thomson had not *based* his religious apologetic on such arguments, but he had come perilously close at times, using one set of scientific ideas (physics) to undermine (bits of) another (biology) that he felt were less conducive to his religious commitments. 'Modern biologists are coming once more to a firm acceptance of something beyond mere gravitational, chemical, and physical forces,' he claimed.[31] The weight thus placed could not help but undermine his broader concordist arguments. If Faraday's peace was private, Maxwell's public but nimble and Stokes's confident, Thomson's verged on the hubristic.

And yet, in spite of his tendency to over-confidence, Thomson was still typical of the general concord that reigned in the relationship between religion and the *physical* sciences in the later nineteenth century. If there was war between science and religion, no one had told the physicists.

'An irreversible doom': war

Gabriel Stokes was elected to the Royal Society in 1851, the same year as Thomas Henry Huxley. Both served as society secretaries through the 1870s. Huxley was elected president in 1883; Stokes succeeded him two years later. The two men admired each other's abilities, scientific and administrative, and worked hard for the Royal Society, but they disagreed vehemently on matters religious and political.

Huxley was notorious for his readiness to maul clerical trespassers on scientific grounds. Even after we have brought some context and nuance to the events of Oxford in 1860, he emerges as a ferociously loyal pugilist, exuberantly 'extinguish[ing] theologians' from around evolution's cradle. Not without reason was he known as 'Darwin's bulldog'.

Nor was this mere anti-clericalism. Huxley was a self-proclaimed scientific naturalist. Throughout human history, he wrote in a late essay, 'naturalism and supernaturalism have . . . competed and struggled with one another'. As 'natural knowledge' has 'gained in precision and in trustworthiness', so the supernatural has 'shrunk'. Whether this amounted to progress or regress, Huxley wrote, was 'a matter of opinion'.[32] His opinion was clear. If Stokes stands as a symbol of the concord between religion and the (physical) sciences in the later nineteenth century, Huxley stands for the conflict.

And not just Huxley. In 1864, animated in part by the hostile reaction to *The Origin of Species* among evangelicals and establishment Anglicans, Huxley drew together a small group of friends to form a dining club that would meet monthly, except in the summer, for the next thirty years. The

X-Club comprised young but increasingly eminent scientists who agitated for academic professionalism and freedom. Members were, in the words of one of their number, the mathematician Thomas Hirst, devoted to science 'pure and free, untrammelled by religious dogmas'.[33] It was a cause that necessitated simultaneously reforming British science and undermining the cultural authority of the Anglican clergy.

The club was successful, after a fashion. Five of its nine members held the presidency of the British Association for the Advancement of Science between 1868 and 1881. Five members also held office in the Royal Society, and three – William Spottiswoode, Joseph Hooker and Huxley himself – occupied the presidency continuously between 1873 and 1885. Given that a generation earlier both presidencies had been the provenance of earls and reverends, the X-Club appeared to have succeeded in its ambitions.

Members didn't have it all their own way, however. Hooker was succeeded as president of the BAAS by Stokes, who was succeeded by Huxley, who was succeeded by Thomson. The X-Club's domination of the Royal Society presidency was followed by a rather different trinity of presidents – Stokes followed by Thomson followed by the surgeon and devout Quaker Joseph Lister. The heights of British science were held alternately by religiously hostile and religiously sympathetic parties.

One of the X-Clubbers who held the presidency of the BAAS, though not the Royal Society, was the Irish physicist John Tyndall, and it was his presidential address in Belfast in August 1874 that marked perhaps the most acrimonious moment in the late Victorian conflict between science and religion, a public scandal that dwarfed the Oxford debate. The address was British science's big night out, attracting as much media attention as 'a budget speech from Mr. Gladstone', at least according to the *Manchester Guardian*.[34] Tyndall used the opportunity not only to advocate uncompromisingly for Darwin's theory, but also to put forward a comprehensive materialist explanation of reality, stretching from atoms to thought. As if this weren't enough, in terms of which Huxley would have been proud, Tyndall pilloried religion's historic involvement in science and warned it off future entanglement:

> Grotesque in relation to scientific culture as many of the religions of the
> world have been and are – dangerous, nay, destructive, to the dearest

privileges of freemen as some of them undoubtedly have been, and would, if they could, be again . . . [religion is] mischievous, if permitted to intrude on the region of *knowledge*, over which it holds no command.

The reaction in Britain was volcanic. One London merchant, a Mr C. W. Stokes, wrote to Richard Cross, the home secretary, suggesting that Tyndall may have made himself liable to the 'penalty of persons expressing blasphemous opinions'.[35] The reaction in Ireland was, if anything, worse. The Catholic bishops condemned the lecture. The Revd Michael O'Ferrall, writing in the *Irish Monthly*, accused Tyndall of delivering a 'message of death' under 'the name of Science'.[36] Most impressively, according to an article in the *New York Times*, the ageing Pope Pius IX wrote to Cardinal Cullen, the archbishop of Dublin, praising the bishops for their response and declaring that nothing is to be so dreaded as 'those spiritual pirates whose trade is to despoil the souls of men'.[37]

That the uproar was covered, in some detail, in the *New York Times* underlines not only its volume but also how this sense of conflict was not limited to the UK. Tyndall's lecture and wider thinking had been influenced by John William Draper's 1862 book *The Intellectual Development of Europe*. Draper, with whose egregious history of science–religion warfare we began, had emigrated to the US aged twenty and taken up a position as a professor of chemistry. He excelled and by 1876 he was the first president of the American Chemical Society. He also considered himself a historian, albeit one with an aversion to footnotes, and from the 1860s he was writing his history of European intellectual life and then, shortly after its conclusion, of the American Civil War. Firmly embedded in American university life, he also remained in touch with British science. Indeed, it was Draper who reduced the Oxford audience to a frustrated stupor on that hot Saturday afternoon in June 1860, before Wilberforce and Huxley took to the stage to respond – in theory – to his lecture.

Draper brought a scientific commitment to – indeed an obsession with – lawfulness to his history. History was 'completely determined by external conditions'.[38] Life was material and, like any good chemist, Draper knew that the material followed natural laws. The result was a kind of historical determinism, whether in the form of human lives, societies, civilisations or even entire peoples, such as Native Americans: 'the race, like each individual of it, submits in silence to an irreversible doom.'[39] Societies matured and aged like bodies. The three social stages of Comte became five physiological ones for the American chemist, and Draper's intellectual history of Europe was premised on the 'analogy between [mankind's] advance from infancy, through childhood, youth, manhood, to old age'.[40]

The following decade he applied the same principles to the history of science and religion, in his *History of the Conflict Between Religion and Science*, each entity neatly reified into a discrete package, the historical relationship between them framed as one of unrelieved 'conflict'. Draper's argument sounded like Huxley on super/naturalism filtered through Boyle's law of gases. This was a narrative 'of two contending powers, the expansive force of the human intellect on one side, and the compression arising from traditionary faith and human interests on the other'.[41] His view of history was bracingly positivist. Like Dickens's Thomas Gradgrind, he believed that 'every page should be ... glistening with facts', which should be clear, impartial and free of the 'pretensions of either party'.[42] Polished, objective facts plus inexorable material laws equalled a story of continuous conflict.

The initial reception of Draper's history of science and religion was rather lukewarm, in part because the book covered much the same ground as his earlier intellectual history of Europe and in part because it was, in effect, a self-fulfilling narrative. As one reviewer complained, 'the word "religion" is to [Draper] no more than a symbol that stands for unenlightened bigotry or narrow-minded unwillingness to look facts in the face.'[43] The book was effectively 'A History of the Conflict between A Very Bad Thing and A Very Good Thing' that showed conclusively that the very good thing was much better than the very bad thing.

Draper, however, had been commissioned to write it by Edward Youmans, an entrepreneurial and hugely successful science populariser, whose *International Scientific Series* (in which Draper's book stood) and *Popular Science Monthly* energetically took the message of conflict to a wider audience. *Popular Science Monthly* published the preface to Draper's *Conflict*, gave it a glowing review and then gave over numerous editorials to defending it from its critics. Youmans adopted a similarly Manichean view of the story. Either there was conflict or there was harmony, and as there clearly wasn't complete harmony, there must have been conflict. The approach worked: *Conflict* went through fifty printings in the US and twenty-four in the UK and was translated into at least ten languages.

Conflict's success was enticing, and twenty years after Draper published his book, Andrew Dickson White wrote an even larger and seemingly equally combative volume called *A History of the Warfare of Science with Theology in Christendom*. White *was* a historian and co-founder of Cornell University, and he had been lecturing about 'The Battle-Fields of Science' since the late 1860s. Just as Draper liked law, White favoured war, and military metaphors dominated his conceptualisation of science and religion, from his early 'Battle-Fields' lecture to his 800-page magnum opus. Participants were combatants, encounters conflicts and the controlling narrative one long, bloody campaign to the ends of intellectual liberation. His history also found its way into Youmans's *Popular Science Monthly*, which published chapters from his book for nearly a decade before it was finally printed in full in 1896.

In among his narrative of conflict, Draper was quite positive about Islam. He called it the 'Southern Reformation' and praised it not only for preserving the embers of classical science but for a theology that seemed to orient it away from corrupting supernaturalism. 'Christendom believed that she could change the course of affairs by influencing the conduct of superior beings. Islam rested in a pious resignation to the unchangeable will of God.'[44]

Not everyone was so positive about Islam, however. Draper made a brief reference to Ernest Renan, a French historian and linguist, whose fame rested on a scandalously naturalistic *Life of Jesus*, published in 1863, which depicted the Son of God as merely a noble teacher and purifier of corrupt, legalistic Judaism. Twenty years after his *Life of Jesus*, Renan delivered a lecture at the Sorbonne, that did unto Islam what Draper and White had done unto Christianity. Indeed, he made Draper and White look positively emollient. Anyone 'with even the slightest education in matters of our time' could clearly see 'the current inferiority of Muslim countries', he assured his audience. This inferiority was primarily a religious issue. 'The decadence of states governed by Islam [and] the intellectual sterility of races' comes from a culture and education that derives from 'religion alone'. Anyone who took the time to travel through the

Orient would be struck by the 'narrow-mindedness of a true believer', religion acting like an 'iron ring around his head, making it absolutely closed to science, [and] incapable of learning anything or of opening itself up to any new idea'.[45] In a debate that would last well into the twenty-first century, and be picked up by, among others, Steven Weinberg and Jamil Ragep (as we saw in chapter 2), Renan argues that the so-called 'Golden Age' of Islamic science shone only because Greek philosophy had been translated by Syriac Christians and had taken root in the former Sassanid empire, and even then it only did so until it was overcast by Islamic theological clouds in the twelfth century.

Renan wasn't critical only of Islam. He damned Catholicism, predictably for silencing Galileo. But even here his knife was being sharpened against the Muslim world for, he reasoned, 'to honour the Islam of Avicenna, Avenzoar, Averroes, is like honouring the Catholicism of Galileo. Theology hampered Galileo; it was not strong enough to stop him.' Such, in microcosm, was the story of Islam and science, at least according to Renan. As he said in his conclusion, 'Islam tolerated philosophy [simply] because it could not prevent it.'

Renan's verdict on science and Islam was about the most severe in an age already inclined to such severity. It – along with White's *Warfare*, Draper's *Conflict*, Tyndall's lecture and Huxley's club – underlines that however much Faraday, Maxwell, Stokes and Thomson might have modelled, with different levels of success, the harmony between science and religion, others saw fervid conflict.

And yet even the warfare between science and religion was not as straightforward as these high-profile examples might suggest. Just as peace came in different kinds, so warfare was not always what it seemed. It is a little known but instructive fact that, despite innumerable throwaway comments to the contrary (including mine above), Huxley was never known as 'Darwin's Bulldog' in his lifetime. The famous nickname was first recorded by the American palaeontologist Henry Fairfield Osborn, two years after Huxley's death, who reported Huxley's own words when escorting the aged Darwin round a laboratory in 1879.[46] 'He said afterwards: "You know I have to take care of him – in fact, I have always been Darwin's bull-dog".' Huxley was, at least in his own reckoning, a defender rather than an aggressor. This, it could fairly be claimed, is a distinction without a difference. After all, it was Huxley himself who coined the metaphor and bulldogs defend through aggression. But the distinction should encourage us to examine the martial lore of this period with particular care. As with all wars, there is fog, confusion and some mythologising to peer through.

Huxley, as we have seen, was poor, gifted, professional and pugilistic. But he was also, in his own way, religious. 'True science and true religion are twin sisters,' he wrote in an essay on 'Science and Religion' in 1859.[47] When his first child Noel died the following year, he wrote to the Revd Charles Kingsley, who had drawn close to him in support, 'more openly and distinctly . . . than I ever have to any human being except my wife'. 'Science seems to me to teach in the highest and strongest manner the great truth which is embodied in the Christian conception of entire surrender to the will of God,' he said. There was a telling coincidence between Christian morality and what made for a good and happy life. 'The gravitation of sin to sorrow is as certain as that of the earth to the sun.' And the Church had a role to play in all this, he believed, providing it could wrench itself from the wrong hands.

[If] the Church of England is to be saved from being shivered into fragments by the advancing tide of science – an event I should be very sorry to witness, but which will infallibly occur if men like Samuel [Wilberforce] of Oxford are to have the guidance of her destinies – it must be by the efforts of men who, like yourself, see your way to the combination of the practice of the Church with the spirit of science.[48]

A third of a century later, a few years before he died, even as he wrote about naturalism driving out the supernatural, Huxley was still sounding a similarly emollient note. 'The antagonism between science and religion about which we hear so much,' he wrote in *Science and Hebrew Tradition*, 'appears to me to be purely factitious.' It was 'fabricated' either 'by short-sighted religious people who confound a certain branch of science, theology, with religion', or 'by equally short-sighted scientific people who forget that science takes for its province only that which is susceptible of clear intellectual comprehension'.[49]

Huxley's objection was to arrogant theological dogma, which abounded in late Victorian Britain. It was, after all, *theologians* who lay extinguished around the cradle of science. His reading of Thomas Carlyle's *Sartor Resartus* had taught him 'that a deep sense of religion was compatible with the entire absence of theology'.[50] The 'exquisite beauty and symmetry' revealed by science filled him 'with that *Amor intellectualis Dei* [intellectual love of God], the beatific vision of the *vita comtemplativa* [contemplative life], which some of the greatest thinkers of all ages' – among them Aquinas – 'have regarded as the only conceivable eternal felicity'.[51]

Even those weapons in Huxley's scientific armoury that seemed most dependable in his anti-religious crusade were more equivocal. Thus, scientific naturalism, he wrote, 'leads not to the denial of the existence of any Supernature; but simply to the denial of the validity of the evidence adduced in favour of this, or of that, extant form of Supernaturalism'.[52] Properly speaking, the naturalism to which science was committed should be silent on bigger philosophical and existential questions. It was, to use today's terms, *methodological* naturalism rather than *metaphysical* naturalism. Such methodological naturalism might have been anathema to Samuel of Oxford and his peers, but it was no more wedded to atheism than it was to religion. Rather, it was most compatible with agnosticism, the term that Huxley famously coined although, as it happens, rarely used.

It was a similar story with Darwinism. In Huxley's view, evolution, which did so much to obliterate the teleological world of William Paley, could in fact be reconciled with the idea of there being ends in creation. 'There is a wider Teleology, which is not touched, by the doctrine of Evolution,' he wrote in an 1869 review of *The Natural History of Creation* by Darwin's leading German advocate, Ernst Haeckel. Indeed, this wider teleology 'is actually based on the fundamental proposition of evolution'; not in the 'coarser form' of Paley's examples but rather in the manner of Maxwell, in the form of 'the mutual interaction, according to definite laws, of the forces possessed by molecules'. 'Teleological and the mechanical views of nature are not,' he concluded, 'necessarily, mutually exclusive.'[53]

Such comments show that Huxley was a long way from being the rabidly aggressive anti-religious canine of legend. His science was deployed selectively against those elements of dogmatic, clerical theology that he considered a threat. In his mind, and that of many of his peers, the greatest such threat emanated from Rome. The later nineteenth century saw the Vatican double down on its dogmatic authoritarianism, following the trauma of the French Revolution and the advent of biblical criticism. This was particularly true of Pope Pius IX, who came to the throne of Peter in 1846 and remained there for an unprecedented thirty-one years. His papacy began liberally enough but Pius was badly shaken by the revolutionary events of 1848, and rapidly lost his reforming fervour. In 1864, he issued the *Syllabus of Errors*, which drew on several decades of papal documents to define eighty heresies judged incompatible with Catholic belief. The syllabus began, unsurprisingly enough, with a condemnation of atheism, naturalism and 'absolute' rationalism. However, it then proceeded to condemn 'moderate' rationalism,[54] and any form of secularism, and concluded, famously, by denying that 'the Roman Pontiff . . . ought to reconcile himself . . . with progress, liberalism and modern civilization.' The Catholic response was mixed; the Protestant one universally hostile. Gladstone spoke for many when he wrote that no one could now convert to Catholicism 'without renouncing his moral and mental freedom', though he subsequently apologised to his 'Roman Catholic fellow-countrymen' for the 'seeming roughness' of his tone.[55]

There was more. Five years later, Pius convened the First Vatican Council to deal with many of the problems identified in the *Syllabus* but, in the process of formulating the Church's response to these challenges, the Council was compelled to grapple with the vexed question of authority and, in particular, papal authority. The idea of papal infallibility – that in certain matters the pope spoke without possibility of error – was hardly new and few bishops disagreed with it in principle. But many felt it was unwise, or at least untimely, to state it definitively. A liberalising Europe would not take kindly to what would invariably be seen as a quasi-medieval assertion of authority. Such reservations were ignored, and Archbishop Cullen of Dublin, honoured by the pope for his response to Tyndall's lecture, drafted the final decree which stated that 'The Roman Pontiff, when he speaks *ex cathedra* . . . [to define] a doctrine concerning faith and morals . . . is possessed of that infallibility with which [Jesus] wished his Church to be endowed.'

The decree confirmed Protestants' worst fears, which were already exercised by the *Syllabus* and by large-scale Catholic migration to historically Protestant countries. Many Britons were alarmed by the growing number of Irish families in Britain; indeed, it was this that lay ultimately behind Darwin's ugly comments about careless, squalid, unaspiring Irishmen. Many others were anxious about the re-establishment of the Catholic hierarchy in England in 1850. In a similar way, Americans fretted over increased levels of immigration from Ireland and southern Europe, provoking fears of 'Rome or home' that lasted well into the 1960s. Intellectual freedom, including but not limited to the freedom of scientists, was, many thought, under mortal threat.

Whether this was in fact the case is rather dubious. The loss of papal patronage, the suppression of the Jesuit order and the reaction to French *philosophes* and Revolution had all chastened a once rich scientific tradition in Catholic countries. But these lands were hardly a scientific desert, still less a prison. The Vatican Observatory had been re-established in 1839, and had flourished under Pius IX, who, in his early years, had also founded the Pontificia Accademia dei Nuovi Lincei, modelled on the Academy of Lynx of which Galileo had been a member. Galileo's works, like all those advocating heliocentrism, had long been removed from the Index and Catholic universities (and monasteries) still taught and sometimes pioneered scientific research, as exemplified by Gregor Mendel, the Augustinian monk whose patient garden experiments laid the ground for genetics. For all that the *Syllabus* and the First Vatican Council seemed to signify severe intellectual repression, scientifically speaking this was, ironically, a period of a slow but steady thaw.

Nevertheless, it was the intellectually oppressive Catholicism of the *Syllabus* that loomed before scientists and historians of science in the final decades of the century. As far as Huxley was concerned, one of the greatest merits of evolution was that it stood in 'complete and irreconcilable antagonism to that vigorous and consistent enemy of the highest intellectual, moral and social life of mankind – the Catholic Church'.[56] He poo-pooed attempts, such as that of the Catholic anatomist St. George Jackson Mivart, to reconcile ideas of evolution and creation, and stated baldly that the antagonism between 'theology and science' was grounded in the unavoidable fact that theology, as illustrated by the idea of papal infallibility, entailed 'belief in a proposition, because authority tells you it is true' (or conceivably 'because you wish to believe it'), which was complete anathema to the scientific mind.[57]

There was a similar pattern behind Tyndall's bellicose address. Tyndall, like Huxley, had been inspired by Thomas Carlyle's atheological religion; indeed, he paid tribute to Carlyle in the closing lines of his presidential address. Like Huxley, he believed that ultimately no 'atheistic reasoning' could 'dislodge religion from the human heart' because religion lived not 'by the force of and aid of dogma, but . . . is ingrained in the nature of man'.[58] And like Huxley, he thought religion still had a role – not, of course, in the region of knowledge but rather in 'the region of emotion', albeit only when 'guided by liberal thought'. But also like Huxley – indeed more so given his Irish heritage – Tyndall was hostile to Catholicism in its various intellectual, social and political forms. Shortly before his Belfast address, the Irish bishops had thrown out a plan to include physical science in the curriculum of the Catholic University. It epitomised the worst kind of reactionary anti-scientific religious closed-mindedness against which Tyndall, Huxley and indeed the X-Club railed.

It would be a mistake to see this bellicosity as simply and narrowly anti-Catholic or indeed anti-anything. Huxley and his peers were *for* naturalism. They were *for* intellectual freedom, up to a point (women, for example, were no more welcome as scientists than the clerics and amateurs the X-Club sought to replace). And they were *for* a cleansed, reformed, enlightened religion, grounded in science, and unencumbered by hierarchy, doctrine, or, one is tempted to add, substance. In Bernie Lightman's reckoning, 'the scientific naturalists were not just aiming to reform scientific theories and institutions. They were also interested in transforming British culture as a whole.'[59] It was no accident that X-Clubbers (and the approach they advocated) could sound like Reformed Protestants of the early seventeenth century energetically burying Aristotle and the schoolmen. The veritable renaissance of their times, Huxley explained in an essay on 'The Progress of Science', could only be explained by the physical sciences and this was 'due to the fact that men have gradually learned to lay aside the consideration of unverifiable hypotheses; to guide observation and experiment by verifiable hypotheses'. It could have been Francis Bacon speaking.[60]

This was also the underlying theme in much of Draper and White's histories. Indeed, as James Ungureanu has recently shown, 'much of the narrative found in Draper and White was drawn from centuries of

Protestant Christian Polemic.'[61] Both had their own axes to grind. Draper's father was a convert from Catholicism who became a Wesleyan Methodist minister; the son inherited his father's antipathy to Rome. Early in his history, Draper records how he would have 'little to say respecting the two great Christian confessions, the Protestant and Greek churches' which, he wrote, rather hopefully, have 'observed a reverential attitude to truth, from whatever quarter it might come'.[62] Roman Catholicism, by comparison, became for Draper the very inverse of science. As he wrote in his Preface:

> Science . . . has never sought to ally herself to civil power. She has never attempted to throw odium or inflict social ruin on any human being. She has never subjected any one to mental torment, physical torture, least of all to death, for the purpose of upholding or promoting her ideas. She presents herself unstained by cruelties and crimes. But in the Vatican – we have only to recall the Inquisition – the hands that are now raised in appeals to the Most Merciful are crimsoned. They have been steeped in blood![63]

Science and society needed another Reformation to free them finally from the shackles of Catholic cruelty. White was less hysterical, but had more reason to detest orthodoxy in general, having come under sustained criticism from ecclesiastics for his role in founding Cornell University as a non-sectarian institution. White's war was not so much between science and Catholicism, as was Draper's, but between science and 'dogmatic theology'. Whereas Draper wanted a New Reformation modelled on the old, White sought a lower-case reformation, in which a pure and undefiled religion based in moral conscience and sentiment could be allowed to grow.[64]

Not all of the anti-religious scientific polemic of the end of the century can be explained in quite this way. Renan's attack on Islam, for example, was clearly not about Christianity. Indeed, in some regards it was not even about religion. Renan was under no illusion about the deleterious effect, as he saw it, that Islam had on intellectual freedom but there was also, he argued, a racial dimension to all this, for 'As long as Islam remained in the hands of the Arab race . . . no intellectual movement of a secular character developed within it.' Even here, though, there was a Reformation undertone, as Renan argued that science and philosophy were only kept alive in the early Muslim centuries because Islam was 'tempered by a kind of Protestantism (the so-called *mu'tazalism*)' which was less fanatical, and less organised, than mainstream Islam. Even in ninth-century Baghdad, science had the Reformation to thank for its survival.

So it was that the war between science and religion, as preached by Huxley, Draper, White, Renan and others like them was, in many respects, the last spasm in the wars of religion; science (or a version of it) being the dagger plunged into the still-twitching body of dogmatic (and primarily Catholic) religion.

And yet the real victim was not religion or even Catholicism but history, with the complex, colourful, ambiguous and hopelessly entangled histories of science and religion reduced to a single narrative of uniform conflict. The pre-eminent centuries of Islamic science became a mistake, their long tail dismissed, the Middle Ages reduced to an era of superstition, the Copernican revolution to a confrontation with clerical obscurantism, Bruno into a martyr for science, the complex theological grounds for the Scientific Revolution ignored, the Royal Society's aversion to sectarianism turned into outright irreligiosity and, of course, the great Galileo transformed into an icon of science, his trial into a cosmic battle of good and evil, and his final house arrest into a testimony, as Milton had put it 250 years earlier, to 'the servile condition into which learning [among Catholics] was brought'.

PART 4

THE ONGOING, ENTANGLED HISTORIES OF SCIENCE AND RELIGION

Clarence Darrow (standing right) cross-examines William Jennings Bryan (seated) on the lawn outside the courthouse in Dayton, Tennessee, on the climactic last day of the 'Scopes Monkey Trial'. Bryan was to die a week later, some claimed because of the stress of the trial. Darrow was not persuaded: 'Broken heart nothing. He died of a busted belly.'

FOURTEEN

THE TRIAL OF THE CENTURY

'They are true parasites': America before Scopes

The Scopes 'Monkey' trial, held in Dayton, Tennessee, in the summer of 1925, would receive huge media attention, and the issue of evolution would go on to acquire iconic status in the (American) story of science and religion. But up until about 1900, the theory was not that big a deal in the US.

Acceptance took some time and was uneven. The influential Princeton theologian Charles Hodge wrote in 1874 that Darwinism 'is atheism [and] utterly inconsistent with the Scriptures', but neither his opinion, nor its bluntness, were typical of the time.[1] Most mainstream Church leaders were happily reconciled to the theory by the end of the nineteenth century, as were virtually all religious scientists. Evangelical ministers were more hostile and, in as far as it is possible to judge, ordinary churchgoers less receptive still. Premillennialists, for whom the expectation of Jesus' imminent return was foundational, were particularly resistant, unwilling to cede any figurative reading to Genesis, for fear of what it would do to the other end of history.

Nevertheless, however much resistance the theory provoked, rejection was a campaigning issue for only a small minority. The reasons for this were scientific as much as religious. Darwin was eclipsed in the final decades of the century as biologists chose to modify and qualify his theory of natural selection. There were coherent reasons for doing so. No one had yet discovered a mechanism for the kind of transmission for which Darwin had argued. Absent the knowledge of genes, only (re)discovered in Gregor Mendel's work in 1900, it was impossible to see how mutations weren't simply diluted away over generations. No one had come up with a convincing response to William Thomson's argument about the age of the earth. The fossil record remained frustratingly incomplete and ambiguous, and before the theory of continental drift had been established, there seemed little pattern to its distribution. All in all, the random nature of natural selection seemed unable to bear the biological, let alone the metaphysical, weight that some wanted to put on it, and scientists were increasingly proposing some kind of Lamarckian or guided evolution, which was apparently easier to reconcile with God.

This was the kind of compromise at which Darwin's friend and correspondent Asa Gray had arrived. Gray was America's leading botanist and the country's earliest and most important advocate for Darwin's theory. He was also a pious Presbyterian and his correspondence with Darwin shows the two of them wrestling with the theological implications of such an apparently arbitrary mechanism for life. Darwin was no more persuaded by Gray's theistic evolutionism than were many American evangelicals, but the very fact that religious scientists of Gray's stature could accommodate the theory with their faith commanded some respect.

The result was that many made conciliatory noises and came to a kind of modus vivendi. Some finessed their reading of Genesis. The Scofield Reference Bible, which was to become the fundamentalists' most respected guide to the scriptures, cautiously plumped for the gap theory of creation. Others acknowledged the truth of evolution by claiming they could trace a kind of special divine guidance in the process. Still others accepted the theory but drew a firm line at human beings.[2] Evolution was an irritant rather than a threat.

As in England in 1860, when *Essays and Reviews* eclipsed *The Origin of Species* for controversy, it was theological liberalism rather than evolution that animated religious sensibilities, and it was theological liberalism that would catalyse the movement that was to become synonymous with anti-evolutionism in the twentieth century. Fundamentalism originated in a series of essays, published from 1910, in which leading evangelicals put their house in order, setting out 'The Fundamentals' of the faith in the face of the liberal threat. Biblical criticism was the real enemy. Biblical credibility was the goal.

Evolution was not much of a concern for the series. Two of the eventual ninety essays tackled it head-on. A few authors understood Darwinism as a cause of religious doubt but most were indifferent to the theory, and a number were openly positive about it. Benjamin Warfield, professor of theology at Princeton, described himself as a 'Darwinian of the purest water', while the Congregational minister and amateur geologist George Frederick Wright insisted that the purpose of Genesis was to undermine polytheism rather than to teach science.[3] Oddly, given what was to happen later in the century, evolution was no more a red flag for fundamentalists than it was for mainstream ministers.

The truce was an uneasy one, however, and it began to unravel in the second decade of the century for a number of reasons. One was scientific. The discovery of genetics and radioactivity within a few years of each other strengthened the credibility of natural selection. Darwin's theory now had the means and the time necessary to do its work, and biologists could jettison Lamarckian or theistic accretions. Evolution in 1915 *felt* more atheistic than it did in 1885.

A second reason was educational. Over this same period, discussion of Darwinism migrated from the academy to wider public debate and thence to the curriculum. Textbooks such as G. W. Hunter's *A Civic Biology*, on which the Scopes trial was based, taught evolution, including of humans, without apology or reservation. Debating the arguments and implications of evolution among consenting adults was one thing; teaching it as unambiguous truth to the children was quite another, not least when it was usually parents rather than schools who were responsible for purchasing the textbooks.

Hunter's textbook illustrated a third problem. Virtually no one – whether adamant fundamentalist or evolutionary atheist – thought Darwinism was 'mere' biology. Understanding the path to life meant understanding the path through it, in particular how society should cope with its problems. In his chapter on heredity and variation, Hunter explained how sexual immorality, alcoholism, epilepsy, feeble-mindedness and other social problems were fundamentally *biological* issues.

Hundreds of families such as those described above exist to-day, spreading disease, immorality, and crime to all parts of this country. The cost to society of such families is very severe. Just as certain animals or plants become parasitic on other plants or animals, these families have become parasitic on society. They not only do harm to others by corrupting, stealing, or spreading disease, but they are actually protected and cared for by the state out of public money. Largely for them the poorhouse and the asylum exist. They take from society, but they give nothing in return. They are true parasites.[4]

The waspish, sceptical journalist H. L. Mencken, one of the many to cover the Scopes trial, wrote that Dayton, the town in which the trial was staged, had pleasantly surprised him. He had been expecting to find the usual 'squalid Southern village, with darkies snoozing on the houseblocks, pigs rooting under the houses and the inhabitants full of hookworm and malaria'.[5] It was not just alcoholics and epileptics that were parasitic on society, it seems.

A Civic Biology helpfully illustrated the fourth reason for the eruption of evolution as a religious issue from the 1910s. If biology redefined the social problem, it also suggested 'The Remedy', as the relevant section in Hunter's book titled it.

If such people were lower animals, we would probably kill them off to prevent them from spreading. Humanity will not allow this, but we do have the remedy of separating the sexes in asylums or other places and in various ways preventing intermarriage and the possibilities of perpetuating such a low and degenerate race. Remedies of this sort have been tried successfully in Europe and are now meeting with success in this country.[6]

Disproportionately from poorer, less well-educated backgrounds, fundamentalists could still tell the difference between social Darwinism and eugenics, and Christian ethics. Squaring Genesis with geology was nothing compared to reconciling the Sermon on the Mount with the Survival of the Fittest.

A fifth reason located these moral and social problems in an international perspective, one that was being distorted by Germany, the source of the 'higher' biblical criticism that was the fundamentalists' nemesis. Although not involved as a nation until 1917, Americans watched the

Great War closely and paid particular attention to the idea that German aggression had been based on the Darwinian ideas. Benjamin Kidd's *Science of Power*, written before the war but revised during it and published after it, traced the links between evolution, in particular that of Darwin's German disciple Ernst Haeckel, and German militarism. More influentially, the evolutionist Vernon Kellogg's *Headquarters Nights* offered first-hand evidence for this link. Based in Brussels through the war and originally a pacifist, Kellogg had multiple conversations with German officers in which they revealed to him the role that Darwinian struggle for survival had played in justifying their belligerence. Kellogg abandoned his pacifism. His ensuing book was widely read.

So it was that scientific, educational, social, moral and international reasons conspired to tip evolution from being an awkward problem for (some) American Christians to being the problem incarnate. Important as they all were, however, they paled into insignificance in comparison with the sixth reason.

'Civilisation is on trial': the road to Dayton

William Jennings Bryan was a populist. Born in 1860, he was the leading progressive politician of *fin de siècle* America: a brilliant orator, three times a Democratic presidential nominee, though never a president, and leader of the short-lived Populist Party in the 1890s.

Not necessarily *of* the people – his parents were wealthy – Bryan was very much *for* them. Optimistic, reformist, devout, charismatic, he fought for the common man: for the unions that were seeking protection for workers in an age of unbridled capital; for the farmers in his home state of Nebraska impoverished by agricultural tariffs; and for the millions of American debtors crippled by deflationary policies, for whom he campaigned against the gold standard, famously declaring to the Democratic Convention in 1896, 'you shall not crucify mankind upon a cross of gold.' He took on the abuses of the liquor trade, spoke out against American imperialism after the Spanish-American War in 1898 and resigned from Woodrow Wilson's government in 1916 over US policy during the Great War. He was known as the 'Great Commoner' and he revelled in the title.

Bryan was in lockstep with co-religionists over evolution. He never believed in the theory but nor was it a campaigning issue for him. 'I do not carry the doctrine of evolution as truth as some do', he said in a speech to a religious gathering in 1904, but 'I do not mean to find fault with you if you want to accept the theory.'[7] He was prepared to countenance it – 'evolution in plant life and animal life . . . might . . . be admitted' – albeit with the rather large caveats that there would need to be 'proof' and that it could only be countenanced 'up to the highest form of animal'.[8] 'Animal life' clearly did not mean human life.

Like the ordinary rural folk he championed, Bryan's worry was the impact the theory had on morality. 'Our chief concern is in protecting man from the demoralization involved in accepting a brute ancestry,' he wrote.[9] For him, evolution was synonymous with the doctrine that might, ultimately, was right. It meant abandoning the law of Christ for the law of the jungle. 'How can teachers tell students that they came from monkeys and not expect them to act like little monkeys?' he often asked his rapt audiences.[10]

Already vexed by German aggression in the Great War, he was deeply troubled by Vernon Kellogg's book, as well as by the increasingly powerful eugenics movement in America. Indiana had been the first to pass a forced sterilisation law in 1907 and dozens of states followed. They sterilised criminals, drunks, promiscuous women, 'morons' and 'imbeciles' ('scientific' definitions of intelligence), as well as a number of poor, unemployed, disabled and black citizens. Some states resisted the movement – Tennessee never passed a sterilisation law, though bills were proposed – but the great Oliver Wendell Holmes, who sat on the Supreme Court for thirty years, nonetheless spoke for his era when he famously declared, in the case of the sterilisation of the 'mentally defective' Carrie Buck, two years after the Scopes trial, that 'three generations of imbeciles are enough.'

Bryan saw imperialism and eugenics as the bitter fruit of Darwin's theory, along with immorality and unbelief. A survey in 1916 by psychologist James H. Leuba revealed how few college graduates remained Christians. Bryan worried. 'What shall it profit a man if he shall gain all the learning of the schools and lose his faith in God?' he lamented.[11] America was on the broad road away from Christ. Darwin was to blame. Bryan would save his flock.

He started speaking against Darwinism in 1921, turning his oratorical gifts and popular credibility against this root of all evil. His authority was the people – 'Have faith in mankind', he declared, 'mankind deserves to be trusted'[12] – and so, in its own way, was his cause. 'Man is infinitely more than science', he proclaimed, and 'science, as well as the Sabbath, was made for man.'[13] Within a year, fundamentalist leaders across the South were calling for a ban on evolution in schools. Teachers, they insisted, must teach what parents judged acceptable, and that did not include evolution or evolutionary eugenics. In 1922, the Kentucky legislature narrowly rejected a motion to ban evolution but its failure did not deter other states and in early 1925 Tennessee passed its own bill, together with a $500 penalty for breaking the law. Bryan approved, although not of the fine, which he thought might make a martyr of his enemies.

The decision split state opinion, energising fundamentalists and eliciting howls of Galileo from their opponents, including most mainstream ministers. 'We . . . must teach science in our colleges [and] this must be done by scientists,' proclaimed one such minister in Memphis. 'Neither priest nor prophet nor apostle, nor even the Lord himself, ever made the slightest contribution to our knowledge of the natural sciences.'[14] The act drew still greater scorn across the country, prompting the American Civil Liberties Union (ACLU), then only five years old and with little legal experience and less success, to advertise that it would defend anyone charged with breaking the Tennessee law. Important intellectual, educational and legal principles were at stake, not to mention commercial opportunities.

Dayton was a small town in Tennessee without much history or industry. Its population had dwindled from a high point of three thousand in the 1890s to around eighteen hundred at the time of the trial. George Rappleyea, manager of the Cumberland Coal and Iron Company which operated in the area, disliked the law but smelt a prospect and persuaded local dignitaries to host an evolution trial in order to boost the local economy. Not everyone was convinced. Some even thought the trial might end up humiliating the town and its state. Nevertheless, the prospect of a show trial, covered by the country's – possibly the world's – media, was too tempting to pass up. Rappleyea wrote to H. G. Wells asking him to present the case for the defence and for evolution. Wells politely declined. The town's civil association formed a Trial Entertainment Committee and arranged accommodation for the press and interested spectators. Dayton prepared for visitors.

The man they chose to test the law was John T. Scopes, football coach and head of science at Rhea County High School in Dayton. Scopes taught physics and maths, though not biology. He was young, co-operative and likeable, but without deep roots in Dayton, and had no problem with the fact that he was bound to be found guilty despite not actually having broken the particular law in question. He was, in effect, a ghost at his own trial, overshadowed by the men, the organisations and the social forces gathering around him.

Scopes admitted to teaching the banned theory. Rappleyea then alerted the local justice of the peace, who issued a warrant for the teacher's 'arrest'. He then wired the ACLU and alerted the local press. The media caravan started moving. The ACLU's plan had been for a narrow test case concerning academic freedom but this was soon derailed. Local prosecutors, afraid of what was being unleashed, drew on the support of the World Christian Fundamentals Association (WCFA) to invite Bryan on to their team, despite the fact that he hadn't practised law since the 1880s. Bryan's acceptance prompted Clarence Darrow to volunteer himself to defend Scopes. Darrow was similar to Bryan in many ways: a bullish man, a first-rate speaker, a vigorous defender of labour rights and a Democratic Party candidate, though for Congress rather than the presidency (he had even campaigned for Bryan in 1896). He was also, however, very different. A brilliant criminal lawyer, Darrow had gained national fame (and infamy) for successfully defending Nathan Leopold and Richard Loeb against the death penalty for their apparently motiveless kidnap and murder of a fourteen-year-old boy. Darrow's defence was based on his belief in psychological determinism and he was well known for repudiating free will as an illusion.[15] This epitomised everything against which Bryan thought he was fighting, as did Darrow's support of evolution (which he believed made mankind gentler and more humane), and his contempt for Christianity, which he regarded with Nietzschean disdain as a 'slave religion'.[16]

With Bryan and Darrow on board, the trial caught fire. Questions of academic freedom and freedom of speech, let alone whether John Scopes had actually broken Tennessee state law, were eclipsed by a ferocious culture war *avant la lettre*. 'Had Mr. Bryan's ideas of what a man may do towards free thinking existed throughout history,' Darrow told the press, 'we would still be hanging and burning witches and punishing persons who thought the earth was round.'[17] 'Darrow is an atheist, I'm an upholder of Christianity,' Bryan shot back. 'That is the difference between us.'[18]

Every exchange upped the ante, as Bryan and Darrow, together with fellow attorneys, addressed audiences and press in the weeks leading up to the trial. By the first day, Friday 10 July, Dayton, Tennessee and America were in a state of frenzied anticipation. For all their bitter differences, Darrow spoke for Bryan, and indeed millions more, when he remarked just before he arrived in town, 'Scopes is not on trial [here]. Civilisation is on trial.'[19]

'This is a battle between religion and science': the trial

On the day the trial opened the courtroom was packed, an audience of five hundred crammed into every possible space. There were over two hundred reporters and photographers present, as well as radio technicians – this was the first trial to be broadcast on US national radio – and a film crew. Journalists wired over two million words from Dayton and the events would dominate front pages across the country for a week. In the courtroom itself, the jury box was relocated to make space for three microphones that relayed the proceedings to the audience outside, an interior redesign that was rich in symbolism.

The weather hovered around 100 degrees Fahrenheit (38 degrees Celsius); the ceiling fans in the courtroom broke. As in Oxford in 1860 people fainted, including, at one point, a member of the prosecution team. The judge repeatedly fretted that the courtroom floor would give way. Although early hopes for visitor numbers proved grossly over-optimistic, Dayton still managed a carnival atmosphere. The town struck a commemorative coin. Shops were festooned with pictures of monkeys. One enterprising soul offered bystanders the chance to have their photograph taken with a chimpanzee, for a fee. Gospel preachers did battle with rationalist speakers on street corners until the police were forced to separate them. 'Whatever the deep significance of the trial,' one newspaper commented, 'there is no doubt that it has attracted some of the world's champion freaks.'[20]

The trial began with a long and pointed prayer from the Revd Cartwright, a local fundamentalist minister. Prayer was to become a running side battle over the next few days, as first Darrow and then the state's modernist ministers, Unitarians and Jews objected to the prayers, or at least the fundamentalists' exclusive right to deliver them. Eventually a compromise was reached, and fundamentalists and modernists took turns.

No such compromise was possible over the crucial issue of expert witnesses. Bryan and his team had reached out to the 'geologist' George McCready Price as a witness to refute evolution. Price, however, had zero scientific authority beyond the fundamentalists and was, in any case, lecturing in England. No other scientific witness, Christian or sceptic, was prepared to take the stand for the prosecution. Bryan sought counsel and was advised 'to exclude all discussion by experts or otherwise on the subject of evolution.'[21] For all his crusading fervour, he was forced towards a more narrow and strictly legal approach.

Darrow and the defence team took the opposite view. They argued that the case could not be decided without considering the *content* of the Tennessee law; in particular, whether evolution was true and whether it was compatible with Christianity. They drip-fed the media names of eminent scientists on whom they would call to build their case, though only a few came through. Bryan objected, insisting that the people in this country had had enough of such 'experts'.[22] 'No specialists from the outside are required to inform the parents of Tennessee as to what is harmful.'[23]

The objection applied just as much, indeed more, to theologians as it did to scientists. 'Why should these experts know anything more about the Bible than some of the jurors?' reasoned one of the prosecution team, to loud 'Amens' from the courtroom.[24] The very foundation of Protestantism located the authority to interpret the scriptures in the individual believer, and the fundamentalists were not about to cede that to a priesthood of academic theologians. Bryan and his team also knew, however, that a cloud of expert witnesses testifying to the compatibility of evolution and Christianity would effectively destroy his argument. If evolution and Christianity were compatible, the fundamentalists' objection was merely a little, local irrelevance. The defence 'will only call those who ... harmonise evolution with [religion]', Bryan feared, and they will 'present a very one-sided view of evolution'.[25]

He was right. In one of the counterintuitive twists of the trial, it was the broadly sceptical, broadly secular defence team that made the most cogent and persuasive case for Christianity. 'We shall show by the testimony of men learned in science and theology that there are millions of people who believe in evolution and in the stories of creation as set forth in the Bible and who find no conflict between the two.'[26] The defence readily acknowledged that there were tensions between evolution and the Old Testament, but they insisted that 'there is no conflict between evolution and Christianity.' It was also Scopes's defence team who, on Day 4 of the trial, finally managed to call an expert witness to the stand, the eminent zoologist Maynard Metcalf, who explained the theory of evolution, testified to its strength, and – because Metcalf was a lifelong Congregationalist who taught Sunday School classes – attested its compatibility with Christianity.[27]

The battle over witnesses persisted. Indeed, it was coming to be *the* issue of the trial. At times, it threatened to kill the drama. The Great Commoner's son, William Jennings Bryan Jr, explained, at considerable length and with much evidence, the accepted rules on the admissibility of expert witnesses in such cases. It had a calming, indeed soporific, effect on proceedings. But it did not change the course of events because, by that stage, it could not. The trial had a momentum of its own and no matter what the protagonists spoke about, they ended up speaking about the big questions.

On Day 2, Darrow discussed the legitimacy of the Tennessee law. He argued that the law established a particular religious viewpoint in public schools. But he went on to note how Christianity was fragmented into hundreds of sects, how such division had left a legacy 'of hatred ... war ... [and] cruelty' throughout the world, how fundamentalism was unleashing 'bigotry and hate' across America, how the Bible was not a book of biology and how most intelligent Christians had not thought it necessary to give up their faith because a literal six-day creation had been found to be nonsense. This was theatre, not law. The court was rapt.

Three days later, Bryan gave an equally barnstorming hour-long performance, in theory about evidence admissibility but in reality fixed squarely on his parallel themes of mankind's divine nature and his trustworthiness. 'The Christian believes man came from above, but the evolutionist believes he must have come from below,' he proclaimed to an admiring, if apparently rather fickle, audience.[28] Bryan was confident about this fact and neither he nor any ordinary Christian needed a theologian to confirm it. 'The one beauty about the Word of God is, it does not take an expert to understand it.'[29]

The oratorical flights of the main characters inspired the minor ones. Dudley Malone responded to Bryan, now almost shouting, 'We are ready. We feel we stand with progress. We feel we stand with science. We feel we stand with intelligence. We feel we stand with fundamental freedom in America.'[30] Following him, again in theory talking about expert evidence, the prosecutor Tom Stewart raised the stakes about as far as they could go. 'They say this is a battle between religion and science. If it is, I want to serve notice now, in the name of the great God that I am on the side of religion.'[31]

It was ferocious, entertaining stuff. The legal teams were angry and exhausted. The crowds loved it. National newspapers printed speeches in full. Over 2,300 papers in the US covered events. International interest was keen. The *New York Times* called it 'the greatest debate on science and religion in recent years'.[32] But the drama was grinding to a halt. On Friday morning, Day 6 of the trial, Judge Raulston ruled that Metcalf was to stand alone and that the court would admit to hearing no other expert witnesses, though he did acquiesce to written submissions. Scopes's students had long since testified that he had taught them from the offending textbook – he had, in fact, coached them in their answers, though it became clear to everyone that the students knew next to nothing about evolution itself. Without expert witnesses to broaden the focus, the trial was all but over. Darrow exploded at the judge, who was widely judged to have been biased towards the prosecution, and narrowly missed being charged with contempt (though he was a few days later). The court rose for the day and journalists started heading for home knowing that the last day of the trial would be a formality.

Prevented from calling any witnesses who might make his case for him, over the weekend Darrow hit upon the idea of calling the prosecution, in the form of Bryan himself, to the stand. Rumours spread that the defence had one last trick up its sleeve and on Monday morning the courtroom was full again. Having heard excerpts from the written testimonies of the scientific witnesses, Judge Raulston then decided to move the proceedings outside on to the platform that had been erected on the courthouse lawn. The heat inside was almost unbearable and Raulston had been told that cracks had appeared in the ceiling of the room directly below the courthouse. Five hundred men and women left the courtroom to join the 2,000 or so outside. Darrow then announced that he was calling Bryan as a witness. There was no precedent for this and the prosecution lawyers, smelling foul play, protested. Bryan, however, did not. Sensing his time had finally come, he took the stand.

For the next two hours, Darrow chipped away at the Great Commoner and his literal interpretation of the Bible. From the outset, Bryan acknowledged that scriptural language could be figurative. When Jesus told his disciples 'ye are the salt of the earth', I would not, Bryan said, 'insist that man was actually of salt'.³³ The Bible was inspired by God but 'he may have used language that could be understood at that time'.³⁴ But he fought hard against any implication that he needed to *interpret* the Bible – that was the broad way to higher criticism, modernism, theistic evolution and spiritual death – and stuck to the plain meaning of the text whenever he could, even at the expense of appearing incurious, ignorant and, at times, inane.

'Did a whale swallow Jonah?' Bryan believed it was actually a big fish. 'Was it the ordinary run of fish, or [was it] made for that purpose?' The Bible doesn't say so I'm not prepared to say. 'Did Joshua make the sun stand still?' I believe what the Bible says. 'But do you believe the sun went round the earth then.' No. 'Did the authors of the Bible?' I don't know what they thought. 'Did you ever discover where Cain got his wife?' No, sir; I leave the agnostics to hunt for her. 'Were there other people on the earth at that time?' I cannot say. 'Did that ever enter your consideration?' It never bothered me. 'There were no others recorded, but Cain got a wife.'

That is what the Bible says. 'Don't you know that there are many old religions that describe the flood?' I don't. 'Have you ever read anything about the origins of religions?' Not a great deal. '[Do] you believe . . . God made the serpent to go on his belly after he tempted Eve?' I believe the Bible as it is. 'Have you any idea how the snake went before that time?' No, sir. 'Do you know whether he walked on his tail or not?' No, sir.

As the afternoon proceeded, the interaction became ever more heated. Other members of the prosecuting team repeatedly interrupted, objecting to the questioning, but Bryan refused to come down from his stage. Bryan accused Darrow of insulting the Tennessee audience and calling them 'yokels'. Darrow denied it and shot back that Bryan 'insulted every man of science and learning in the world because he does not believe in your fool religion'.[35] By now, the two men were shouting at each other. 'Your honour, I think I can shorten this testimony,' said Bryan, turning to Judge Raulston. 'The only purpose Mr. Darrow has is to slur at the Bible . . . I want the world to know that this man, who does not believe in a God, is trying to use a court in Tennessee—' Darrow cut in. 'I object to that.' Bryan persisted. ' . . . to slur at it . . .' Darrow shouted over him. 'I object . . . I am exempting you on your fool ideas that no intelligent Christian on earth believes.' At which point, Raulston seems to have had enough and adjourned proceedings for the day.[36]

'Between science and Bryanism': aftermath

Of the myths that have gathered around the Scopes trial, as many and thick as those around Galileo and the Wilberforce–Huxley debate, two are particularly popular. The first is that the events of that scorching Monday marked an obvious and universally acknowledged victory for Darrow over Bryan. No one could deny that Darrow's cool logic had eviscerated Bryan's fundamentalist ignorance. The second is that, their lead man so publicly humiliated, American fundamentalists went quiet and retreated into a subculture to lick their wounds, from which they were tempted by Ronald Reagan half a century later, with promises of power and influence. Neither is true.

It was raining on Tuesday morning and the proceedings returned inside. It seemed a fittingly anti-climactic end. Raulston prohibited any more examination of Bryan and struck his testimony from the court record because, he claimed, it shed no light on anything relevant to the trial. Darrow and the defence team had nowhere left to go and duly asked the jury to bring a verdict of guilty against Scopes, thereby depriving Bryan of the opportunity to deliver his concluding remarks. The jury, who had been absent for most of the trial as the legal teams fought over the admissibility of evidence, obliged and Raulston fixed the fine at $100, the minimum required. The Scopes trial was over.

But then, as after any modern presidential debate, the spinning began. Both sides claimed victory, one legal, the other moral. Bryan cited the prosecution's success. He cited Darrow's contempt of court and his contempt for Christianity. He revised what would have been his closing speech and made plans to tour the country. Six days later, he died from apoplexy, passing away in his sleep on Sunday afternoon after addressing the Methodist church in Dayton in the morning. Reports claimed that he had been broken by the experience, and for many fundamentalists he became a martyr for their cause, but that was to underestimate his resilience and optimism. 'Broken heart nothing,' Darrow unceremoniously remarked. 'He died of a busted belly.'[37]

Darrow and his team claimed the moral, and intellectual, victory. They cited Raulston's repeated obstruction of their cause, the chain of expert witnesses whose presence had been confined to paper, and, of course, the incurious, closed-minded ignorance that Bryan had displayed in technicolour on the Monday afternoon. The fundamentalists had won the battle but they had truly lost the war.

The press was divided, usually along party lines. Bryan was stupid but Darrow was malign. The world had been shown how intellectually vacuous fundamentalism was, but it had also seen how condescending secular elites could be. 'The truth is that when Mr Darrow in his anxiety to humiliate and ridicule Mr Bryan resorted to sneering and scoffing at the Bible, he convinced millions ... that Bryan [was] right ... that the contest at Dayton was for and against the Christian religion.'[38] The ACLU, keen not to alienate moderate Christian opinion, tried quietly to drop Darrow from the appeal case, fearing he was now a liability. The appeal itself ground its way to the Tennessee Supreme Court, which found the original law constitutional but set aside the conviction on a technicality, and drew a line under the whole affair by preventing any further appeals.

In the immediate term, the causes of fundamentalism and anti-evolution enjoyed a surge. Donations rose. Church membership continued to grow. The following year, the Southern Baptist Convention voted unanimously that 'this Convention accepts Genesis as teaching that man was the special creation of God.' Anti-evolution laws and regulations were subsequently passed in Texas, Mississippi, Arkansas and Louisiana, although they failed totally in the Northern States. None was repealed for over forty years. Hunter's infamous textbook deleted a six-page section on evolution for the Southern States. It was only towards the end of the decade, when 1928 saw the country's first Catholic presidential candidate, Al Smith, and 1929 the Wall Street Crash, that fundamentalist energies were directed elsewhere.

The trial passed from newspapers to history books and then to entertainment, narrated most often by those who saw it as a straightforward triumph of liberalism over fundamentalism, reason over faith and science over religion. From Frederick Allen's early history of the 1920s, *Only Yesterday*, published in 1931, to *Inherit the Wind*, Jerome Lawrence and Robert Lee's 1955 play that was turned into a successful film five years later, the story was remoulded through the lens of its famous last day. Rappleyea was forgotten, Scopes became an innocent victim of mob action, Bryan emerged as a mindless reactionary and the complex, difficult entanglement of evolution with eugenics, a doctrine that lost some of its shine after the Holocaust, was largely ignored. Retelling smoothed the story, as it did for Galileo, Wilberforce and Huxley.

You could hardly blame the historians or playwrights. After all, had not a member of the prosecution counsel itself said that this was a battle between religion and science? This was the version that would pass into history, not the more accurate one of the defence lawyer, Dudley Malone, who commented before the trial that 'the issue is not between science and religion, as some would have us believe . . . .[but] between science and Bryanism.'[39]

Malone was right about the events of 1925 but, as with Galileo, such details became casualties of the alleged war. Bryan himself may have been humiliated but his rhetoric had inspired millions who had heretofore, like Bryan himself, been rather agnostic about Darwinism. He had, in effect, fused evolution and godlessness in the fundamentalist mind. The century would go on to show that what Bryan had joined together in Dayton, Tennessee, would prove very hard to put asunder.

Albert Einstein talks with the Belgian Catholic priest and cosmologist Georges Lemaître (right). Lemaître was the first to propose mathematically, in an obscure scientific paper, that the universe was expanding and had an origin in time. Atheists smelt a theological rat.

FIFTEEN

ENTANGLED AND UNCERTAIN

'A purely scientific matter': the new physics

The most famous scientist in the world was not invited to take the stand in Dayton but he was asked to comment on the whole affair. According to the *Pittsburgh Sun*, Albert Einstein remarked in June 1925 that 'any restriction of academic liberty heaps coals of shame upon the community which tolerates such suppression', although he added that he didn't want to interfere 'in an American "family squabble" '.[1]

Einstein was not the only eminent physicist to pronounce on the trial – Marie Curie signed a letter of protest about it – but he was undoubtedly the best known. Ever since his *annus mirabilis* of 1905, when he published *four* discipline-transforming papers – on Brownian motion, the photoelectric effect, special relativity and the equivalence of mass and energy – but more so once his later theory of general relatively was confirmed by the British physicist Arthur Eddington's observation of a solar eclipse in May 1919, Einstein had been acclaimed a genius. A world exhausted and demoralised by human barbarity found in the avuncular, brilliant, charismatic and instantly recognisable Jewish-German physicist a figure of hope and wisdom. He had stood on Newton's shoulders and seen further into reality than even the great English mathematician. People sought his opinion: on relativity, on science, on reality, on God.

In 1921, as part of a much-fêted global tour, Einstein visited London and dined with, among others, Randall Davidson, the archbishop of Canterbury. What ramifications, the archbishop enquired during the meal, did his theory of relativity have for religion? Einstein did not pause.

'None,' he replied bluntly. 'Relativity is a purely scientific matter and has nothing to do with religion.'[2]

The archbishop was, presumably, relieved. His question had not come out of nowhere. Einstein's visit had been engineered by Lord Richard Haldane, a politician and philosopher, who had just published a book entitled *The Reign of Relativity*, which attempted to explain the theory and enlist it for wider philosophical ends. Haldane had informed Davidson that relativity had profound consequences for theology, and encouraged the bemused cleric to try to make sense of it. 'The Archbishop can make neither head nor tail of Einstein,' one of his staff confided to the physicist J. J. Thomson, 'and protests that the more he listens to Haldane . . . the less he understands.'[3]

It is hard to blame him. Haldane's was one of many attempts to draw relativity away from 'pure science' into the realm of epistemology and ethics. Einstein's theory lifted the veil on the nature of reality. If his inquiry chose to 'stop short . . . of the general problem of the Relativity of all Knowledge', as Haldane said in his Introduction, that was simply due to his, and science's, stringent 'self-denial'.[4] Philosophers would rush in where physicists feared to tread.

In spite of what he told the archbishop, Einstein did not help his own case. It would be a few years before he started to speak openly about religion, but his 'purely' scientific language still habitually gravitated to theology. That was precisely what caught the press's attention. His recurrent references to 'the Lord' were hardly calculated to persuade listeners that relativity, or indeed any part of his physics, was a 'purely scientific matter'. Indeed, it is doubtful even Einstein believed it was.

Newton, whose towering achievement had cast a shadow from which only now physicists were emerging, was intensely devout, and had let the world know it (or, rather, know some of it). The later nineteenth century had been studded with pious physicists and mathematicians, albeit most rather more circumspect about the religious implications of their work than Newton. And now a new era of physics was revising the nature of reality, undermining ideas of determinism, placing the subjective, human observer back at the centre of things and even positing the outrageous idea that, in spite of empirical evidence, the universe may not have been around for ever but may have come into existence at some point in the past. Who could honestly say these were *only* scientific issues? Who could stop the new physics from bleeding into metaphysics, science into religion? The two, it seemed, were destined to be inextricably entangled. The problem was, no one was really certain how.

'Spooky action at a distance': on not bringing down the house

Physics was grinding to a halt at the end of the nineteenth century, confident in what it had achieved and in what little there was left to be discovered. Albert Michelson, a Nobel-winning American physicist, remarked in a lecture in 1899 that 'the more important fundamental laws and facts of physical science have all been discovered.' Our future discoveries, he stated, 'must be looked for in the sixth place of decimals'.[5]

There was still, however, much to discuss, not least about how recent discoveries might relate to religious belief. Through the later nineteenth century, greater understanding of the laws of thermodynamics had cemented the idea that the universe was winding down, and heading for a long, slow heat death. Some ventured that this was a 'purely scientific matter' with no great significance for religion. Others disagreed, interpreting the universe's ever-growing entropy as yet another scientific blow to the religious, meaning Christian, belief in a final conflagration and/or resurrection. The universe, it appeared, was destined to end with a whimper rather than an apocalyptic bang. Still others argued it proved that the universe was not as eternal and self-sustaining as the atheists argued and that, if it was winding down, presumably it had also once been wound up. To paraphrase a later American president, thermodynamics had spoken but it was taking a while to discern what it had said.

Thermodynamics was soon to become a sideshow, however. Around the time of Michelson's lecture, Marie Curie and her husband Pierre were investigating a phenomenon she called radioactivity, which appeared to show the release of intense energy from within certain atoms, a discovery that would earn her two Nobel Prizes and result in her death, leaving her notebooks so radioactive that they are still stored in a lead-lined safe. Her breakthrough was extraordinary, though it appeared at first to be at odds with the principle of the conservation of energy and also to challenge the historic understanding of atoms as the literally 'indivisible' building blocks of nature.

Two years after Marie's first Nobel Prize, Einstein published a paper on the electrodynamics of moving bodies. Prepared to doubt Newton's authority (at least in extreme conditions), Einstein and his wife Mileva Marić combined Maxwell's equations with the idea that the speed of light was a constant irrespective of the speed of any observer, to argue that the absolute and uniform nature of space and time inherent in Newton's mechanics was not entirely accurate. Time dilated and mass increased at extreme speeds, and it appeared that there was no 'absolute space' against which objective and definitive measurements could be made. Ironically, one of the strongest pieces of empirical evidence then available for the theory came from an experiment measuring the speed of light conducted by Albert Michelson twenty years earlier.

Over the following decade, Einstein applied these ideas to gravity to form his theory of general relativity, which fused and mathematically described the three dimensions of space and one of time. The theory made predictions about light being bent by the Sun's gravity, and when these were confirmed by Eddington's observations in May 1919, both theory and theorist achieved global fame. The idea of relativity was launched upon a sometimes over-enthusiastic general public.

Another of Einstein's great papers of 1905 would, in the slightly longer term, give rise to yet another transformation of physics, one to which Einstein was never reconciled. 'On a Heuristic Viewpoint Concerning the Production and Transformation of Light' argued that electromagnetic radiation behaved as if it consisted of discrete, independent packages (or quanta) of energy as well as being wavelike, and this idea would, in the hands of Niels Bohr, Werner Heisenberg and Erwin Schrödinger, form the basis of quantum theory in the later 1920s.

Relativity attracted a certain amount of creative commentary, as philosophers and theologians used it to speculate on everything from miracles to morality. Quantum theory attracted even more. The theory threw up ideas of uncertainty, subjectivity, complementarity and entanglement, all for highly specific and technical reasons but all amenable to wider deployment. The uncertainty principle, proposed by Heisenberg in 1927, stated that it was impossible to measure simultaneously both the position and the momentum of certain particles – not because of the inadequacy of observation equipment but because of an inherent indeterminacy that was part of the dual matter-wave nature at the quantum level. At its deepest level, nature simply could not be fully known.

This basic 'ontological' uncertainty was sometimes linked with the observer effect, the idea that (even passively) observing an experiment could change its results. This was (or would become) common sense at a human level, such as when anthropologists embedded themselves among the subjects and in so doing influenced their behaviour. But it was anything but common sense for the physical sciences, which were predicated precisely on the idea that matter ran on iron rails and didn't give a damn who was watching. Physics was *the* objective discipline and the idea that the results you gathered depended in some way on your presence as a subjective observer – even if that observer was actually an electronic detector – was as unsettling as it was counterintuitive.

Complementarity, an idea associated with Niels Bohr, was a response to Heisenberg's ideas about uncertainty. If a system could no longer be described comprehensively by means of a single, definitive and objective account because of the quantum world's fundamental matter–wave duality, physicists would need to make use of complementary descriptions – parallel accounts of the same phenomenon that were *both* true but also incommensurable. As with the observer effect, this was not a new idea – indeed Bohr effectively imported it from other disciplines – but it was new to physics, which aspired to a single, comprehensive account of the physical realm. Reality, it now seemed, could afford two – perhaps more? – equally truthful narratives.

The idea of entanglement was developed by Schrödinger, who coined the word (in German) in response to Einstein's criticism of quantum theory and who reckoned it *the* characteristic trait of quantum mechanics. Entangled particles were those that, having shared immediate physical space, could not thereafter be described independently of one another, irrespective of how far apart they were. Superficially less mind-stretching than the ideas of inherent uncertainty or subjectivity, the idea still seemed to suggest that information – such as about spin or momentum – could be transmitted from one particle to another instantaneously, with no regard for the ordinary rules of local causality or the speed of light. Two such particles might be entangled, separated and then placed on opposite sides of the universe and yet still, it was argued, exchange information instantaneously. Einstein was not persuaded and famously dismissed the idea as 'spooky action at a distance'.

By 1935, when Einstein and Schrödinger were exchanging papers on entanglement, the prediction of Albert Michelson thirty-six years earlier seemed wildly hubristic. Reality was very considerably stranger than physicists of his generation had once thought, although not only those of his generation. Einstein, born twenty-seven years after Michelson, contested and resisted quantum theory for the rest of his life, despite the evidence slowly gathering in its favour and the fact that his own work lay at its origins. The greatest minds of the physics world did not always agree on the science, let alone what it might mean for religion.

In one regard, Einstein's comment to the archbishop was undoubtedly correct. Nobody found or lost God on account of relativity or quantum theory. For all that the new physics adjusted the foundations of reality, few imagined that it brought the religious house down, if only because there was no longer one built above it. Newton had famously claimed that he had had his eye on how his system might work 'with considering men for the belief of a Deity', but after the work of Laplace and the fall of Paley, no one based their religion on Newtonian physics. And that meant that when physics was remodelled in the early twentieth century, there were no religious beliefs or doctrines built on it that could come crashing down. The curse of concordism, which had afflicted science and religion since classical times, and had led to the greatest moments of tension in their history – over Aristotle in the thirteenth century, or Aristotle (again) in the seventeenth, or natural theology in the nineteenth – was finally lifted.

At least it was in *physics*, for that was the other reason why the early twentieth-century revolution failed to spark a science–religion conflagration. It was not, as is sometimes claimed, because the ideas were just too complex to grasp. Not fully understanding science had rarely deterred people from using it for their wider political or religious ends in the past. Rather, it was simply that early twentieth-century physics had a limited amount to say about the nature of the human. The arguments that ricocheted around the streets of Dayton in the 1920s were rich with human significance. The theories of relativity and uncertainty and the observer phenomenon were not. The historic *casus belli* for science and religion felt a long way away. Those who read the still-popular histories by John Draper and Andrew White and watched this space in the hope of further fireworks, left disappointed.

'A conscious and intelligent mind': Einstein and friends

No fireworks did not mean no discussion, however. Indeed, it was precisely because there was so little, religiously speaking, resting on the new physics that the conversations around it could be so (religiously speaking) interesting. The new physics opened up plenty of questions without offering definitive answers.

One was about the openness of the universe itself. In place of a Newtonian universe – mechanistic and apparently deterministic – the new physics found a reality that was undetermined at its most fundamental level. A universe of uncompromisingly closed causation, into which God had to break and enter to get what he wanted, moved aside for one that was looser at the joints. Process theology, the idea that God coaxed rather than coerced creation into co-operation, became popular.

Whatever an open universe might mean for divine action, it appeared to leave more space for human action. Some seized on this. Reviewing the developments of modern science in his 1928 book *The Nature of the Physical World*, no less a scientist than Eddington claimed that 'religion first became possible for a reasonable scientific man about the year 1927.'[6] He went on to explain that if 'our expectation should prove well founded that 1927 has seen the final overthrow of strict causality by Heisenberg, Bohr, Born and others', it meant that the world was more open to all kinds of things, including 'common activities (e.g. falling in love)', which the 'consistently reasonable man' had heretofore been compelled to deny as impossible within a strictly deterministic universe. The indeterminacy of the universe brought back freedom and love, and religion, as credible options.

In reality, much of Eddington's argument was tongue-in-cheek. He readily acknowledged that people had always 'managed to persuade themselves that they had to mould their own material future notwithstanding the yoke of strict causality',[7] and many, including Eddington himself, would point out that however much quantum theory might challenge strict 'Newtonian' causality at a subatomic level, it did nothing as grand as establish the reality of free will. Nonetheless, for all Eddington's playful provocation, his wider point – that 'the compartments into which human thought is divided are not so water-tight that fundamental progress in one is a matter of indifference to the rest' – remained valid.[8]

Hence the fascination the new physics had for so many people, vanishingly few of whom understood it in any detail. However specific and technical they were in origin, relativity, uncertainty, the observer effect and complementarity looked and felt like philosophical, even existential ideas – as did the idea of beauty. The conviction that beauty was important to physics was hardly new. Indeed, it was precisely the elegance with which Newton's work unified and explained disparate phenomena in heaven and earth that recommended it to admirers. But the idea now received renewed attention.

Paul Dirac was an English theoretical physicist who held the Lucasian Chair of Mathematics (like Newton and Stokes), shared the Nobel Prize for Physics with Schrödinger in 1933 and made brilliantly original contributions to both quantum theory and relativity. He was also very serious, precise, focused and perhaps autistic, with no personal interest in aesthetics. 'I don't see how you can work on physics and write poetry at the same time,' he once berated a young Robert Oppenheimer. 'In science you want to say something nobody knew before in words everyone can understand. In poetry you are bound to say something that everybody knows already in words that nobody can understand.'[9]

And yet, he was adamant that beauty was not just *a* factor in determining the truth of physics but the supreme factor. The foundations of relativity, he wrote, are 'stronger than what one could get simply from the support of experimental evidence', precisely because of their beauty. 'It is the essential beauty of the theory which I feel is the real reason for believing in it.'[10] Indeed, he would later go further still and claim that 'it is more important to have beauty in one's equations than to have them fit the experiment.'[11]

Dirac was unusual in this only for his strength of opinion. Heisenberg once remarked in conversation with Einstein that 'if nature leads us to mathematical forms of great simplicity and beauty . . . we cannot help thinking they are "true", they reveal a genuine feature of nature.'[12] For all Einstein resisted the implications of quantum theory, it was a sentiment he shared. It was a position that had a certain coherence and credibility within a theistic understanding of creation, which claimed that the reality, and correspondence of, beauty, goodness and truth reflected something of a creator. But it was harder to reconcile with a cosmos of accident and chance. Why should beauty matter in a rough, random, material universe?

Such considerations fed into a wider, longstanding debate between materialism and idealism. Materialism had long been associated with atheism and, in the popular mind, irreligiosity. A universe comprised solely of matter left little space for spirit, in spite of what Margaret Cavendish or David Hartley had argued two centuries earlier. With idealism, by contrast, mind came first, the human mind reflecting something of the divine one. The new physics seemed to steer the world picture towards idealism.

Eddington led the way, putting forward his ideas in a book entitled *Science, Religion and Reality*, but it was James Jeans, another Cambridge cosmologist and physicist, who best captured the argument when he wrote, in his popular science book *The Mysterious Universe*, that the contemporary advance of physics made the universe 'look more like a great thought than like a great machine. Mind no longer appears to be an accidental intruder into the realm of matter . . . [and] we ought rather hail it as the creator and governor of the realm of matter.'[13]

Jeans's book, like Eddington's comparable *Nature of the Physical World*, was successful, and particularly popular with preachers and Christian apologists, in spite of the fact that Jeans himself was no religious believer but merely an agnostic with a mystical bent. His god was the god of the mathematicians rather than of the philosophers, let alone of Abraham and Isaac. Indeed, his god *was* a mathematician, at least according to the newspapers.[14]

Not all mathematicians nor, indeed, philosophers or theologians were persuaded. English philosophers had drifted away from the high point of idealism in the late nineteenth century, and were not convinced that the new physics was enough to return them to it. Bertrand Russell, the country's pre-eminent philosopher at the time, rejected the idealist attempt to find meaning behind matter. He was one of several sceptics who pointed out that there was something unconvincing about the way in which preachers had once claimed to have discovered God in Newton, and now claimed they had discovered him again in the thing that replaced Newton.

For their part, theologians were rather lukewarm too. They had only, in the last generation or two, finally freed themselves from the natural theological corpse to which they had eagerly shackled themselves when it had been the picture of rude health, centuries earlier. They were not eager to reattach themselves. Under the influence of the Protestant theologian, Karl Barth, the energy was directed forcefully away from natural theology. The God of the authentic Christian tradition was radically different from his creation. There existed an unbridgeable gap between God and the world. Knowledge of God was possible only in as far as God had revealed himself to his creatures. Believers could not and should not try to read anything of God from his creation. Natural theology was a mistake, and a dangerous one at that.

Many theologians were influenced by Barth's arguments, even if they did not share them with the same vigour. Frustratingly, however, the language and ideas of the new physics kept on slipping back into a theological register. 'Einstein keeps talking about God,' someone remarked during an evening discussion at the fifth Solvay Conference in 1927. 'What are we to make of that?'[15]

The Solvay Conferences had been, since 1911, the pre-eminent gathering of the world's physicists and chemists, and the fifth, which met to discuss the emerging theory of quantum mechanics, was probably the most famous. Pretty much everybody who mattered in this world was there, including Einstein, Schrödinger, Heisenberg, Dirac, Marie Curie, Wolfgang Pauli, Max Born, Niels Bohr and Max Planck. One evening, a group of delegates spent a few hours ruminating on the relationship between science and religion. Although absent himself, Einstein was the spark that got the whole thing going.

Max Planck, a German physicist who had first identified energy quanta and won the Nobel Prize for his efforts, emerged as having one of the more sympathetic views. A lifelong Lutheran, he believed that science and religion effectively operated in different spheres. 'Science deals with the objective, material world. It invites us to make accurate statements about objective reality and to grasp its interconnections. Religion, on the other hand, deals with the world of values. It considers what ought to be or what we ought to do, not what is.' Although no naïve materialist – he would later write that 'all matter originates and exists only by virtue of a force which brings the particle of an atom to vibration . . . we must assume behind this force the existence of a conscious and intelligent mind' – he nonetheless was clear that religion and science were non-overlapping magisteria, to use the modern phrase.[16]

Heisenberg was not persuaded. He was not more sceptical than Planck – indeed he too was a practising Lutheran – but he was simply less sure the two could or should be kept separate. 'Although I am convinced of the unassailability of scientific truth in its own sphere,' he would later say, 'I have never been able to dismiss the content of religious thinking simply as a stage in human consciousness which we have superseded.'[17] The religious approach to the world and the scientific understanding of it could not be separated so easily. 'I doubt whether human societies can live with so sharp a distinction between knowledge and faith,' he remarked that evening at Solvay.

Wolfgang Pauli, another quantum pioneer and (future) Nobel Laureate, agreed with Heisenberg. He felt that the 'sharp clash' between science and religion had been the result of 'the idea of an objective world running its course in time and space according to strict causal laws', which the new physics was busy undermining. But he also felt that the relationship between science and religion would not necessarily be improved by the new physics. Society, he argued, was in particular danger whenever fresh knowledge threatens to explode the old spiritual forms, and he feared that 'the parables and images of the old religions will [lose] their persuasive force' and that 'the old ethics [would] collapse like a house of cards' leaving 'unimaginable horrors [to] be perpetrated'.

Dirac emerged as the most sceptical of the party, complaining that they were even talking about religion. 'If we are honest – and scientists have to be – we must admit that religion is a jumble of false assertions, with no basis in reality'. The very idea of God was 'a product of the human imagination'. Religion itself was 'a kind of opium that allows a nation to lull itself into wishful dreams'. Its continued existence was simply down to the fact that 'some of us want to keep the lower classes quiet'. Religions contradicted one another. Religious belief was an accident of birth. And religious ethics taught subservience, distracting humans from the real task of living. 'Life,' he said, 'is just like science: we come up against difficulties and have to solve them.'

Niels Bohr, not present at the conversation but whose views were latterly reported by Heisenberg, exhibited the most interesting combination of these disparate views. He was sympathetic to Dirac's position, the idea of a personal God being entirely foreign to him. He also, however, recognised that Dirac had a very narrow view of language, and that 'religion uses language in quite a different way from science', it being 'more closely related to the language of poetry than ... of science'.

Bohr was also prepared to relate the religious ideas, in which he didn't believe, with the new science. Subjective and objective were now, after relativity and quantum mechanics, no longer the absolute concepts they once had been. 'It is no longer possible to make predictions without reference to the observer or the means of observation.' Complementarity underlined how there could be alternative descriptions of the same thing. There was more than one way of understanding the world.

As we have seen, Einstein was not present during the Solvay conversation, but his influence was palpable. 'It is extremely difficult to imagine that a scientist like Einstein should have such strong ties with a religious tradition,' one of the group observed, correctly. His transparent genius meant that everyone wanted him on their side, but any attempt to turn the God-bothering physicist – Bohr once berated him for telling God what he could and couldn't do – into an orthodox believer was destined for failure.

Einstein certainly *talked* about God a lot but, in reality, he had jettisoned any formal religious beliefs in his teens. He never attended religious services and never prayed. He disliked mysticism. He could not conceive, let alone admire, a God who punished and rewarded people, not least because he was a thoroughgoing determinist. He repeatedly distanced himself from the idea of a personal God. He refused a traditional Jewish burial. 'The word God is for me nothing more than the expression and product of human weaknesses,' he wrote years later to the Jewish philosopher Eric Gutkind, who had sent him his book *Choose Life: The Biblical Call To Revolt*. 'The Bible is a collection of honourable, but still primitive legends which are nevertheless pretty childish.' This was not the voice of a believer.

Nor, however, was it the voice of an atheist, in any conventional sense of the word. Einstein could, and frequently did, use the God-talk in simply idiomatic ways. After a concert in April 1930, he rushed over to the soloist Yehudi Menuhin in raptures, embraced him and said, 'Now I know there is a God in heaven.'[18] This was God as aesthetic bliss; not philosopher, let alone saviour.

And yet, his idea of God was more than a mere figure of speech. 'I'm not an atheist and I don't think I can call myself a pantheist,' he once said, when asked to define God.[19] 'I believe in Spinoza's God,' he told Rabbi Herbert Goldstein of the Institutional Synagogues of New York, 'who reveals himself in the orderly harmony of what exists.'[20] Beauty mattered, and not just in a merely subjective sense, even assuming the subjective could ever, now, be mere. All the finer speculations in the realm of science 'spring from a deep religious feeling', he remarked in 1930. In the order, beauty and intelligibility of creation, he found signs of the 'God' he also heard in music.

This was emphatically not the personal God of the Abrahamic faiths, but nor was it the invented God of the fearful human imagination. Indeed, Einstein could be as withering about atheism as he was about religion. There are still people, he remarked at a charity dinner during World War Two, who say there is no God. 'But what really makes me angry is that they quote me for support of such views.' He had as little time for monomaniacal unbelievers as he did for the obsessively religious. 'There are fanatical atheists whose intolerance is of the same kind as the intolerance of the religious fanatics,' he said in 1940.[21]

There was, then, a (pleasing) uncertainty about Einstein's religious position: the closer you got to it, the harder it was to establish. In as far as it is possible to say anything definitive about his beliefs, they appear to be a kind of cross between deism and Spinoza's pantheism, impersonal but present in the equations, perhaps even breathing fire into them. But to make anything authoritative of such beliefs would be a singular mistake. As he knew only too well after 1919, Einstein was treated as a figure of great authority, both intellectual and moral. But that was an irony, indeed a curse. It was only as a punishment for his own contempt for authority, he once joked, that Fate chose to make him an authority himself.

'Therefore, there is a Creator': a new beginning

Even great authorities could get things wrong. During the fifth Solvay Conference, Einstein was introduced to a relatively young physicist and cosmologist who, although not a delegate, was working in Brussels where the conference was being held. Georges Lemaître was thirty-three years old, had recently been awarded his PhD from MIT, and a few months earlier had published a paper on 'A homogeneous universe of constant mass and increasing radius accounting for the radial velocity of extragalactic nebulae'.

The paper, which had appeared in an obscure Belgian journal, had argued, on the basis of Einstein's own theory of general relativity, that the universe was expanding, and indeed that the speed of the galaxies was in proportion to their distance – the further away, the faster they travelled. Einstein had been alerted to the young cosmologist's paper and walked with him through Brussels' Parc Léopold discussing his ideas. He could not fault the young man's mathematics. Technically, it was flawless. But from the point of view of physics, he thought his ideas were 'abominable'.[22]

Einstein no more liked the idea of the expanding universe of Lemaître's calculations than he did the indeterminate one of quantum theory. The immutability of the universe, like its determinism, was an article of faith for him. But there was another aspect to Lemaître's theory that concerned Einstein, as it would do many others when his ideas seeped into wider scientific discussion in the early 1930s. Georges Lemaître was a devout Catholic, indeed a priest, and his idea of an expanding universe, with the implications of a beginning in time, seemed to many to be a bit suspect.

Nesting beneath the flawless mathematics, they could smell a theological rat.

Lemaître was educated by Jesuits and studied civil engineering before fighting in World War One. He went on to study mathematics and physics, while also training for the priesthood into which he was ordained in 1923. His 1927 paper had been first published in *Les Annales de la Société Scientifique de Bruxelles*, the journal of a society of Catholic scientists, and so it was only when the paper was translated into English, four years later, and Arthur Eddington published a commentary on it, that the idea gained traction. Lemaître was invited to lecture in London, during which time he coined the phrase 'Primaeval Atom' as a way of describing the universe's point of origin. The fact that the British Association meeting at which Lemaître was speaking was focused on the physical universe and spirituality did not aid his cause. Many were instinctively suspicious of the idea, if not of the jocular and likeable priest himself.

A static and eternal universe was not synonymous with atheism, in spite of the longstanding tensions around this Aristotelian idea throughout Christendom. Siger of Brabant had proposed something similar in the thirteenth century, albeit his ideas were among those condemned in 1277. That recognised, there were many eminent minds of the time – among them the philosopher Bertrand Russell, the mathematician and future president of the Rationalist Press Association Hermann Bondi and the blunt-talking English astronomer Fred Hoyle – who found in the idea of an eternal universe confirmation of God's non-existence, or at least his irrelevance. The idea of a cosmos that had come into existence was bad enough. That it had been identified by a Catholic priest was intolerable.

Hoyle was particularly animated. Religion, as far as he was concerned, was 'but a blind attempt to find an escape from the truly dreadful situation in which we find ourselves'.[23] Faith was a cop-out and religious reason nothing but enforced dogma. 'If the facts of the case should disagree with the dogma then so much the worse for the facts.'[24] He went to war against Lemaître's 'primaeval atom', working with Bondi on formulating a coherent 'Steady State' theory and coining the phrase 'Big Bang' in a post-war radio broadcast, in a move that was widely considered to be derisive of Lemaître's theory, although Hoyle denied this.

Einstein had similar, if less forthright reservations – 'it suggests too much the (theological) idea of creation', he told Lemaître – but the observational evidence slowly tilted in the Belgian's favour, and by the time cosmic microwave background radiation was detected in 1964, not long before Lemaître died, the consensus was for a Big Bang. Science had confirmed that the universe had a moment of creation. And therefore, presumably, a creator?

That, at least, was what Pope Pius XII tried to argue. Addressing the Pontifical Academy of Sciences in 1951, the pope drew on these new ideas to claim that 'present-day science . . . has succeeded in bearing witness to the august instant of the primordial *Fiat Lux* [let there be light].' The conclusion was unavoidable. 'Hence, creation took place. We say: therefore, there is a Creator. Therefore, God exists!'[25]

Lemaître was not impressed. He had long been clear that the Bible was simply not interested in such matters. 'Neither St Paul nor Moses had the slightest idea of relativity,' he told an interviewer for the *New York Times* in 1933. 'The writers of the Bible were illuminated more or less – some more than others – on the question of salvation. On other questions they were as wise or as ignorant as their generation.'[26] His theological training had also alerted him to the important longstanding distinction there is in Thomist theology between creation and a beginning. Aquinas had been able to entertain the possibility of a universe that was both eternal and created because 'created' in this sense did not mean having a beginning in time so much as being dependent on another for its existence. Creation was, in effect, a metaphysical claim; the Primaeval Atom a physical one. That being so, the pope's jubilant theological conclusion from scientific premises was a howling category error.

Lemaître interceded. He contacted the pope's scientific adviser and encouraged him to dissuade the pope from further pronouncements on such matters, which were theologically naïve not to mention scientifically unhelpful. He was successful, although papal reticence on such matters would not stop many believers from seeing in the Big Bang a demonstration of God's existence, just as many non-believers took the apparently immutable permanence of the universe as demonstration of his non-existence.

The fact that it was a Catholic priest who first established that the universe had a beginning – not by faith, not by dogma, but by flawless mathematics – is an attractive detail within the complex histories of science and religion. But that he then went on to warn the pope himself *away* from drawing any theological conclusions from his work is a still more appealing twist in this particular tale. In the strange world of the new physics, nothing was as it seemed.

Sir James George Frazer, whose monumental book *The Golden
Bough* would endeavour to explain scientifically the origins of
religion. Ironically, given Frazer's purely literary appreciation
of the Bible, *The Golden Bough* would have a greater impact
on the literature of the twentieth century than on its science.

SIXTEEN

INFANTILE DELUSIONS

'The natural history of religion': primitive religion

Paul Dirac's view, at the 1927 Solvay Conference, that religion originated among 'primitive people, who were so much more exposed to the over-powering forces of nature than we are today, [and who] personified these forces in fear and trembling' was not mere prejudice.[1] It was science.

Sceptics had inferred such ignorant and naturalistic origins since the eighteenth century, but it was only in the later nineteenth that theories of this nature acquired the mantle of science. Darwin and the missionaries again stood at the source. 'There is no evidence that man was aboriginally endowed with the ennobling belief in the existence of an Omnipotent God', Darwin wrote in *The Descent of Man*, citing as his witnesses not 'hasty travellers, but ... men who have long resided with savages'.[2] However, he continued, if by religion we mean belief in 'unseen or spirit-ual agencies', it was clear that such beliefs were 'almost universal with the less civilised races'. Such an expansion of the category of religion – away from the coherent, doctrinal model inherited from post-Reformation Europe and towards a more capacious, more ritualised, if not necessarily less cognitive pattern exhibited by 'savage' peoples – would be crucial for the new science of religion. But it would be an anthropologist whom Darwin cited in *The Descent*, rather than Darwin himself, who would pioneer the field.

Edward Burnett Tylor was born in 1832. Raised a Quaker, he spent much of his twenties travelling abroad on account of his fragile health. The experience awakened an interest in unfamiliar cultures and provided the raw material for his first book, on ancient and modern Mexico.

Anahuac was published in 1861 and although neither a commercial nor an academic success, it did mark the start of Tylor's scientific quest for the origins of religion.

There were moments in which Tylor could sound uncannily like Voltaire or Denis Diderot's *Encyclopédie*:

> The great influence of the priests in Mexico was among the women of all classes, the Indians, and the poorer and less educated half-castes. The men of the higher classes, especially the younger ones, did not appear to have much respect for the priests or for religion, and, indeed, seemed to be sceptical, after the manner of the French school of freethinking.[3]

Elsewhere, he was pure John William Draper:

> The guilt of thus bringing down Europe intellectually and morally to the level of negro Africa lies in the main upon the Roman Church [as] the records of Popes Gregory IX and Innocent VIII, and the history of the Holy Inquisition, are conclusive evidence to prove.[4]

At the time of *Anahuac* this was Quakerism speaking, with its dislike of priesthood and ritual, rather than outright scepticism. Tylor would, however, soon leave his Quaker faith, but not the distrust of Catholicism it had instilled in him. Such prejudices notwithstanding, his work would strive to be more than mere polemic, systematic rather than sceptical and discerning rather than dismissive. He was appointed as Oxford's first reader in anthropology in 1884 and its first professor in 1895, and was knighted in 1912. His was, in effect, the first sustained attempt to offer a 'natural history of religion', to borrow the title of his 1889 Gifford lectures, which Tylor himself borrowed from David Hume.

Tylor had suggested, in his 1865 work *Researches into the Early History of Mankind*, that 'dreams may have first given rise to the notion of spirits' (an idea that Darwin had quoted approvingly in *Descent*) and in his following book, *Primitive Culture*, he set out this theory in detail. As with so many thinkers of his age, Auguste Comte lurked in the shadows. History was progressive, with a predictable three stages discernible. Progress was not smooth or uniform, however, and a number of beliefs and behaviours somehow survived into later, more developed, stages of civilisation.

Religion was one such 'survival'. Dreaming was ubiquitous among primitive humans, as was the practice of oneiromancy, 'the art of taking omens from dreams by analogical interpretation' (exhibit A: the story of Joseph in the book of Genesis).[5] A savage knows his friend or enemy is dead, Tylor reasoned, and yet, when he sees him in a dream or vision, he naturally takes this 'spectral form' to be 'a real objective being', which confirms the existence of the dead man in some ethereal dimension. From such a category error, he infers the existence of a soul, which becomes 'the gateway into a complex region of belief'.[6] The result is a kind of 'animism', a word that Tylor coined or, rather, repurposed for anthropology to mean the belief in the existence of other spirits, from local sprites right up 'to the rank of powerful deities'.[7]

Tylor claimed that animism characterised 'tribes very low in the scale of humanity', but argued that it was also present, albeit much modified, even 'in the midst of high modern culture'.[8] It reflected well on 'savage philosophy' to recognise that a similar kind of animism still survived in 'civilized countries', in the form of popular belief in ghosts, souls, spiritualism and the like. Indeed, institutional religious beliefs might seem more sophisticated than primitive animism or contemporary spiritualism but, on closer inspection, they were really all the same beast under the skin.

This was the main argument of Tylor's (unpublished) Gifford lectures on the *Natural History of Religion*. The spiritual landscape of the Bible was, *mutatis mutandis*, the same as that of animists the world over, even those as 'barbarous' as indigenous Tasmanians, the so-called savages who had been systematically killed by civilised white settlers. There were 'deluge legends' and demons and angels and souls the world over. The idea of spiritual beings 'consort[ing] sexually with men and women', as he put it delicately in *Primitive Culture*, was equally universal.[9] Even something like the eastward orientation of the Catholic priest at Mass, recently reintroduced into Established Church practice by the Oxford Movement, had its roots in primitive, animistic sun worship.

Tylor was able to marshal an impressively wide range of sources for his theories. His brief discussion of sexually adventurous spirits in *Primitive Culture*, for example, took in the Antilles, New Zealand, the Samoan

Islands, Lapland, Hinduism, St Augustine and Pope Innocent VIII. His range and creativity impressed many readers, not least the man who was to become anthropology's (anti-)Christ to Tylor's John the Baptist.

James George Frazer wrote, in the preface to the first edition of *The Golden Bough*, his ever-expanding magnum opus on the origins of religion, that his interest in early societies had been stimulated by Dr E. B. Tylor. Tylor's work, he claimed, in entirely appropriate language, 'opened up a mental vista undreamed of by me before'.[10] Twenty years his junior, Frazer learned much from Tylor (although latterly fell out with him), just as he had from William Thomson, who had given him, at the University of Glasgow, 'a conception of the physical universe as regulated by exact and absolutely unvarying laws of nature expressible in mathematical formulas'.[11] Thomson had judged this conception as wholly compatible with, indeed as evidence for, the truth of Christianity. Frazer did not.

The first edition of *The Golden Bough* published in 1890, then still a mere two volumes, was subtitled 'A Study in Comparative Religion'. It put forward the idea, which Frazer acknowledged taking from his friend, and the book's dedicatee, W. Robertson Smith, that 'the conception of a slain god' was central not only to Christianity but to all religions. Religions had their origins in fertility cults, which celebrated, and attempted to secure, the eternal, life-giving cycle of death and rebirth. Frazer traced how many of these cults involved a sacred priest-king in their ritual, by whose sacrificial death future life in all its abundance could be secured.

This basic framework expanded, both in terms of evidence and structure, as the volumes grew. In a second edition, Frazer explained how humans developed through a certain number of stages – no prizes for guessing how many – beginning with magic and ending with science. Magic attempted direct control over nature but when that repeatedly failed, humans turned to an intermediary, exercising power through gods, or eventually a god, with whom they had to bargain, giving agricultural produce, or blood, or life to ensure the future. Eventually religion, no more successful at manipulating reality than magic, gave way to an age of science.

This was hardly a novel schema for the time and what distinguished Frazer's work was the sheer amount of material he accrued – *The Golden Bough* grew to twelve volumes plus a supplement by the time Frazer died

– *and* the unapologetic way he applied his ideas to Christianity. This became more noticeable between the first edition in 1890 and the second in 1900, during which time the subtitle also changed from 'A Study in Comparative Religion' to 'A Study of Magic and Religion'. The first edition had left his audience to read between the lines, at least when it came to foundational Christian ideas of incarnation, atonement, resurrection and Eucharist. Ten years later – aided by the fact that his mentor Robertson Smith, who had been an accomplished Old Testament scholar and a minister in the Free Church of Scotland, was no longer alive to challenge and correct him – Frazer was more forthright. The message was inescapable: the religion of the Bible was the religion of primitive men.

By means of innumerable examples, Frazer showed how the stories and rituals and creeds that Jews and Christians thought were quintessentially theirs hardly differed from those of primitive tribes and cults through history and across the world. The wholesale conversion of the heathen, to which missionaries dedicated so much energy, was really little more than a change of robes. There were ideas of holiness, pollution and purification; of priestly kings, sacred names, holy water and royal blood; of sacrificed gods, resurrections aplenty and god-eating ceremonies that recapitulated it all. It was impressive stuff, except that it all felt rather self-fulfilling.

Frazer could not but view such cultures from his position within the civilisation in which he lived. Sceptic as he was, he was nonetheless a man of his solidly Christian times. So much was unavoidable. More problematic was the fact that, for all that he wrote about other cultures, Frazer, unlike Tylor, never actually visited any of them, let alone did any original fieldwork. He knew no other lens through which to view 'primitive' peoples and so translated most of what he wrote about into a familiar lexicon. Much of what he said in *The Golden Bough* and in his later *Folk-Lore in the Old Testament* was effectively a version of Judaeo-Christianity.

Thus, the Njumas of East Africa were the equivalent of 'the Levites of Israel'.[12] The Grand Lama of Lhasa was the 'Good Shepherd who laid down his life for the sheep'.[13] A ritual among the Wachaga of East Africa was 'a solemn league and covenant'.[14] 'Scapegoat', a word coined by William Tyndale in his aborted translation of the Old Testament to describe a particular ritual within the book of Leviticus, became the focus of an

entire 400-page volume of *The Golden Bough*, which covered the ideas of
the 'transference' and 'expulsion' of evil in every imaginable culture and
time. Frazer's very language – of Adam and Noah, of being born again, of
baptism, sacrament, holiness and resurrection – could not help but make
parallels, even when they were tenuous at best. The cumulative result of all
this was to generate a powerful sense of unity across cultures, at the
expense of detail and difference. It was intended to reveal 'the savage
beneath the sacred', which it did in technicolour, but it also left the lingering suspicion that there was only really one kind of religion, and that there
was nothing whatsoever original about Hebrew culture.

Frazer himself insisted that he had come to praise the scriptures rather
than to bury them, and in his 1895 *Passages of the Bible*, he wrote of how
the book was 'noble literature ... fitted to delight, to elevate and to
console'.[15] The irony is that this became the main reason why his own work
would be remembered and honoured: its contribution to literature.[16] T. S.
Eliot would famously acknowledge his debt to Frazer in his notes to *The
Waste Land*, and however mischievous those notes were, no student of the
poem can ignore *The Golden Bough*. James Joyce would draw on Frazer's
ideas in *A Portrait of the Artist as a Young Man*, *Ulysses* and *Finnegans
Wake*. W. B. Yeats would reference him in 'Sailing to Byzantium'. D. H.
Lawrence wrote to Bertrand Russell in 1915, after reading Frazer, that now
'I am convinced ... that there is another seat of consciousness than the
brain and the nerve system: there is a blood-consciousness which exists in
us independently of the ordinary mental consciousness, which depends
on the eye as its source or connector'.[17] Robert Graves, John Millington
Synge, Wyndham Lewis, even John Buchan, were shaped by his ideas. The
literary power of Frazer's anthropology far exceeded its scientific merits.
As Frazer might have said of the Bible itself, whatever scientific truth there
might have been in *The Golden Bough* – and later scholars would come to
the conclusion that there was not much – there appeared nonetheless to
be a deeper, more primeval, mythical truth within the book.

'A whole climate of opinion': infantile religion

The association of 'primitive' cultures with childhood was a familiar trope
among anthropologists. Magic and religion were, after all, remnants of

humanity's infancy. Children delight to play with rattles and drums, Tylor had reasoned, and so it is telling that 'the savage magician clings with wonderful pertinacity to the same instrument.'[18] The comparison of such savages to 'grown-up children' may be 'trite', he admitted, but it is 'in the main a sound one'.[19] There was more than one way of being childlike, however. You could, for example, actually be a child. Perhaps religion was not so much a survival of humanity's childhood but of childhood itself. Or perhaps the two were the same.

Sigmund Freud read a lot of anthropology. He adopted Tylor's ideas about dreaming and animism and mentioned or referenced Frazer nearly a hundred times in his book *Totem and Taboo*. But he was not an anthropologist. He was a doctor, an anatomist turned neurologist and psychotherapist. Above all, he was a scientist, determined to bring the principles of scepticism, rigour, hypothesis, evidence and verification to the study of the mind, or so he claimed. He was also rationalist, materialist, positivist and determinist. Science was salvific. 'Our best hope for the future is that intellect – the scientific spirit, reason – may in process of time establish a dictatorship in the mental life of man,' he wrote in 1932.[20]

As science rose, religion must fall. Although Jewish by birth and upbringing, Freud appears never to have believed in God or immortality, and was indifferent to any emotional or existential pull religion might have exerted on him. 'I was always an unbeliever,' he wrote to a friend, although he did at least credit his cultural Jewishness – with its years on the margins of European intellectual life – for giving him the perspective and disposition to make his great breakthrough.[21] 'Why did none of the devout create psychoanalysis?' he asked his friend, the Swiss pastor Oskar Pfister, rhetorically, in 1918. 'Why did one have to wait for a completely godless Jew?'[22] Whatever inadvertent intellectual advantage it gave him, religion was nonetheless always something he looked *at* rather than *through*.

For a rationalist, he was very interested in irrational behaviour. From the late 1880s, he became convinced that psychologically disturbed patients could be cured by taking them back to moments in their childhood, often under hypnosis, in order to deal with unresolved trauma. These powerful, often sexual, subconscious forces, many derived from infancy, swayed human behaviour long after adulthood and supposed rationality was reached. And this was the key to religion. 'The ultimate basis of man's need for religion,' he confided his 'flash of inspiration' to Carl Jung in 1910, 'is *infantile helplessness*, which is so much greater in man than in animals.'[23]

Freud had been developing these thoughts on religion for nearly a decade by that stage. In 1907, he had put forward the idea that religion was an obsessional neurosis. In the same way as neurotics exhibited compulsive traits to deal with their phobias, so the religious performed rituals and ceremonies to calm conscience and ward off anxieties. In the year he wrote to Jung, Freud introduced a more explicitly sexual dimension to this picture. The personal God, he wrote in a study of Leonardo da Vinci, 'is psychologically nothing other than an exalted father'.[24] That father was both provider of security and repressor of instincts (particularly sexual ones), loving but authoritarian and intimidating. The result was a powerful mix of need and guilt that worked its way out as religious piety.

These ideas reached maturity in his 1913 book *Totem and Taboo*. The title nodded to Frazer's *Totem and Exogamy* published three years earlier, while the subtitle – 'Some Points of Agreement between the Mental Lives of Savages and Neurotics' – introduced Freud's own fresh contribution. The four essays in *Totem and Taboo* did not set out a systematic understanding of religion but they did provide the raw material from which one could be assembled. Religion's origins lie in the infant's ambivalent attitude to his – notoriously it is *his* – father: dependent but resentful. This is not only the origin of man, however, but of mankind. Reworking a Darwinian ideal of the 'primal horde', Freud posited how sons, ruled over and then driven out by the dominant father, who claimed exclusive sexual rights over females in the tribe, came together to kill the tyrannous father and consume him in a commemorative feast. In doing so, they acquired the father's strength and simultaneously achieved fraternal unity. 'The totem feast, which is perhaps mankind's first celebration, would be the repetition and commemoration of this memorable, criminal act with which so many things began, social organization, moral restrictions and religion.'[25]

This primal murder, together with its guilt and consequences, became embedded in the human psyche and was transmitted through generations as part of the collective human unconscious. For this, Freud borrowed the 'law', developed by Ernst Haeckel, that 'ontogeny recapitulates phylogeny'.

This argued that the development of the individual organism followed the development of the species as a whole. Thus, it was believed, a human foetus went through stages – fishlike to reptilian to avian to mammalian – that 'recapitulated' evolution. On the basis of this model, individual human Oedipal instincts were a recapitulation of the primeval and universal act of killing and consuming the father. Religion became the personal and communal attempt to atone for primeval guilt.

Freud's union of historical, biological and psychoanalytic ideas in his scientific description of religion in *Totem and Taboo* was strikingly original, and it remained central to his understanding even as he integrated more familiar theories. The origins of religion were tied up with the conventional belief in human stadial development, as societies progressed from animism through religion to science. In *The Future of an Illusion*, published in the year of the fifth Solvay Conference, he added explanations that could have come from Dirac's mouth. Religion was born of our 'wretched, ignorant and downtrodden ancestors', was dependent solely on 'instinctual wishes' and 'inner experience', and full of 'contradictions, revisions and falsifications'.[26] It was no more than an illusion, albeit an illusion born of – and perversely capable of satisfying – our deep psychic needs. Faced with the 'majestic, cruel and inexorable' forces of nature, vulnerable humans personified them, turning them into gods that could be appeased or bribed.[27] The result made life manageable. 'Over each one of us there watches a benevolent Providence which is only seemingly stern and which will not suffer us to become a plaything of the over-mighty and pitiless forces of nature.'[28]

This was not religion's only function. Freud reiterated how it was also, indeed still primarily, a response to inner or social threats, as well as external, natural ones. The religious attitude, complete with ritual and ethical code, was the way in which civilisations coped with the necessary repression of primal, violent and sexual instincts, all of which could be traced back to the trauma of infancy. As he wrote in *Civilisation and its Discontents* in 1930, 'the derivation of a need for religion from the child's feeling of helplessness and the longing it evokes for a father seems to me incontrovertible.'[29]

All this still left open the problem of monotheism, however. Granted, the religious urge came from human(ity's) infancy, and found some satisfaction in animism and magic and ritual and the gods. But whence

God? Freud tackled the question head on in *Moses and Monotheism*, completed on his deathbed. The book proposed a somewhat novel reading of the book of Exodus. According to Freud, Moses was not Hebrew but an Egyptian follower of the monotheistic pharaoh Akhenaten. Following Akhenaten's death, Moses led a band of these committed monotheists out of Egypt but they subsequently rose up and murdered him and abandoned his religion for that of the volcano god Yahweh. Guilt predictably followed, passing down into the collected Jewish subconscious and emerging in the 'wish-phantasy' of a returning messiah, who would lead his people to 'salvation and the promised sovereignty over the world'.[30]

Moses and Monotheism contorted the biblical story into the Procrustean bed of *Totem and Taboo*. The result is a wish-phantasy as wrong as it is absurd, as critics were not slow to point out. It failed to dim Freud's star, however, which was still burning brightly when he died in September 1939. Whatever the future would make of his ideas about the origins of religion or his claim to be uncompromisingly scientific – and in neither regard did he escape unscathed – it was impossible to ignore them or indeed not to be shaped by them. As the poet W. H. Auden wrote in his memorial poem, Freud was less an individual than 'a whole climate of opinion'.

'Superstition, like belief, must die': revaluing religion

With typical modesty, Freud considered his work to have effected a Copernican revolution in the way humans understood themselves, supplying a new and robustly scientific perspective on existence. In a strange way he was right, although only because, as we have seen, the Copernican revolution was not really that revolutionary.

The problem was, as his multiplying critics pointed out, he was so often wrong, drawing indefensible conclusions from partial reasoning based on flimsy evidence. Often wrong did not mean always wrong. Ideas of repression as a form of psychological self-protection and of religion as a kind of cultural regulator fared well. However, those ideas on the genesis of religion that were original to him, and representative of his approach, turned out to be unsubstantiated to the point of untenable.

This was obviously the case for *Moses and Monotheism*, but it was no different in his earlier work on the primal horde, its original sin of parricide, the consequent totem feast and the mysterious embedding of ensuing guilt in the collective psyche. Primatologists would show that primates lived and bonded in a variety of different ways, few of which corresponded to those necessary for Freud's ideas. Anthropologists would similarly deny any definitive archetypal primitive human 'horde'. There was simply no evidence for any foundational act of parricide. Freud's schema was criticised for being a relentlessly male interpretation, seemingly oblivious to the agency, or sometimes even the existence, of women.

Sexual envy and violence were – are – common enough among primates, including 'primitive' and 'civilised' humans, contrary to Margaret Mead's 1928 classic of anthropology, *Coming of Age in Samoa*, which purported to reveal a simple island of peaceful promiscuity. However, Freud's attempt to build the entire edifice of religion on such sexual aggression and guilt was as unrealistic as the picture Mead drew, which also turned out to be wildly inaccurate. Totemism – the beliefs and rituals centred around spiritually significant plants, animals and natural objects – emerged as nothing like as universal as Freud believed; the totem meal he posited even less so.

Most damagingly, there was just no mechanism for the passing on of Freud's key ideas. In the 1940s, Darwinian evolution finally found a life partner in Mendelian genetics, and natural selection became credible as the main engine of evolution. Lamarckianism, the idea that organisms passed on acquired characteristics to their offspring, which had been creeping back into biology for nearly fifty years, now fell out of favour. Freud was an unapologetic Lamarckian. His ideas about the collective psyche's origins of religion needed Lamarck. As a result, they slowly sank beneath the waves, weighed down with acquired characteristics.

Underlying all these errors that spread across Freud's theories like cracks over thin ice was the simple problem of bad science. Freud was as committed to bringing the study of the mind into the realm of scientific rigour as he was to anything and yet all too often, at least when it came to his ideas about religion, evidence was forced to fit the theory. His unwavering commitment to the inheritance of acquired characteristics; his indifference to the problem of primal guilt transmission; his positing of entities, like the id and the ego, and of complexes, like the Oedipal, for which there was at best tenuous evidence: all this revealed a mind so convinced of its own theories that it just *knew* that the evidence supported them, or it made the evidence support them, or it ignored it if it didn't. Theory usually precedes evidence in scientific practice, but in Freud's case with religion, it got so far ahead that it lost sight of the evidence altogether.

What may be said in his defence is that Freud simply followed the footsteps of his anthropological forerunners. Tylor and Frazer may have put the anthropology of religion on an academic footing, but their armchair analysis, and the colonial, positivist and progressive prejudices that came with it, seriously weakened the empirical strength of their work, permitting elaborate and/or patronising theories to run far ahead of the evidence. This was beginning to change by the second quarter of the twentieth century, and so it is no surprise that early criticism of Freud's religious ideas often come from anthropologists who *had* spent years among the people about whom Freud and Frazer pontificated. 'All that one can say of the theory [of a totem meal],' Edward Evan Evans-Pritchard wrote dryly in 1965, 'is that, whilst eating of the totem animal could have been the earliest form of sacrifice, and the origin of religion, there is no evidence that it was.'[31]

Evans-Pritchard knew whereof he spoke. His ground-breaking *Witchcraft, Oracles and Magic among the Azande* was published a matter of months after the final, supplementary volume of Frazer's *Golden Bough* but, in the historian Timothy Larsen's words, it demonstrated an 'athletic . . . leap' in terms of 'method, theory, practice and personal convictions.'[32] Evans-Pritchard had spent years among the Azande in the Upper Nile region, and then years more among the Nuer, on whom he published in 1940. Unlike his predecessors, for whom religious belief, like the people who held it, was primitive in the most loaded sense of that word, he not only engaged in detailed and immersive participant-observation fieldwork but sought to understand the subjects of his study on their own terms, rather than those of the self-evidently superior, scientific-minded West.

This did not entail wholesale relativism. Evans-Pritchard was clear that 'witches, as the Azande conceive them, cannot exist', that the Azandes' oracles 'can tell them nothing' and that the Azandes' magic was completely ineffective.[33] That did not, however, mean that the Azande were simply ignorant or irrational. Their spiritual explanations did not supplant natural ones so much as supplement them. They didn't deny natural causality but nor did they believe natural causality ruled out spiritual explanations. Moreover, shockingly, given how anthropologists had treated such beliefs beforehand, Evans-Pritchard argued that, within their own closed system, the Azandes' beliefs *made sense*. 'Their mystical notions are eminently coherent, being interrelated by a network of logical ties, and are so ordered that they never too crudely contradict sensory experience but, instead, experience seems to justify them.'[34] Such ideas could (and would) be used to justify the kind of relativism that would place Azande magic on the level of European science, but that was not Evans-Pritchard's intention. Rather, he was initiating the move away from the default position of understanding 'primitive' people and religious beliefs as simply infantile and delusional.

In a similar vein, when he came to write on Nuer religion in the 1950s, Evans-Pritchard claimed that it had 'features which bring to mind the Hebrews of the Old Testament'.[35] This, of course, was precisely the kind of comparison that Tylor, Frazer or Freud might have made, but unlike them Evans-Pritchard meant it as a compliment. Accordingly, he not only made comparisons, and analysed Nuer beliefs by means of Latin theological terms like *ex nihilo* or *deus absconditus*, but ended the book on an instructive note:

Nuer religion is ultimately an interior state. This state is externalized in rites which we can observe, but their meaning depends finally on an awareness of God and that men are dependent on him and must be resigned to his will. At this point the theologian takes over from the anthropologist.[36]

Evans-Pritchard's work on the Azande (and the Nuer) was seminal in reorienting the early anthropological presuppositions about religion's simplicity and childishness, but the light he shone on his own tribe was hardly less important. His eminence was beyond question. When he returned to Oxford after the war, he took up a Fellowship at All Souls along with the Chair in Social Anthropology, which was his for nearly a quarter of a century. Shortly before doing so, however, he converted to Roman Catholicism, and was received into the Church at Benghazi Cathedral in Libya. Never out of the Christian fold – his father had been an Anglican clergyman – Evans-Pritchard's conversion was nonetheless striking, as was his insistence that post-Enlightenment attempts to make Christianity more palatable to modern man were destined to failure. Willingness to abandon prophecies, miracles, dogma, theology, ritual, tradition, clericalism and the supernatural amounted to nothing more than a 'desperate effort to save this ship by jettisoning its entire cargo', he wrote. It was religion that had dismantled religion, not science. 'Why blame the geologists, the anthropologists, and the biblical scholars,' he asked rhetorically, 'when the dykes had been breached centuries before?'[37]

That did not mean that the anthropologists had nothing to answer for. These remarks were made in his 1959 Aquinas lecture, 'Religion and the Anthropologists', in which he showed quite how systematically anti-religious sociology and anthropology had long been. 'All the leading sociologists and anthropologists contemporaneous with, or since, Frazer were agnostics and positivists,' he said.

> If they discussed religion they treated it as superstition for which some scientific explanation was required and could be supplied. Almost all the leading anthropologists of my own generation would, I believe, hold that religious faith is total illusion, a curious phenomenon soon to become extinct and to be explained in such terms as 'compensation' and 'projection' or by some sociologistic interpretation on the lines of maintenance of social solidarity . . . Religion is superstition to be explained by anthropologists, not something an anthropologist, or indeed any rational person, could himself believe in.[38]

Religion was, almost by definition, an infantile delusion. It was anthropology's job to explain why it persisted.

Evans-Pritchard's critique of the anti-religious assumptions of religion-studying anthropologists was sharp. But his opinions were also fed by his own experience as a Catholic convert. He once remarked how having converted he 'had to face some personal difficulties'.[39] According to one more recent account, 'his conversion to Catholicism was regarded as a kind of defection from the mainly rationalist, agnostic or "humanist" principles of his pre-war friends at LSE.'[40]

In actual fact, he was not alone in his beliefs. Mary Douglas, one of the greatest post-war British anthropologists, was also a Catholic, though in her case from the cradle, and the institute over which Evans-Pritchard presided in post-war Oxford boasted a number of other believers, including the Catholics Godfrey and Peter Lienhardt, the Hindu M. N. Srinivas and the Jew Franz Steiner. Moreover, for all the hostility Evans-Pritchard's religion may have provoked, there was no doubt that the unreflectively anti-religious cloud under which the anthropology (and sociology) of religion had been born was beginning to thin. The ideas of progress – let alone of inevitable progress – that had been so foundational to the work of Frazer, Tylor and Comte had taken a bit of a knock in the first half of the twentieth century, as had the assumption of Western moral superiority, and the necessarily positive link between science and civilisation. An era of decolonialisation made the ethnocentric superiority of early anthropologists, on the back of which had ridden similarly superior attitudes to religion, a little harder to sustain.

And yet, for all this, the climate of the age could still be breathless with scientific excitement as humanity reached for the moon. The default view, at least among intellectuals, was that religion was, at heart, little more than a kind of primitive superstition which was destined for history, leaving only ruins in its place. As Philip Larkin wrote of the decaying churches he visited on quiet Sundays:

> superstition, like belief, must die,
> And what remains when disbelief has gone?
> Grass, weedy pavement, brambles, buttress, sky,

One of many anti-religious posters from the Soviet era that juxtaposed science (here represented by the fresh-faced young girl yearning for technology and education) and religion (the ugly, gnarled grandmother dragging her to church). It declares 'Religion is poison.'

SEVENTEEN

STORMING THE HEAVENS

'I'll give you my halo, you give me your helmet': religious science

However much they may have disagreed about the details or even the desirability, physicists like Dirac, anthropologists like Frazer, psychoanalysts like Freud and poets like Larkin were in accord: religion's days were numbered. The only real question was *how* it would go.

The default answer was 'naturally'. Freud spoke for many when he wrote, in *The Future of an Illusion*, that 'a turning away from religion is bound to occur with the fatal inevitability of a process of growth.'[1] More science just meant less religion. It was a simple zero-sum game.

There were moments, however, when things appeared more complex. Frazer had hinted that religion might disappear only if it were purposefully rooted out, though he deplored revolutionary attempts to do so.[2] Freud had used a particularly telling word when he stated that 'our best hope . . . is that . . . the scientific spirit, reason – may in process of time establish a *dictatorship* in the mental life of man.'[3] He wrote this in 1932, when the word dictatorship had rather lost the right to be interpreted loosely or idiomatically. Indeed, this was already fifteen years into a dictatorship that would generate the most vicious confrontation that science and religion had engaged in, or ever would: an explicit, sustained, heavily theorised confrontation between the two that dwarfed anything that had happened in Rome, Oxford or Dayton.

Scientific knowledge 'slowly but surely undermines the authority of all religions', wrote the revolutionary Marxist philosopher Nikolai Bukharin in 1920.[4] The conviction underpinned much of what led to and emerged from the Bolshevik Revolution three years earlier. Communism is 'founded on the scientific worldview' and has 'no room for gods, angels, devils or any other fabrications of human fantasy', proclaimed Yemelyan Yaroslavsky, the historian and author of the book *How Gods and Goddesses Are Born, Live, and Die*.[5]

Science had disproved religious claims, showing belief to be no more than primitive superstition, or the manifestation of human need and alienation, or perhaps the projection of human pride on to a spiritual canopy. Whichever it was, science had shown it false and revealed a world empty of gods and spirits. There was more to this confrontation than just rehashed Comte or Frazer or Freud, however. Soviet communism was based on a scientific theory of history, as well as of religion. History was the process of material development in which traditional factors like great men, big ideas or divine providence were rendered irrelevant by collective endeavour and the material conditions of existence, in particular the production, exchange and consumption of goods. Understanding the dynamics of these relationships allowed humanity to discern the course of history, heretofore visible only to God himself.

Science also enabled humanity to redirect the world, orienting history towards 'scientific socialism'. A scientific approach to the past enabled a scientific approach to the future, facilitating 'scientific government' that would render the authority structures of the past redundant. Science now governed the future as God had once done through his providential sovereignty.

Underpinning this scientific approach to religion, history and government there was a scientific understanding of mankind. Religious talk of the soul was metaphysical nonsense, eternal life was a false prize rendered appealing only by injustice on earth and holiness was no more than an ecclesiastical confidence trick. Viewed through the clarifying and objective lens of science, undistracted by the yearnings foisted on desperate humans by economic injustice or unscrupulous priests, humans were just material beings, destined for comfort in life (once liberated by communism) and then oblivion in death. When the Church of St Seraphim was razed in 1927, it was replaced by the Donskoi Crematorium, which had no space – literally – for consideration of the soul or an afterlife.[6] It was the scientific concept of humanity in bricks and mortar.

The logic and coherence of all this appeared faultless. It was only religion's historic dominance that prevented it from withering away now that the future had arrived. The spiritual vine was long dead, grubbed up by the hard truths of science, and it only remained upright because it was tethered to the trellis of political authority. Once separated, religion would fall and die.

So it was that the Bolsheviks, once in power, set about creating a secular state. Shortly after the Revolution, they moved to separate the (Orthodox) Church from the state, passing the innocuous-sounding 'Decree on the Freedom of Conscience and on Church and Religious Associations', which appeared to guarantee freedom of both religion and irreligion. A subsequent decree nationalised all monastic and Church land. Another on civil marriage removed control over registration of weddings from the Church. Church-run schools were transferred to the Commissariat of Enlightenment. Religious instruction of the young was forbidden outside the home and then, three years later, inside it. Clergy were reclassified as 'servants of the bourgeoisie' and had all financial support withdrawn.

The effect was devastating on the Church but less so on religion, which failed to disappear as anticipated. As Lenin remarked, 'we have separated the church from the state, but we have not yet separated the people from religion.'[7] Intellectuals became impatient, and at its Twelfth Congress the Party passed a resolution stating that although religion would collapse when the economic restructuring of society had finally been completed, in the meantime '[we need] intensified, systematic propaganda, graphically and convincingly revealing to every worker and peasant the lie and contradiction to his interests of any religion'.[8]

The chosen vehicle for anti-religious work was the Soviet League of Militant Godless, which organised articles, posters, demonstrations, lectures, discussion circles and plays, all with the intent of separating the people from religion. The League adopted various lines of attack – undermining theology, mocking clerics, caricaturing the Church – but one of the most popular was the celebration of science as the solvent of faith. Slogans were blunt and punchy. 'Science instead of religion.' 'Destroying religion, we say: study science.' 'Reason against religion.' Electricity was a favourite. One poster compared how God electrified the village (with lightning strikes of terrifying vengeance) with how the village electrified itself (with a picture of orderly pylons powering tidy homes). Another depicted a tubby priest futilely wielding a saw-toothed cross against the base of a pylon, pathetically trying to topple the threat to his authority.[9] Stories were told of peasants who, having once believed, fell on their knees and exclaimed 'electricity totally destroyed my faith in God.'[10]

Industrialisation was a similar motif. Posters showed elegant smoke stacks towering over crumbling church cupolas. Space was a third front. The Moscow Planetarium opened in 1929, the first in Russia. Symbolically placed next to the zoo to underline the materialist basis of heaven and earth, the institution was intended to inspire a fascination with the skies, while simultaneously capturing them for science.[11]

As with the effect of the early secularising decrees, the impact was not as decisive as the authorities had hoped. The League of the Militant Godless continued its work through the 1930s but became less important as Soviet atheism turned more aggressive. School curricula became officially anti- rather than non-religious. Churches were closed by the thousand and clergy were forbidden to reside in cities, and then singled out as counter-revolutionary threats during collectivisation and dekulakisation.[12] Although Russia boasted around 54,000 churches, chapels and monasteries on the eve of the Revolution, Orthodox churches and communities could be numbered in their hundreds by 1940.

The war effort precipitated a thaw in Soviet anti-religious campaigns and for a decade or so following 1943 it seemed that religion might establish a vastly reduced and heavily policed, yet ultimately stable, niche in this most self-consciously scientific of societies. It was not to be. Nikita Khrushchev replaced Stalin in 1953. His early noises on religion were revisionist, and he told the French newspaper *Le Figaro* that 'the question of who believes in God and who does not is not [one] that should give rise to conflicts – it is the personal affair of each individual.'[13]

The underlying message had not changed, however. The thaw had seen an increase in clergy, if only because many were returned from prison, as well as a rise in religious publishing. But it was too risky. In 1954, two Central Committee departments published a report entitled 'On Major Insufficiencies in Natural-Scientific, Antireligious Propaganda', and a few months later the Party launched another, albeit shorter lived, anti-religious campaign.

Over thirty-five years after political secularisation, and around twenty since the enforced closure of the country's churches, it was now harder to take the political line in propaganda. The Church had long since lost any formal authority. Targeting it for its abuse of power rang hollow. Accordingly, the new campaign was more ideological than political, and science took a big role in 1954 and then an even bigger one in the more sustained anti-religious campaign that restarted four years later.

This was all in line with Khrushchev's own views. 'I have long ago freed myself from [the idea of God],' he told a French journalist in 1958. 'I am an advocate of the scientific worldview. Science and belief in supernatural forces are incompatible and mutually exclusive views.'[14] More importantly, it was in line with recent national achievement and pride. The USSR had become the world's second nuclear power in 1949. It detonated a hydrogen bomb only four years later, considerably earlier than the West had anticipated (albeit with the help of a little espionage). Its military hardware matched that of the West, as did its pace of industrialisation, while the Soviet economy grew faster than anywhere else in the world in the 1950s, bar Japan. Scientific socialism worked.

The impressive progress of Soviet technology, industrialisation and state planning in the forties and fifties lent a powerful legitimacy to the scientific premises of the new anti-religious campaigns. But it was space that offered the finest frontier for this war. In October 1957, the Soviet Union successfully launched the world's first artificial satellite, *Sputnik I*. The following month, *Sputnik II* carried a living animal, Laika the dog, into orbit, albeit only briefly. Two years later, *Luna I* became the first spacecraft to approach the moon, and two years after that, *Vostok I* took the cosmonaut Yuri Gagarin into an orbit from which, unlike Laika, he returned safely.

These were the miracles of the age, wrought by man not by God. Scientific materialism had enabled humans to transcend their earthly realm and penetrate the heavens. The potential for anti-religious symbolism was endless. Space was open for propaganda. Accordingly, the intensification of anti-religious campaigning from 1958 played heavily on scientific, and particularly cosmic, themes. Znanie, another association of intellectuals tasked with enlightening the masses, had been founded in 1947, and had taken up the work of the defunct League. Somewhat subdued in its first decade, it launched a new atheist journal, entitled *Science and Religion*, in 1959, which ran innumerable articles on how Soviet man was now successfully 'storming the heavens'.[15] Five years later, the academic Institute of Scientific Atheism was founded in Moscow. The Moscow Planetarium had by then become the headquarters for the co-ordination of scientific anti-religious work, and by the early 1950s there were about a dozen other planetariums across the country. Under Khrushchev they spread and by the early 1970s the country boasted over seventy. They put on lectures and 'debates', organised youth astronomy clubs and the like. Many were popular and successful.

Posters multiplied. In the most famous, a smiling cosmonaut floats in the skies above the churches, surveying the prospect and cheerily proclaiming 'There is no God'. In another, a smiling child shows his picture of an ascending rocket to his incredulous, aged, angry, cross-wearing grandfather. In a third, entitled 'Seventh Heaven', God is knocked off his cloud by a speeding rocket with CCCP scrawled on its side. A fourth shows an aged deity floating in the heavens next to a cosmonaut, gripping his space suit and pleading 'Let's swap: I'll give you my halo, you give me your helmet.'

The superstars in all this were the cosmonauts themselves, who dutifully informed their expectant audiences how their achievements had finally dispensed with God. Khrushchev had declared that the cosmonauts had been dispatched to space to investigate religious claims of a heavenly paradise and had found nothing. 'The fact that a human being had flown to the cosmos was a bitter blow to the church,' Gagarin wrote in his *Autobiography*. 'I was delighted to read accounts of how believers, under the influence of scientific achievements, had turned away from God, assented to the fact that there was no God and that everything associated with his name was rubbish and nonsense.'[16] State publications concurred. An editorial in *Izvestia* crowed about how Gagarin had given 'a terrible headache' to believers on account of not running into anyone during his orbit.[17] The letters page echoed the verdict. 'Yuri Gagarin overcame all belief in heavenly powers that I had in my soul,' read one. 'Now I am convinced that God is Science, is Man.'[18]

Gagarin's successor in space, Gherman Titov, was even more outspoken. He informed the Seattle World's Fair in 1962 that he had not seen any gods or angels on his many orbits of the earth.[19] His and Gagarin's rockets had been built by the people, not by God, he told them. Titov believed not in God but in 'man, his strength, his possibilities and his reason'.[20] His remarks in Seattle turned him, briefly, into the face of Soviet scientific atheism, and on his return he was given an editorial in *Science and Religion*, in which he tackled the question 'Did I meet God?' (his answer was no): he observed, rather literally, that 'the prayers of believers will never reach God, if only because there is not air in that place where he is supposed to exist.'[21]

The harmony between science and scientific atheism was not quite as perfect as the antics of the cosmonauts made out. Over the previous thirty years, the Soviet authorities had thrown their considerable weight behind the biology of Trofim Lysenko, Stalin's director of biology, which rejected Darwinian evolution and Mendelian genetics on the grounds that they contradicted Marxist–Leninist ideology. Natural selection had, after all, legitimised the liberal capitalism of imperial Britain, against which scientific socialism had long struggled. It was, by definition, unsound. Lysenkoism became the only acceptable theory and its opponents were denounced and sometimes killed. The result was to retard Soviet biology for a generation and severely reduce crop yields for an already hungry population.

There were two ironies here. The first was that it was the same materialistic view of humanity that supposedly underpinned the Soviets' triumphs in the space race that simultaneously caused their failures on earth. Materialism could cut both ways, it seemed. The second was that the constrictive effect that scientific atheism had on the life sciences was rather like that which Christian theism, at least according to its critics, had exercised on Darwinism nearly a century earlier. To an objective observer, it might almost look as if it were ideology in general, rather than religious ideology specifically, that had a deleterious effect on science.

These two ironies concerning Soviet biology, however, paled before a third, which attended the entire Soviet scientific war on religion. The temptation for atheism to become a religion had been present from the earliest revolutionary days. Socialist thinkers like Anatoly Lunacharsky and Maxim Gorky had sought to invest Marxist materialism with a religious aura, creating secular saints: 'examples worthy of being followed by the generations of the future'.[22] Their so-called 'God building' project was not popular with the Party's leaders. 'The Catholic priest who seduces young girls,' Lenin wrote, 'is far less dangerous . . . than . . . a democratic priest . . . who preaches the building and creating of god.'[23] The idea was dropped.

Yet the activities of anti-religious campaigns repeatedly gravitated to religious ritual and dogma, if only because 'a sacred space never remains empty'.[24] People were encouraged to replace religious holidays with secular ones. Church rituals were turned into atheist ones. There were 'red weddings' and 'civilian funerals'. Icon corners in houses were replaced by godless or Lenin corners. Christening became 'Octobering'. None of this was ever very successful but when Soviet man stormed the heavens in the late 1950s, a new hope emerged. The promise of a new humanity had been wrapped up with ideas of revolution for decades but the achievements of science, particularly in space, now appeared to deliver the goods. Cosmonauts were the prototype of this new man, liberated from the past, from ignorance and superstition, from humanity's historical earthbound existence. The space age heralded a new stage of history, a true New Age. Humanity finally transcended the earth. Man was revealed as the true creator. The heavens were conquered. Science offered salvation.

And as with all salvific messages, there was a powerful missionary zeal about what now was to be done. 'It is man who, without the help of God, builds a new and joyous life,' wrote the former priest Nikolai Rusanov. 'The believer should not wait for heavenly paradise, because it does not and will not exist.' The true kingdom of heaven was at hand, now, through the power of science. Earthly paradise was not a dream to wait for but something 'which will be built in the next fifteen to twenty years, here, in our godless Soviet country'.[25]

'People think of these men as not just superior men but different creatures': scientific religion

And this was precisely what others feared. The uneasy wartime truce between America and the Soviets did not last long and within a year

American policy, indeed American identity, was being formed in opposition to Soviet godless techno-utopianism. As Dwight Eisenhower said during his election campaign in 1952, the Cold War was at heart a moral and spiritual struggle. 'What is our battle against communism if it is not a fight between anti-God and a belief in the Almighty?'[26]

American scientific superiority was not in doubt, at least at first, but many were worried that the nation lacked the spiritual vigour to resist atheistic communism. The nation's religious vulnerability was often juxtaposed with its scientific strength. President Truman warned the Federal Council of Churches that if the civilised world were to survive, Americans had to acquire 'spiritual strength' that was 'of greater magnitude' than the 'gigantic power [of] atomic energy'.[27] A few years later, Omar Bradley, US Army chief of staff, sounded the same note when he warned that his countrymen had grasped the mystery of the atom but rejected the Sermon on the Mount.[28] What shall it profit a nation to gain the scientific world, if it lost its soul in the process?

The fears precipitated a programme of state-sponsored spiritual stimulation. America had enjoyed numerous religious revivals throughout its history but they had tended to be bottom-up affairs: organic, democratic, innovative, messy. In the fifteen or so years after the end of World War Two, the state took an active interest in the nation's spiritual health, or at least its spiritual identity. In 1952, the government established a National Day of Prayer and, two years later, it added 'under God' to the Pledge of Allegiance. 'In God We Trust' was put on postage stamps and the following year on currency, and by 1956 it had formally replaced *E pluribus unum* as the national motto. Over the same period, numerous national faith drives – the Religion in American Life campaign, the Committee to Proclaim Liberty, the Foundation for Religious Action – sought to reawaken and fortify the nation's religious commitment.

The campaigns worked, after a fashion. During the 1950s, a record number of Americans attended religious services and church membership rose from 88 million in 1951 to 116 million a decade later. But it failed to calm post-war anxiety, as the USSR kept pace in the arms race and surged ahead in space.

It was in this context – of spiritual renewal mixed with political anxiety – that America launched its own space programme. For all that the rhetoric of transcendence swirled around its mission, NASA was not a religious enterprise. Astronauts were permitted to carry small religious items into space as part of their personal kit but they were not selected for their religiosity; almost the opposite in fact. Norman Mailer described the typical astronaut as 'powerful, expert, philosophically naïve, jargon-ridden and resolutely divorced from any language with grandeur to match the proportions of his endeavour'.[29] They were test pilots and engineers, not visionaries. The lengthy process of selecting astronauts was specifically designed to weed out any who might allow the experience of space travel to obscure their technical judgement and cool. Even Freud got a look-in, potential space men being examined for any signs of the Freudian hypothesis 'that a love of flight had its origins in feelings of sexual inadequacy, reflecting a narcissistic drive or a wish for self-destruction'.[30] It is not clear how many prospective astronauts fell at this particular hurdle.

None of this meant that NASA was irreligious, however, and a good number of prominent figures were publicly pious. In a strange way, these exceptions became the perceived rule. Some senior figures – such as Hugh Dryden (NASA's deputy administrator from 1958 to 1965), Major General John Medaris (who headed the army's missile development programme in the late 1950s before studying theology and being ordained) and Eugene Kranz (NASA's second chief flight director) – were very devout. Most famously, at least among those who stayed on terra firma, was Wernher von Braun, who ran the Marshall Space Flight Center in the 1960s and oversaw the development of the Saturn V rocket, and who was openly pious. He had had a life-changing religious conversion not long after he arrived in the US (though cynics claimed that this might have had something to do with his need to wipe clean the memory of his willingness to work on rocket technology for the Third Reich).*

There could have been a similar religious enthusiasm at the political level. Aviation had long had associations with 'touching the face of God' and now Americans were heading into space, theo-political rhetoric had the excuse to soar with them. When Gordon Cooper, one of the early

---

* I find it hard to think of Wernher von Braun without the lyrics of the brilliant pianist and satirist Tom Lehrer coming to mind: ' "Once the rockets are up,/ Who cares where they come down?/ That's not my department," says Wernher von Braun.'

Mercury astronauts, revealed that he had recorded a prayer during his orbital flight on the last Mercury space flight, the heads of the Astronaut Office at NASA suggested that he use it to close the speech before a joint session of Congress.[31] A few years later, Richard Nixon, who occupied the White House when Apollo 11 landed on the moon, enthusiastically greeted the astronauts on their return by saying that their journey had marked 'the greatest week in the history of the world since the Creation'. Not everyone agreed. Billy Graham gently suggested to the president that the week in which the Son of God had given his life for humanity and been resurrected from the dead might, at least from a Christian point of view, have greater claim to the title.[32]

It was the astronauts, however, in whom the public was really interested and whose spiritual beliefs, remarks and actions were most closely scrutinised and celebrated. Some spoke of them as if they were the kind of new, transcendent man that the Russian cosmonauts were supposed to epitomise. 'You get a feeling,' remarked one news correspondent just before Apollo 11 left for the moon, 'that people think of these men as not just superior men but different creatures.' They are, he continued, like people 'who have gone into the other world and have returned, and you sense they bear secrets that we will never entirely know, that they will never entirely be able to explain'.[33]

However hyperbolic that feeling might have been – American authorities rarely spoke of astronauts with the same kind of eschatological fervour that Russians did of their cosmonauts – the religious words and gestures of these 'creatures' certainly caught the imagination of the public. John Glenn, another of the early Mercury astronauts and the first to orbit the earth, spoke explicitly about the spiritual impact of his journey. An elder in the Presbyterian Church, he confirmed, many years later, that space flight deepened and strengthened his faith. 'To look out at this kind of creation and not believe in God is to me impossible.'[34] Somewhat more publicly, during their orbit of the moon, on 24 December 1968, the crew of Apollo 8 read the opening ten chapters of the book of Genesis, in the King James translation, to an audience of around a billion people. The act was approved by NASA's Office of Manned Space Flight and remains strangely moving more than fifty years on. The following year, 'Buzz' Aldrin, another Presbyterian elder, privately celebrated communion during the first lunar landing (although he kept this secret at the time) and then read from Psalm 8 – 'When I consider the heavens, the work of Thy fingers, the moon and the stars which Thou hast ordained, what is man that Thou art mindful of him?' – on the voyage home. Jim Irwin, who was part of the Apollo 15 mission in 1971, had lost his faith as a child but had an epiphany on the moon's surface – 'I felt the power of God as I'd never felt it before' – and returned to found the High Flight Foundation, an interdenominational evangelical organisation. He subsequently led two expeditions to find evidence for Noah's Ark on Mount Ararat in Turkey, although both were unsuccessful.[35]

When such 'superior men' sensed God in space, who could disagree? The novelist and Christian apologist C. S. Lewis for one. Lewis died in the early years of the space race but remarked in a late essay that whether or not an astronaut found God in space was primarily a function of the mind he brought with him. 'To some, God is discoverable everywhere; to others, nowhere. Those who do not find Him on earth are unlikely to find Him in space . . . much depends on the seeing eye.'[36]

Lewis was surely right, but space travel still had the potential to challenge and change people's existing views, whether faithless or faithful. A number of pastors remarked on how their flock often still clung to literal cosmologies, which were shaken by humanity's incursion into the heavens. Men on the moon could provoke as much discomfort as jubilation. Where exactly did God live now?

Others, less wedded to such biblical literalism, were still concerned about what the heavens said about the earth. It was now clear that the universe boasted other galaxies that would have other stars that would have other planets, at least some of which might bear life. Such a possibility had, of course, long been postulated in theological thought experiments going back to the fourteenth century. William Derham had written in his *Astro-Theology*, in 1721, that 'Myriads of Systems are more for the Glory of God, and more demonstrate his Attributes than [just] one.'[37] But that did not necessarily make it any easier to come to terms with the prospect, and awkward questions were left hanging in the air. What now of human uniqueness? What now of God's special care for the earth? What now for the incarnation and God's mission of salvation? Theologians wrestled and worshippers wriggled.

Astro-theology was not their biggest worry, however. All the while that NASA was opening up the heavens for Americans, the earth was being reshaped around them. The post-war faith crusades petered out. The number of churchgoers began to plateau then fall. John F. Kennedy's bid for the White House inspired the same fear as had the Catholic Al Smith's in 1928. Kennedy's insistence that he would favour home over Rome was intended to calm Protestant fears but it could just as easily be read as indifference to Christian ethics, an accusation that his personal lifestyle did little to help. The Supreme Court prohibited school prayers at one end of the decade, and at the other end the organisation American Atheists filed a lawsuit claiming that the Apollo 8 crew's reading of Genesis 1 had violated the First Amendment. The case was a failure and their appeal to the Supreme Court was dismissed, but it could nonetheless be read as a sign of the times. It was one of the reasons why Buzz Aldrin kept his lunar communion service private.

Whatever the scientific successes, spiritual epiphanies and theological questions provoked by the NASA programme, it was secularising pressures like these that really stirred religious anxiety. In 1964, the discovery of cosmic microwave background radiation finally tilted the scientific community away from a steady state universe towards an expanding one. As noted two chapters ago, this could (although need not) be taken as affirmation of a creator but it also gestured in the direction of a long, slow and rather uneventful demise for the universe. Where was God's final act of consummation to be found? The same earthly fragility that pulled some into soft eco-spirituality pushed others to hard eschatology. Jim Lovell had observed, during the Apollo 8 mission, that no trace of man's existence on earth could be discerned from outer space. What indeed was man that God was mindful of him, as Buzz Aldrin has asked? Not much, at least without God, the images seemed to say. Human civilisation had reached for the stars, just as it was crumbling on earth. God was preparing to act. End-times literature – most famously *The Late Great Planet Earth*, ghost-written by Carole Carlson for Hal Lindsey – became wildly popular.

It was no coincidence that all this was happening in the wake of 1959, when America, like the rest of the Western world, had celebrated the centenary of the publication of Darwin's *Origin of Species*. The eugenics movement that had walked in lockstep with Darwinism in the 1920s had stumbled out of World War Two disgraced and silenced, although not dead. Evolution was now less tarnished by association and easier to accommodate within the education system, which is precisely what the Biological Sciences Curriculum Study, founded in 1958, recommended. The creation–evolution ceasefire was over. In 1961, John Whitcomb and Henry Morris published *The Genesis Flood*, which offered a 500-page defence of a universal flood within a Young Earth Creationist time frame. The book was to prove immensely popular and influential and helped resurrect the fundamentalist anti-evolutionary cause which had been dormant for a generation.

The book stood four-square on biblical literalism. Whitcomb was a professor of Old Testament studies and the first chapter begins with his and Morris's declaration that 'our conviction [is that] the Bible is the infallible Word of God'.[38] However, its interest and significance lies not in its biblical but its scientific framing. Morris had a PhD in hydraulic engineering and was professor of civil engineering at Virginia Polytechnic Institute. The book argued that the Fall had precipitated the entropy inherent in the second law of thermodynamics and that a worldwide flood had laid down the various layers that geologists had been unearthing for two centuries. It used the language of 'sedimentary strata', 'Eternal Oscillation' cosmology, rubidium and potassium methods and 'geochronology'. An entire appendix was dedicated to 'palaeontology and the Edenic curse'. Despite the fact that the authors insisted that 'evidences of full divine inspiration of Scripture are far weightier than the evidences for any fact of science', it was clear that the argument of the book rested on its (alleged) scientific credentials.[39]

It was typical of the movement that was to become Creation Science. The Creation Research Society was founded in 1963, the Creation-Science Research Centre in 1970, the Institute for Creation Research in 1972. Future attempts to place creationism and its offspring, Intelligent Design, on the curriculum would hinge on scientific, not religious, reasoning. For all that biblical literalism powered the anti-evolution movement, just as it had during the time of the Scopes trial half a century earlier, it was science that now legitimised it. Social authority had shifted. In much the same way as Soviet science had appropriated religious language and ideas in order to legitimise itself in the post-war period, so American fundamentalism dressed up in the language of science to do the same. Paradoxically, for each to secure its authority in the tense, strange decades after World War Two, science and religion effectively swapped clothes.

The American palaeontologist and evolutionary biologist, Stephen Jay Gould, who took the term 'magisteria' from Catholic thought and applied it to science and religion. His attempt to bring peace to what, at the time, was a fractious relationship by separating the two into a magisterium of facts and one of values was thoughtful, well meaning but ultimately wrong.

EIGHTEEN

IRREDUCIBLY COMPLEX

'Scratch an altruist': ending the mystery of human nature

American creationists were not the only ones to worry about Darwinism. Others, who knew rather more about the theory and harboured no doubts as to its truth, had their concerns too. One such was George Price.

Born in 1922, Price was a physical chemist turned theoretical biologist. He worked on the Manhattan Project and at Harvard University and Bell Laboratories before being successfully treated for cancer, receiving a substantial insurance payout, abandoning his wife and children and moving to London. He was an atheist and a materialist, and abrasive and dogmatic with it.

In London, Price met the brilliant young British biologist W. D. Hamilton, who was developing a gene-centred approach to evolution that helped explain examples of apparent altruism in nature. Why would any animal give its life for another, particularly those to which it was not directly related? Many did, such as the eusocial insects that formed hyper-cooperative colonies, where labour was divided and innumerable creatures laid down their lives so that others could reproduce. What was the possible evolutionary benefit of this?

Hamilton's idea was that the problem dissolved if evolution was centred on preserving the gene rather than the organism. Altruistic self-sacrifice made sense if it improved the chances of genes being passed on, even if that was not via direct offspring. Although Price was only an amateur biologist, he was captured by the idea and did the maths to prove Hamilton right. The equation appeared to show that altruism was nothing of the kind and that what humans thought was self-giving love was in fact an unconscious, self-interested calculation, a trick played on us by our genes. As journalist Andrew Brown put it, tongue-in-cheek: 'through algebra, George Price had found proof of original sin.'[1]

His success drove Price to despair. In June 1970, four years after coming to London, he experienced a powerful religious conversion, becoming convinced not only of God's existence and love but of the necessity of Christ's teaching. Price was inspired by the story of the Good Samaritan. He quarrelled with his priest for being insufficiently zealous. Jesus' example of complete self-sacrifice became his light. He proceeded to give away his money and possessions, at one point writing to his friend, another eminent biologist, John Maynard Smith, and telling him that he had 15p left in the world. He invited homeless men to live in his flat. They behaved so badly, he ended up sleeping in his office. When that became untenable, he moved to a squat near Euston station. Increasingly exhausted and unstable, he cut his carotid artery with a pair of nail scissors in January 1975. His funeral was attended by Hamilton, Smith and the four homeless men who lived in his flat. 'The trouble with George,' the priest remarked at the service, 'was that he took his Christianity too seriously.' Hamilton, no believer, demurred. 'I think George felt that if it was good enough for St Paul, it was good enough for him.'[2]

The death of George Price was a world away from the anti-evolutionary movement that was then spreading through the US population like a successful genetic mutation. Price knew what he was talking about. But the fears were curiously similar. Fundamentalist Christians rejected Darwinism. On the surface, that was because, they said, it contradicted Genesis. No doubt that is what they believed. But beneath that particular fear, as in the 1920s (and indeed today), believers within the emerging religious right worried that evolution threatened to do away with human uniqueness, dignity, morality, even rationality; indeed, that it erased the category of 'human' altogether. In his own informed but unbalanced way, Price feared the same.

He had reason. A few months after Price died, the American biologist E. O. Wilson published a book entitled *Sociobiology: The New Synthesis*. It was to become a landmark text, bringing together two key ideas: that evolution could explain social behaviour just as successfully as it could, say, physical form; and that evolutionary explanations centred on the survival of the gene rather than the organism. Altruism, co-operation, parental care, communication, aggression: all could be explained by reference to genes. The book's final chapter, entitled 'Man: From Sociobiology to Sociology', made it clear that our species was not exempt from this reasoning. In the guise of sociobiology, evolution by natural selection laid claim to pretty much everything.

Wilson's book was special for its size and ambition rather than its argument. It was, after all, a self-declared 'synthesis'. In a book published the previous year, on *The Economy of Nature and the Evolution of Sex*, the American biologist Michael Ghiselin had taken the same evolutionary scalpel to society and morality. 'No hint of genuine charity ameliorates our vision of society, once sentimentalism has been laid side,' he wrote. 'What passes for co-operation turns out to be a mixture of opportunism and exploitation.' Where it is in its interest, an organism will co-operate or even serve another. However, 'given a full chance to act in his own interest, nothing but expediency will restrain him from brutalizing, from maiming, from murdering – his brother, his mate, his parent, or his child.' This was a serious argument, rather than the plotline of a *Saw* film, and it concluded with perhaps the most famous aphorism from this new scientific turn: 'Scratch an altruist and watch a hypocrite bleed.'[3]

Ghiselin's writing was vivid and colourful, which was part of the problem. The new sociobiological synthesis attracted some brilliant writers, whose purple prose popularised population genetics for an audience that might otherwise not be gripped by the subject. But it came at a cost. Supreme among the new writers was the British zoologist Richard Dawkins, whose book *The Selfish Gene*, published the year after Wilson's *Sociobiology*, became a classic in the field. Dawkins' language could be as lurid as Ghiselin's. His book, he explained early on, argued that humans, like other animals, are 'machines created by our genes. Like successful Chicago gangsters, our genes have survived, in some cases for millions of years, in a highly competitive world.'[4]

The image of geriatric Chicago gangsters was an odd one. The machine metaphor was more serious. It pointed to a certain conception of the human, underlined by Dawkins' description of genes surviving 'in huge colonies, safe inside lumbering robots', i.e. us. When criticised for that use of the word 'robot', he replied that 'part of the problem lies with the popular, but erroneous, associations of the word "robot" ', before going on to tell people what they should understand from the word.[5]

It was a similar case for the book's controlling metaphor. There were moments when the book seemed to suggest that genes made people self-ish: 'gene selfishness will usually give rise to selfishness in individual behaviour . . . let us try to teach generosity and altruism, because we are born selfish.'[6] There were others when the author simply countersigned Ghiselin's vision of reality: 'I think "nature red in tooth and claw" sums up our modern understanding of natural selection admirably.'[7]

And yet sometimes the vehicle was thrown into reverse. Dawkins never actually meant *people* were selfish; it was their genes. Moreover, genetic selfishness often worked by making people co-operative and even altruistic. Selfish and ruthless were only, it seems, 'figures of speech.'[8] As he wrote at the end of one chapter of the same name, even with selfish genes at the helm 'nice guys can finish first.'[9] Dawkins admitted, many years later, that the book could just as easily have been called *The Co-operative Gene* or *The Altruistic Vehicle*, but at the time he was more robust (or perhaps more slippery) in his defence of the metaphor. When the philosopher Mary Midgley took him to task for his use of the word, he replied, 'In effect I am saying: "Provided I define selfishness in a particular way an oak tree, or a gene, may legitimately be described as selfish".'[10] The defence recalled Humpty Dumpty in Lewis Carroll's *Through the Looking-Glass*. ' "When I use a word," Humpty Dumpty said in rather a scornful tone, "it means just what I choose it to mean – neither more nor less". Either way, if selfish could actually mean co-operative or altruistic, it made pinning down what was actually at stake in this debate about human nature rather difficult.

However slippery the terms could be, and however selfish or co-oper-ative humans and their genes actually were, what was clear was that socio-biologists were making sweeping claims about life, humanity and moral-ity, a vast land grab that displaced the other disciplines; religion (obviously) among them. 'When you are actually challenged to think of pre-Darwin-ian answers to the questions "What is man?", "Is there a meaning to life?", "What are we for?", can you, as a matter of fact, think of any that are not now worthless except for their (considerable) historical interest?' Dawkins asked in a footnote. 'Before 1859, all answers to those questions were [simply wrong]. The sociobiological vision swept aside the detritus of intellectual history. Indeed, it was cosmic, almost religious, in its scope. 'The only kind of entity that has to exist in order for life to arise, anywhere in the universe,' *The Selfish Gene* ended, 'is the immortal replicator.'[11] Heretofore, talk of immortality or necessarily existing entities had been the provenance of theologians and metaphysicians. No longer.

Not everyone who had been walking in this terrible darkness was convinced they had seen a great light. Sociologists were not as enthusiastic about sociobiology as were biologists, and indeed not all biologists were enthusiastic. The American palaeontologist Stephen Jay Gould was one of sixteen scientists who wrote a letter to the *New York Review of Books* in 1975 that was published under the title of 'Against "Sociobiology" '. The authors protested against Wilson's 'biological determinism', his tendency to oversimplify complex phenomena like intelligence, his speculative reconstruction of human prehistory and his willingness to invoke genes favouring homosexuality, creativity, entrepreneurship, drive and mental stamina, despite there being no evidence for their existence.[12] They criticised him for stumbling into a moral quagmire of his own making – 'for Wilson, what exists is adaptive, what is adaptive is good, therefore what exists is good' – and for deploying the get-out-of-jail-free card idea of 'maladaptive' behaviour whenever it suited him. And if their antipathy wasn't quite clear enough, they also drew attention to evolution's repeated history of 'explaining' human (mis)behaviour, such as criminality and alcoholism, its consistent (genetic) justification of the status quo and of existing privileges for certain social groups and its historical links to 'the eugenics policies which led to the establishment of gas chambers in Nazi Germany'. *Sociobiology*, they concluded, should be taken seriously, not for its scientific credentials 'but because it appears to signal a new wave of biological determinist theories'.[13] The letter drew predictable return fire, and over the next few decades the debate over the moral, social and religious implications of Darwinism degenerated into acrimony. However much the different Darwinian factions squabbled with one another, though, they could at least agree on their common enemy.

'Breath-taking inanity': anti-evolutionism 2.0

The fundamentalist Christian rejection of evolution long predated the new sociobiological, gene-centred 'synthesis' of the 1970s, but that synthesis did prove extremely useful in rallying foot soldiers to the cause.

By the time *Sociobiology* was published, culture lines were hardening in the US. Evangelicals and Catholics had started to overcome their longstanding mutual distrust, and turned their collective fire on the secular humanists, liberals and atheists who were destroying their country. As half a century earlier, much of the blame could be pinned on the Darwinian conception of human nature that was allegedly rotting America from within. The charge was not straightforward. The calculating and competitive world of selfish genes was as recognisable to (many) Americans as it had been to British liberals a century earlier. An ethos of struggle and survival was hardly alien to a culture of rugged individualism, let alone to 'Reaganomics'. Here at least, the selfishness of genes found fertile ground.

What bothered many Christians was not so much the competitiveness of nature as the moral relativism implicit in evolution, and the undermining of human agency altogether. Dawkins, for example, relentlessly emphasised how genes were the masters and organisms mere emissaries. 'A body is really a machine blindly programmed by its selfish genes,' he wrote in *The Selfish Gene*.[14] 'The fundamental truth [is] that an organism is a tool of DNA rather than the other way around,' he said in *The Extended Phenotype*.[15] The 'individual organism [is] not fundamental to life [but] . . . a secondary, derived phenomenon', he explained in *Unweaving the Rainbow*.[16]

The sociobiologists denied that their theory implied any such moral deracination but they often tied themselves in knots when doing so. Dawkins claimed that for all that our genes 'instruct us to be selfish . . . we [humans] are not compelled to obey them all our lives'.[17] 'We, alone on earth, can rebel against the tyranny of the selfish replicators,' he exhorted at the end of one chapter.[18] This was an odd reversion to pre-Darwinian beliefs: humans were somehow qualitatively different from the rest of creation. Why humans were so completely unlike any other species in their ability to rebel, how they were immune to the iron rule of the gene and how exactly this meat-robot was supposed to mutiny against its circuitry was unclear. Indeed, precisely who the person was, other than his or her all-controlling genes, was far from obvious.

However (un)convincing this defence was, it was the message of moral relativism and reduced agency that hit home with American Christians. Children were being corrupted at school, streets were no longer safe to walk, people no longer took responsibility for their actions, the moral certainties of the past were ignored and God was firmly pensioned off by evolution. The ultimate fault lay with Darwin.

It could all have come from the 1920s except for the fact that this time, with the scientific turn of creationism still ringing in fundamentalists' ears, the tactics were more sophisticated. In 1981, the governor of Arkansas signed into law an Act that mandated that public schools in the state 'shall give balanced treatment to creation-science and to evolution-science'. The law was, predictably, challenged in a legal case, *McLean v. Arkansas Board of Education*, and the judgment found unequivocally that 'creation science is not science', and that the Arkansas law was really about 'the advancement of religion'.[19] A few years later, a similar case, *Edwards v. Aguillard*, ended up before the Supreme Court. The court heard from seventy-two Nobel laureates, seventeen State Academies of Science and seven other scientific organisations, which together submitted a written statement in support of the appellants, who were challenging the constitutionality of a Louisiana State law that mandated the equal treatment of creation science and evolution science in the classroom. The court found for the appellants again. The creationists were running out of road.

They changed direction. The phrase 'Intelligent Design' had had a long history – Darwin had used it in his letters and Tyndall in his Belfast address[20] – but it wasn't until the late 1980s, when US courts effectively roadblocked creation science, that 'ID' was repackaged as a coherent (pseudo)scientific theory and deployed against evolution. *Edwards v. Aguillard* had heard that 'creation scientists' liked to point to the 'high probability' that life was 'created by an intelligent mind'. The ID movement focused on the probability and the mind, and began to argue that there were some things in nature that simply could not have evolved naturally. For example, the flagellum, a whip-like structure by means of which bacteria are propelled, was supposedly 'irreducibly complex'. This meant that it comprised too many discrete, working parts to be amenable to evolution's trademark blind and gradual process. The obvious conclusion was that it was designed, although ID advocates were rather opaque about how and by whom.

The first question – how – had no answer precisely because ID left a place holder where science normally put a theory. Instead of positing a natural mechanism to explain a natural phenomenon, it concluded that no natural mechanism was up to the job. Mutation and selection were questioned, found wanting on inadequate grounds ('irreducible complexity' is effectively impossible to define) and then replaced with a vague supernatural answer.

But only a vague one, for the answer to the second question – by whom – had to be kept sufficiently distinct from the God of Christianity in order to avoid the fate of creation science. ID advocates preferred to speak of an intelligent agent or mind or designer than of anything, or anyone, more obviously religious. For those who had eyes to see, the designer was clearly the God of creationism, having swapped his robes for a lab coat, but the fiction persisted.

From the publication of its first major text, *Of Pandas and People*, in 1989, the movement flourished in the US. It attracted some intelligent minds. The law professor Phillip Johnson found his life's cause after reading Richard Dawkins' *The Blind Watchmaker* and became its leading light. The movement picked up where creation science had fallen and attempted to get ID into schools alongside evolution. The result was an all-too-predictable return to court. In 2005, a Pennsylvania district court adjudicated between the Dover Area School District and eleven parents of children studying biology there. *Of Pandas and People* stood at the heart of the hearing, which inevitably came to be called the Dover Panda Trial, although there were distinctly fewer pandas on show than there had been monkeys eighty years earlier.

The question was essentially the same as that of *McLean v. Arkansas* and *Edwards v. Aguillard*, only this time with ID taking the place of creation science. The result was the same too. After a forty-day trial, the judge found that ID was a religious movement, little more than relabelled creationism, wholly unscientific and completely inadmissible on the school curriculum. 'The breath-taking inanity of the Board [of Education]'s decision [to admit the teaching of *Of Pandas and People*] is evident when considered against the factual backdrop which has now been fully revealed through this trial', the judge concluded witheringly, adding that 'those who disagree with our [decision] will likely mark it as the product of an activist judge.'

He was right. Judge John Jones III was personally lambasted (despite being a churchgoing Republican, and so hardly an obvious anti-ID culture warrior), but in spite of – or perhaps because of – the repeated failure and demolition of creation science and Intelligent Design in the classroom and the courts, the anti-evolutionary movement remained stubbornly popular. When the Pew Forum asked Americans their view of evolution in 2005, around the time of the Panda trial, 48% said they thought humans had evolved over time (26% due to natural processes, 18% guided by a supreme being, 4% not knowing how), while 42% said they thought humans had always existed in their present form.[21] In a similar question, asked nearly a decade later, 33% of Americans believed that humans had evolved 'due to natural processes', 25% said they had evolved 'due to God's design' and 34% said they had 'always existed in their present form' (the remaining 8% either said 'evolved [but I] don't know how' or simply 'Don't know').[22] The numbers were sobering but at least, many thought, unique to America. Indeed, Stephen Jay Gould remarked towards the end of his life that the world need not worry about creationism because it was a peculiarly American phenomenon, 'a local, indigenous, American bizarrity'.[23] He was wrong.

As discussed in chapter 12, the Muslim world – or, rather, the Muslim worlds, as there was sometimes as much difference between different Muslim cultures as there was between them and the rest of the world – had initially received Darwinism wrapped in a Western colonial mantle. This, plus the absence of any central authority and therefore any 'official' Islamic position on the theory, meant that the response to it was localised, inconsistent and heavily coloured by wider social considerations. Evolution never quite attained the totemic status it did in religious America.

Nonetheless, by the end of the twentieth century, evolution rejection had become almost as widespread in Islamic countries as it was in America. According to one survey, the percentage of people who thought that evolution 'could not possibly be true' exceeded 50% in Turkey,

Indonesia, Pakistan, Malaysia and Egypt.[24] A different Pew Forum study found that in nine of the twenty-two Islamic countries surveyed, fewer than half of the population believed that humans and other living things had evolved over time. The figures were contested and could be read in different ways. The Pew Study also noted that in only four of the countries did more than half of the people say that humans have remained in their present form since the beginning of time.[25] The Darwinian glass could be half empty or half full. But either way, it clearly wasn't completely full. Religious rejection of evolution was not solely a fundamentalist Christian problem.

Moreover, some Muslims could be as willing to campaign on the issue as their Christian counterparts. In 2007, French public schools, American politicians and many others, including your author, received the unsolicited gift of an 850-page *Atlas of Creation* produced by Harun Yahya, otherwise known as Adnan Oktar, a Muslim creationist working in Turkey. Yahya had come to prominence in the 1990s, just as ID was maturing, with his book *The Evolution Deceit: The Scientific Collapse of Darwinism and Its Ideological Background*, during which time he made common cause with American creationists. Mustafa Akyol, a journalist and follower of Yahya, built links with the Discovery Institute, which reciprocated by sending some of its fellows to what was billed as the first ID conference in Turkey in 2007. The word was spreading.

As it happened, Islamic anti-evolutionism ploughed its own course. It parted company with both creation science, whose literal, Young Earth adherence to Genesis was an irrelevance to Muslims, and with ID, whose designer was too anaemically religious and distant from the Allah of the Qur'an. Muslim creationism was its own species, not simply a variation of the American beast. Yet it shared one key factor with its Christian counterpart, namely the belief that evolution undermined human nature and morality. 'The reason why a special chapter is assigned to the collapse of the theory of evolution is that this theory constitutes the basis of all anti-spiritual philosophies,' Yahya wrote in his book *Nightmare of Disbelief*.[26] Evolution was widely associated, among Muslims, with secularisation and moral decay and, in Europe at least, it was also tied up with issues of migration and identity.[27] As always, science and religion interbred with politics and culture.

'Replete with religious metaphors': irreducible complexity

The combination of a revitalised sociobiology, confident in its ability to explain all human life, and religiously motivated rejection of a well-attested and long-established area of science, provided the perfect ingredients for renewed conflict.

The American neuroscientist Sam Harris began working on his book, *The End of Faith*, on 12 September 2001 and when it was published three years later, it helped catalyse the noisiest anti-religious movement in decades. So-called New Atheism was self-evidently a moral reaction to the events of the time, but religion's moral turpitude could not be divorced from its intellectual backwardness, and so the attack on religion was commonly framed as, at heart, a science vs. religion issue.

This was greatly helped if you were convinced that science and religion were effectively different attempts to do the same thing. Dawkins spoke for many when he claimed that religion was 'a scientific theory',[28] 'a competing explanation for facts about the universe and life',[29] or a straightforward alternative to evolution: 'God and natural selection are . . . the only two workable theories we have of why we exist.'[30] However much this definition might have perplexed theologians, philosophers of religion and a great many believers, it was easier to sustain in an age in which (a) creation science, Intelligent Design and Islamic creationism were actively engaged in positing supernatural explanations in place of evolutionary ones; and (b) evolution, in the form of the sociobiology synthesis, was making ambitious claims to explain all of life – mind, morality and meaning included. In effect, when creationism/ ID made a land grab for science, and sociobiology made a land grab for everything, there could only be conflict.

The zero-sum game thus laid out, the rhetoric soared to levels not seen even in the heyday of Draper and White. One side endlessly replayed the genealogy of Darwin, eugenics and Nazis. The other opined on how science flew people to the moon, whereas religion flew them into buildings. Religion was compared to a virus ('like smallpox only harder to eradicate'), although when challenged about the judiciousness of this particular comparison, Dawkins retreated into one of his trademark HumptyDumptyisms by explaining that his use of the word virus to describe religion was intended to be somehow technical or 'special'.[31]

There was a pleasing clarity to the dispute, which drew its energy from the social, political and cultural circumstances of the moment: terrorism, migration, the role of the religious right in bringing born-again Christian George W. Bush into power, and the ensuing war in Iraq. In effect, Islam and fundamentalist Christianity played the same role in the revived science and religion battle of the early twenty-first century that Ultramontane Catholicism had done in the original battle at the end of the nineteenth century.

As then, however, the full story was not as clear as the polemicists would have it. In the first instance, the religious response to evolution was hardly universally hostile. In 1996, Pope John Paul II, speaking to the Pontifical Academy of Sciences, affirmed the Catholic Church's acceptance of evolution. Building on the 1950 encyclical of Pius XII, *Humani generis*, John Paul II stated how, more than a half-century on from Pius's acknowledgement that there was in principle no conflict between evolution and the faith, the Church now acknowledged 'new findings' that led her to 'the recognition of evolution as more than a hypothesis'. Indeed, he went on, in an affirmation as resounding as in any biology textbook (other than *Of Pandas and People*), 'this theory has had progressively greater influence on the spirit of researchers, following a series of discoveries in different scholarly disciplines. The convergence in the results of these independent studies – which was neither planned nor sought – constitutes in itself a significant argument in favour of the theory.'[32] There was similar affirmation from pretty much every mainstream Christian denomination, for most of whom Intelligent Design and creation science were embarrassments.

An ever-growing body of public opinion research found that attitudes to and, in particular, rejection of evolution were not simply a function of religion. In 2009, as the world celebrated a double-Darwin anniversary (200 years since his birth, 150 since the publication of *The Origin of Species*), polling companies surveyed public opinion on the topic. The results underlined how complex the phenomenon of anti-evolutionism was. Ignorance and uncertainty were the primary winners.[33] In the UK, roughly a quarter of people were *consistent* in their acceptance of evolution by natural selection and roughly a quarter were *consistent* in their rejection of it; the rest exhibited greater or lesser levels of confusion.

Religious believers were only slightly less likely to agree that evolution is a theory 'so well established that it is beyond reasonable doubt'. Conversely, many of those who rejected the theory were not religious. There *was* a correlation between religiosity and rejection of evolution but not a particularly strong one.

There were even nuances in the homeland of the anti-evolution move-ment – evangelical America. Elaine Howard Ecklund, a professor of soci-ology in the US, conducted a series of detailed studies exploring what reli-gious Americans thought of science, scientists and Darwinism. She found that while evolution rejection was vastly greater among evangelicals than it was among mainline Protestants, other religious groups or people of no religion, many evangelicals were more open than the stereotype would have it. 'Religious Americans appear quite flexible in what they are willing to believe regarding creation and evolution,' she concluded, '*but* their underlying schemas motivate them to keep some role for God and respect what they see as the sacredness of humans.'[34] Once again, the idea of the human lay somewhere near the heart of evolution rejection, a fact supported by further surveys that showed how people were more willing to countenance evolution in general than human evolution in particular, and more willing to countenance an evolutionary explanation for human physical form than for human moral sensibility or consciousness.

Moreover, criticism of evolution was not simply a preserve of the reli-gious right. Elements of the secular left had their problems too. This was evident in the initial letter against E. O. Wilson, which, although not overtly political, drew on longstanding Marxist concerns that reifying 'human nature' – whether through God or biology – merely legitimised existing social stratification. Human behaviour was fixed: men were aggressive, women were nurturing, delinquents were immoral and the lower classes were feckless, and there was little any of them could do about it. It was the same criticism – the erosion of human freedom and moral agency – that emanated from the religious right, albeit driven by different social concerns.

There were, therefore, attenuating details within this all-new war of science and religion. Some leading evolutionists denied there was any necessary conflict with religion. One of Gould's final books, *Rocks of Ages*, argued that the tension between science and religion 'exists only in people's minds and social practices, not in the logic or proper utility of these entirely different ... subjects'.[35] Properly speaking science and religion were NOMA: non-overlapping magisteria. Science covered the empirical realm – 'what is the universe made of (fact) and why does it work this way (theory)', and religion the realm of 'ultimate meaning and moral value'.[36] On close inspection, such a clean division was hard to sustain and those who already disliked Gould for his relentless criticism of gene-centred sociobiology were hardly likely to be persuaded by his ameliorative turn in the war against religion. But his arguments were a reminder of how all-out-war wasn't in fact all-out or even necessarily war.

A more subtle challenge came from a growing sense that, in some areas, most notably the spin-off discipline of evolutionary psychology, sociobiological claims to explain all life were rather overambitious. Theories often failed to fit the data or offer convincing explanations or make reliable predictions.[37] Sometimes they were just plain absurd. Apparently, humans found the savannah particularly lovely because they evolved in the African savannah, which afforded them 'expansive views, so predators, water, and paths can be spotted from afar'.[38] And women naturally preferred pinks and reds because they had harvested red berries in the hunter-gatherer phase of human existence. And babies cried at night to prevent their parents from having sex, which would create more babies that would compete for limited maternal resources. This was supposed to be science.

The case for the prosecution was set out in a 2001 volume entitled *Alas Poor Darwin*, the first chapter of which saw the American sociologist of science Dorothy Nelkin dissect what she called the 'Religious Impulse' in evolutionary psychology.[39] According to the American Association for the Advancement of Science, Nelkin noted with a nod to Gould that the differences between science and religion have to do with the kind of questions asked: 'Science is about causes, religion about meaning. Science deals with how things happen in nature, religion with why there is anything rather than nothing. Science answers specific questions about the workings of nature, religion addresses the ultimate ground of nature.' Sociobiology, however, blurred these boundaries by deploying genetic explanations in 'a language replete with religious metaphors and concepts such as immortality and essentialism', and seeing the gene 'as a kind of sacred "soul" '. It was this unwarranted invasion of the territory of religion (and philosophy and the humanities) that lay behind evolutionary psychology's failures.[40]

There was more. One of sociobiology's great thrusts against religion was the sheer randomness of evolution. As Gould himself had written, in a more analogue age, rewind and replay the tape of life and you would get an entirely different picture. There was no plan to life, no order, no direction, just chance and necessity. And yet, mounting evidence appeared to show that chance and necessity repeatedly added up to direction, or rather directions. Blind as evolution was, it was still constrained by the chemical and physical conditions on earth, which meant it repeatedly converged on familiar solutions. There were, after all, only so many ways of seeing, hearing, smelling, eating and moving about on the planet. What this meant for the human beast was complex. As Simon Conway Morris, professor of evolutionary paleobiology at Cambridge wrote, humans were undoubtedly, 'from the present evolutionary perspective', unique, a seemingly accidental event after billions of years of evolutionary chance. However, if we were accidents, it looked as if we might be accidents waiting to happen. 'If we had not arrived at sentience and called ourselves human, then probably sooner rather than later some other group would have done so.'[41] The whole thing felt a bit like an elaborate cosmic card trick – lots of shuffling but you still ended up picking the one you were meant to.

None of these qualifications challenged, let alone undermined, evolution. Indeed, most of them presupposed and depended on the theory. Nor, of course, did any of them amount to proof of religious beliefs. What they did do was to show how the noisy rhetoric around science and religion that characterised the first decade of the new millennium was not a global conflict, but simply just a little, loud, local skirmish.

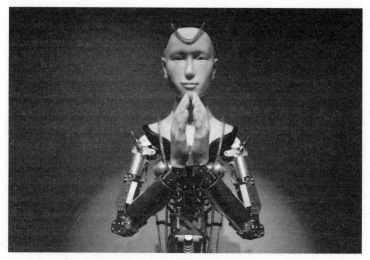

AI invites precisely the kind of questions that have underpinned the histories of science and religion over the centuries – in particular, what/who is the human, how do we tell, and who gets to decide?

NINETEEN

ARTIFICIAL ANXIETIES

'His brain was on fire': abnormal activity

One of the spin-offs from the genetic turn in sociobiology was the idea of memes. Memes were the cultural equivalent of genes; 'units' – like an idea, a tune or a catchphrase – that multiplied, evolved and, if successful, spread through society. They had great (anti-)religious potential. Religious ideas, Dawkins explained, like God, hell or blind faith, were all just effective memes, despite the fact they caused 'great psychological anguish'. Memes were treated, long before they were appropriated by the internet, as a serious scientific idea.

There were a few problems. On closer inspection, memes were not testable, had no empirical support, couldn't be defined with precision, were not quantifiable, had no consistency, elicited no intersubjective agreement and were not subject to repeatable experiments. Invisible, undetectable, unwarranted, unhelpful and unnecessary things, they fulfilled pretty much all the criteria their disciples ascribed to God. Arguably, they demanded more faith too. As the philosopher Anthony O'Hear remarked, 'if memes really existed, they would ultimately deny the reality of reflective thought.'[1]

Dawkins had pictured memes 'propagating themselves in the meme pool by leaping from brain to brain'.[2] The effect was to dismantle the human mind, which became, in Daniel Dennett's words, 'an artefact created when memes restructure a human brain so as to make it a better habitat for memes'.[3] Just as bodies could be reduced to genes, so could minds to memes.

Or to electrical activity. The tight, indeed indissoluble, link between the breathtaking poetry of human thought and emotion, and the prosaic material brain from which they emanated, had vexed thinkers since the eighteenth century. For those wedded to any kind of substance dualism – the idea that humans were made up of two kinds of stuff, material and spiritual – every advance in brain science pushed the spiritual human ever closer to the edge. Those, like Hartley, who saw the spiritual as embedded in and emergent from the material were less threatened. But even they were not totally immune as 'mechanisation' threatened to flatten more than just the human spirit. The more people peered into the brain, the more they puzzled at what exactly they found there. And by the early twenty-first century, they were able to peer a long way in.

The idea that religion was all in the mind was hardly new. 'To say [God] hath spoken to [a man] in a Dream, is no more than to say he dreamed that God spake to him.'[4] So wrote Thomas Hobbes in *Leviathan*. Scoffers had made much of the illusory nature of religious experience and not only scoffers. In his novel *The Idiot*, Fyodor Dostoyevsky described Prince Miskin's quasi-religious experience before an epileptic seizure. 'It seemed his brain was on fire . . . the sense of life, the consciousness of self multiplied tenfold . . . mind and heart were flooded with extraordinary light [and] . . . all anxieties were . . . resolved in a kind of lofty calm, full of serene, harmonious joy and help.'[5]

From the mid-twentieth century, science was able to follow where philosophy and literature had led, and clinical neurology began to confirm the link between (some) temporal lobe epileptic seizures and (some) intense religious experiences. Numerous patients suffering from epilepsy reported mystical 'premonitory symptoms or auras', which were linked to abnormal and intense electrical activity before and during a seizure.[6] The ecstasy of certain intense religious experiences could, it seemed, be traced to simple neural (mal)functioning.

And not only intense ones. Mystical experiences had been reported everywhere and throughout history, and they seemed to be as core to human nature as hunger or lust. One study of Vietnam veterans, comparing those who had suffered a 'penetrating traumatic brain injury' with a healthy control group, found that the former group 'presented markedly increased mysticism' that could not be accounted for by the subjects' history or character. The researchers concluded that their study gave further support to the longstanding hypothesis that 'executive functioning' in the brain – the processes of self-regulation that allow us to focus and plan – 'causally contributes to the down-regulation of mystical experiences'.[7] In other words, rational thought held the mystical in check and when the rational was damaged, the mystical escaped.

This was not to claim that spirituality was the preserve of epileptics or the brain-damaged. As 'neurotheology' flourished, the devout became sought-after lab rats, and Buddhist monks and Catholic nuns, among others, were fed into MRI scanners or probed and measured while at meditation or prayer. The results showed increased bilateral frontal lobe activity and decreased right parietal lobe activity at precisely the time when the monks and nuns reported reaching a state of 'total absorption' or 'oneness'.[8] Decreased activity in the parietal lobes, areas associated with processing temporal and spatial orientation, correlated to the absence of a sense of self, a characteristic of deep meditation and prayer.

Not all spiritual activity showed up in the same way. Nuns, who used words as part of prayer, showed increased activity in the subparietal lobes, areas associated with language processing, particularly when compared with Buddhists, who favoured visualisation over articulation. Islamic prayer practices were assessed via changes in cerebral blood flow, and revealed a decrease in flow in the prefrontal cortex, together with brain patterns that were distinct from ordinary, 'secular' techniques of concentration. Researchers hypothesised that 'the changes in brain activity may be associated with the feelings of "surrender" and "connectedness with God" . . . experienced during these intense Islamic prayer practices'.[9] Union with God, or with the universe, or the loss of the self, or of time, or unqualified submission and surrender, were all simply neural states; indeed, states into which you could train yourself.

Moreover, just as neuroscientists could link religious experiences, whether intense or meditative, to measurable temporal and parietal lobe activity, so it seemed that generic religious interests also had a neurological basis. 'The ongoing belief pattern and set of convictions (the religion of the everyday man [*sic*]) . . . may be predominantly localized to the frontal . . . regions . . . of the right hemisphere.'[10] Religiosity, like religious experiences, was all in the mind.

Such a connection was hardly *sui generis*. One study gave former Mormon missionaries a simulated one-hour religious experience in a controlled environment while their brains were scanned using fMRI. It found, in the deadpan words of the researchers, that 'the association of abstract ideas and brain reward circuitry may interact with frontal attentional and emotive salience processing, suggesting a mechanism whereby doctrinal concepts may come to be intrinsically rewarding and motivate behavior in religious individuals.'[11] This meant, in the rather less deadpan way that the study was reported, that religious and spiritual experiences activated the same brain reward circuits as love, sex, gambling, drugs and music.[12] As far as MRI scanners were concerned, religion was pure sex & drugs & rock 'n' roll.

The implication of all this – that religion was no more than a normal (or sometimes abnormal) neural state – could be tested. If religious experiences could be traced to electrical signals in the brain, presumably artificial brain stimulation could provoke religious experiences. If religion could be deconstructed, then it could also be reconstructed. And lo it was so. Scientists reported that psilocybin, the active ingredient in magic mushrooms, would often produce mystical experiences when administered to religious individuals.[13] Others cut out the middle mushroom altogether and went straight for the brain. The American psychologist Michael Persinger developed a head-worn device that (apparently) disrupted communication between the temporal lobes by use of weak magnetic fields. In doing so, it was claimed, the helmet was able to induce a sense of a 'presence' or even an out-of-body experience among wearers. It was inevitably baptised 'the God helmet', although sadly the notable atheists who tried it failed to meet their maker.

Opinions of what all this meant for the brain varied. For some, it appeared to suggest that there was a kind of 'God module', a specific area dedicated to religious experiences, in which (ab)normal activity could induce the mystical, irrespective of a person's beliefs or background.[14] Others argued that the stimulated experiences – harmony, significance, peace, joy – were generic rather than specifically religious, and that they needed to be passed through the filter of an existing religious worldview in order for them to be interpreted as such.[15] They were, in effect, experiences that could be *viewed* as religious rather than being religious experiences themselves.

Beyond this, however, there was greater uncertainty over what any of it actually meant for religion. In the first instance, the link between temporal lobe epilepsy and religious experiences, on which so much subsequent research was built, came under critical scrutiny. The evidence for particularly intense religiosity among people suffering from temporal lobe epilepsy was, in fact, not particularly strong. There were also difficulties with replicating some 'induced God' experiments, not to mention problematic presuppositions with others. Those who thought that the intense, trippy, mystical experience of psilocybin was typical of religion, clearly had not spent much time reading the Book of Common Prayer.

These were all valid concerns, but they were overshadowed by the bigger questions kicked up by the emerging discipline of neurotheology. That there were strong correlates between localised brain activity and religious experiences or practices was hardly a surprise. There were neural correlates for pretty much everything that people thought or did, whether recognising a face, solving a sudoku puzzle, listening to music or falling in love. That hardly meant such experiences were only 'in the mind'.

That there were some particularly noteworthy neural correlates was hardly earth-shattering either. Practised Buddhist meditators were reported to have a thicker cortex in those brain regions associated with attention, but a similar observation could be made for less obviously spiritual activities. A much-reported study on licensed taxi drivers in London, renowned for their extensive spatial knowledge, found that their posterior hippocampus was significantly larger than those of comparable subjects and even that the hippocampal volume correlated with the amount of time spent as a taxi driver. If you engage in a particular activity every day, all day, with intention and intensity, it apparently changes your brain.

Perhaps the biggest challenge to the idea that neuroscience could simply explain away religiosity, in all its varied forms, was that the same line of attack ended up explaining away a great deal else. In any neuroscientific critique of the spiritual, there was considerable collateral human damage. God and the human were in this together.

Experiments as early as the 1960s had found that there was a build-up of electrical activity in parts of the brain, known as the 'readiness potential', before a subject made a voluntary movement. In a series of famous experiments in the 1980s, Benjamin Libet and collaborators measured in milliseconds the sequence of (a) the moment at which a subject decided to perform a simple action, like flexing the wrist or fingers, (b) the point at which the 'readiness potential' became measurable in the brain; and (c) the moment at which the action occurred. In the ordered, rational world in which humans thought they lived, the sequence should have been a-b-c, but Libet found it was, in fact, b-a-c. The brain seemed to know what was going to happen before the subject did. Libet concluded, as would many subsequently, that our conscious decision to act is not the true cause of the movement.[16]

Libet's conclusions did not go unchallenged and, in particular, there was much debate about whether the moment at which someone consciously decided to perform the action could be measured with the requisite precision. But the general message was still disconcerting. The same approach that showed how neuroscience could explain away spiritual experience and belief also seemed to explain away free will and intentionality.

And morality. In 1848, when working on a railroad construction near Vermont, a foreman named Phineas Gage had a substantial metal bar – 2 cm wide by 1 metre long – blown clean through his head, entering through his cheek and emerging at the top of his skull. Astonishingly he survived, but the damage transformed him from being a conscientious and hard-working employee into a dissolute, unreliable gambler. A hundred and fifty years later, his remains having been exhumed and studied with a CT scan, it was possible to digitally reconstruct the injury and trace the link between neural damage and behavioural change. There was nothing spiritual or ethereal or existential about morality, any more than there was for intentionality. It too could be reduced to signals in the brain.

Shocking as the tightness of this connection was in 1848, it was no longer news by the time Mr Gage's skull was being scanned. By the early twenty-first century, the link between brain and behaviour was well established. In 2000, a schoolteacher was arrested for visiting pornographic websites that featured children, and then making an advance on his stepdaughter. He was convicted, the jury unpersuaded by his claim that he simply couldn't help himself. The day before sentencing, he was admitted to hospital, distraught, complaining of a severe headache. An MRI scan revealed a tumour the size of an egg pressing on his right frontal lobe. It was removed and his paedophilia disappeared. The tumour subsequently started to regrow and the urges returned.[17]

The teacher's story was less dramatic than Gage's but just as indicative. Neurologists often reported cases in which a patient's ability to act freely or morally was severely compromised by cranial abnormalities. MRI and CT scans appeared to confirm what Hobbes and La Mettrie had proposed 250 years earlier. Body, mind, morality, spirituality: all was matter. Peer into the brain and you would not find a soul or a spirit or, for that matter, free will, consciousness or morality. They disappeared in a maze of neural networks.

Except, of course, that anyone who thought they would find something that looked like a soul or a moral compass or conscious free will would simply have been making a massive category error. Neuroscientists looked for patterns of blood flow or electrical activity and found them. They were able to detect material correlates to human beliefs and behaviours. But geneticists could do the same via the genome. Evolutionary biologists and psychologists claimed to explain all this by means of natural selection. Psychoanalysts described it through the interaction of the conscious and the unconscious, anthropologists through the development and norms of culture, sociologists through patterns and structures of human association.

All the explanations carried weight. Most were defensible. Some were right. But being right didn't mean being uniquely right. Humans could be described and understood in different ways, and to reduce them to any single description or explanation was to flatten them and subject them to what is sometimes called 'nothing buttery': humans are 'nothing but' their brain activity, or 'nothing but' their genes, or their evolutionary past, or their subconscious urges. Humans *are* the product of brain activity but they are also the product of their genes, and their evolutionary past, and their education, and their cultural practices, and their socio-economic circumstances and their metaphysical beliefs. The only reason for this multi-layered complexity to be reduced to a simplistic either/or is, it seems, to satisfy the perennial urge for a disciplinary land grab.

Cartography offers a helpful metaphor in all this. A single country or area may be charted by physical, topographical, political, electoral, geological, meteorological, historical, population, resource and road maps. All can be correct, in their own terms, but none can fairly lay claim to be the only true map. There are different ways of looking at the same phenomenon, especially if that phenomenon is a complex one. Humans are fiendishly complex phenomena, beasts characterised by more than one nature, and amenable to more than one kind of description or explanation. Neuroscience stands no more chance of finding morality or the soul in an MRI scan than ethicists or theologians will locate evidence for frontal lobe activity in the Nicomachean Ethics or the Bible. But for either party to claim that, therefore, the other is simply wrong or illusory is simply disciplinary overreach.

That does not mean that these are simply non-overlapping magisteria, as Stephen Jay Gould put it. There are plenty of areas in which that is true, where science and religion don't have much (or indeed anything) to say to one another and don't really overlap. In some places, NOMA makes sense. But the human being is emphatically not one of them. Indeed, it is over the human that science and religion most clearly *do* overlap. Humans are *both* 'material' creatures, which are measurable and explicable according to the methods of science, *and* 'spiritual' ones, who talk about and aspire to things like meaning, significance, transcendence, purpose, destiny, eternity and love, which have always been the building blocks of a religious understanding of reality. Science and religion are partially overlapping magisteria, and they overlap within us.

'Humans are merely tools': AIpocalypse now

Hobbes's dismissal of the religious dreams and visions that plagued the ignorant and the pious was predicated on his conviction, spelled out in the opening pages of *Leviathan*, that 'life is but a motion of Limbs'. The heart, he reasoned, was 'but a Spring, the Nerves, but so many Strings'. That being so, could we not say that 'all Automata . . . have an artificial life', or that human ingenuity might 'make an Artificial Animal'?[18] The human body was, after all, just a complex machine.

Much the same could be said of the human mind. In his *Elements of Philosophy* written a few years later, he reasoned that 'ratiocination' – reasoned thought – was, at heart, 'computation'. What went on in the head was ultimately no more than 'addition and subtraction . . . multiplication and division'.[19]

The implication of these two convictions was momentous. If the body was no more than springs and joints and the mind no more than addition and subtraction, it should, in theory, be possible to build a human from scratch. The idea remained a dream even as, fulfilling Hobbes's prophecy two hundred years later, the logician George Boole firmly established 'the Laws of Thought on Which are Founded the Mathematical Theories of Logic and Probabilities'.[20] A century after Boole's work was published, the computer scientist John McCarthy coined the phrase 'artificial intelligence' at a conference at Dartmouth College, New Hampshire, and helped establish a discipline that promised to turn Hobbes's vision into reality. If neuroscience was able to deconstruct all that was quintessentially human into signals in the brain, there was, in principle, no reason why computer science couldn't reconstruct it, one bit at a time.

It was decades before AI could boast of anything more than an aptitude for chess, but by the time AlphaGo, a programme produced by Google's Deep Mind, beat Lee Sedol, the eighteen-time world champion, at the fiendishly difficult 'Go', it was beginning to look as if the game was up. For some, the prospect positively pulsed with potential. Technology offered the opportunity for the plodding, irrational, limited, analogue human to be augmented and transformed. Medicine (or rather, medicine and public health) had more than doubled life expectancy (in some countries) over the course of the twentieth century. Gene therapy promised to extend it further in the twenty-first and brain–computer interfaces offered the possibility of cheating death altogether by scanning, digitising and uploading the entire brain into a storage facility from which it could be downloaded and re-embodied at some point in the future (if desired).

Whether or not this particularly ambitious goal was achievable, humans might be radically improved in other ways. AI offered the potential to break out from the constraints imposed by evolutionary biology and re-engineer our frail, faulty humanity. Humans could use information technology to improve their memory, their mental processing and even their morality. As Ray Kurzweil, the American futurist who was the St Paul of this transhumanist gospel, claimed, with puppy-dog enthusiasm: 'We're going to get more neocortex, we're going to be funnier, we're going to be better at music. We're going to be sexier . . . We're really going to exemplify all the things that we value in humans to a greater degree.'[21] Amen.

Re-engineered as a kind of super-species that was capable of self-virtu-alisation, humans might then escape the limitations of earth, and take their new form beyond the solar system, spreading out across the cosmos in some kind of interstellar mission. Such a human transformation was part of a wider, cosmic transformation that made the Soviets' 'new man' seem positively pedestrian. Intelligent machines could engineer other intelligent machines. Liberated by ever-faster processing power, AI would grow exponentially until it arrived at 'The Singularity', the point at which the new superintelligence would leave behind humans and their ponder-ous sublunary lives, and reshape itself, us and the earth as it saw fit. 'Humans are merely tools for creating the Internet-of-All-Things, which may eventually spread out from planet Earth to pervade the whole galaxy and even the whole universe. This cosmic data-processing system would be like God. It will be everywhere and will control everything.'[22]

The idea that all this might not work out so well for humans themselves occurred to more than simply the professional dystopians. Stephen Hawking warned that once this singularity was reached, humans would not be able to compete. 'The development of full artificial intelligence could spell the end of the human race.'[23] The idea that AI could be humans' 'final invention' became commonplace.[24] The Astronomer Royal Martin Rees wrote about how genetic modification combined with cyborg tech-nology would lead to a world dominated by 'inorganics – intelligent elec-tronic robots – [which] will eventually gain dominance'.[25] Yuval Noah Harari followed up his hugely popular *Sapiens* with an equally popular *Deus*, which suggested that spirituality, humanity, liberty and morality would be superseded by 'the data religion', in which all experience was reduced to data patterns. Humanity, he wrote, would turn out 'to have been just a ripple within the cosmic data flow', with the ensuing 'Internet-of-all-Things . . . [becoming] sacred in its own right'.[26]

The religious resonances within all this were painfully obvious. The world of AI seemed inexorably to gravitate to religious issues and language: post-mortem existence, immortality, re-embodiment (a kind of secular resurrection), human transformation and improvement, not to mention the possibility of a post-human 'mission' to the rest of the cosmos and indeed wholesale cosmic transformation. Early Jewish and Christian apocalyptic visions were exemplified by three notable characteristics: alienation within the world, desire for the establishment of a heavenly new world and the transformation of human beings so that they may live in a perfect, new creation. More or less the same factors characterised visions of the AIpocalypse. 'Having downloaded their consciousnesses into machines, human beings will possess enhanced mental abilities and, through their infinite replicability, immortality.'[27]

For some, this was just another staging post in the long history of the way in which science and religion repeatedly found themselves in conflict. Science, now in the form of AI, offered humanity redemption, transformation and eternity, an overpowering eschatological vision to replace the old new heavens and old new earth. Once again, religion was pensioned off. As one writer in the *Atlantic* put it, 'AI may be the greatest threat to Christian theology since Charles Darwin's *On the Origin of Species*.'[28]

That there *was* potential for conflict was beyond doubt. One BBC report on whether AI would transform religions presented various priests and believers with examples of roboticised religiosity. The faithful tended to receive them with qualified enthusiasm, but their qualification was a firm one. AI was fine as far as it went but that could only be so far because AI did not and could not have 'a soul'.[29]

There was a certain circularity in the reasoning here: only humans can have a soul which is why robots don't have a soul. Combine this with the popular pseudo-neuroscientific view – that, as Harari expressed it, when 'scientists opened up the [human] black box, they discovered there neither soul, nor free will, nor "self" – but only genes, hormones and neurons'[30] – and all the ingredients for a perfect confrontation were there. The headlines were ready. 'Religious groups fight AI research because "it threatens the soul".' Precisely because they overlap across the human, the *potential* for conflict between science and religion is always a live one, whether talking about algorithms in the twenty-first century or astrology in the fourth.

And yet, if the long history of science and religion has anything to teach us, it is that this conflict is only potential, not inevitable. Indeed, if the main argument of this book is right, and it is the complex, multi-layered and varied natures of the human beast that lie at the heart of so many interactions between science and religion, then it is just possible that the age of AI might open up a space for enriching dialogue rather than closing it down in the face of defensive argument.

One obvious area for such dialogue lies in the intractable ethical questions that now adhere to AI research like its own shadow. That AI 'should be oriented towards respecting the dignity of the person and of Creation', as Pope Francis put it, is almost a truism.[31] Only naïvely techno-positive Pollyannas deny that ethics has a vital, if difficult, path to pick through the thorns of progress. Some naturally gravitate to religious traditions to find it. David Brenner, co-founder of the organisation AI and Faith, observed that those working in AI now find themselves asking some of the biggest questions in life but are doing so 'mostly in isolation from the people who've been asking those questions for 4,000 years'. Rob Barrett, who took a doctorate in applied physics at Stanford before going to work on data storage technology at IBM, found himself wrestling with the question of what was the right thing to do when working on default privacy settings for an early web browser. 'That was when it dawned on [me]: "I don't know enough theology to be a good engineer".' He persuaded his employer to give him a leave of study, and ended up doing a second PhD, in Old Testament theology, and then leaving the industry.[32]

The ethics of AI is an obvious place for science and religion dialogue. But what marks this moment out as especially interesting is that AI demands we go beyond the familiar territory of 'how should we use this new kit?' and enter the realm of 'what even *is* this new kit?', 'how similar is it to ourselves?' and, by implication, 'who, then, are we?'.

Discussion around AI, and in particular its relationship to its inventors, is often predicated on a certain understanding of the human that treats cognitive power – our capacity to select, sort and process information – as the most important thing about us. Humans are thought and thought is calculation. Hobbes may have died but, in the early twenty-first century, Hobbes has risen. Speaking at the launch of the Leverhulme Centre for the Future of Intelligence in October 2016, Stephen Hawking claimed that 'intelligence is central to what it means to be human.' Given that, as he went on to say, 'there is no deep difference between what can be achieved by a biological brain and what can be achieved by a computer', the conclusion was reasonably obvious.[33] Intelligence is what makes us human. Computers will one day be super-intelligent. Therefore . . .

It's a strong argument until you begin to think about it. 'Intelligent human believes that intelligence is what makes us most human' is not a line of reasoning that should go unchallenged. Hawking claimed that 'everything that our civilisation has achieved, is a product of human intelligence, from learning to master fire, to learning to grow food, to understanding the cosmos.' But one could just as easily swap the word intelligence here for co-operation. Nothing we have achieved as a civilisation has been done without some form of co-operation. Even Newton, the archetypal independent intelligence, worked with dead giants. Intelligence without co-operation may not get you as far as co-operation without intelligence.

To be clear, it is not so much that Hawking is wrong about intelligence. Humans *are* uniquely intelligent, and it has been through the exercise of this intelligence that we have managed to transform the planet. (It may even be through the exercise of this intelligence that we transform it back.) Rather, it is that this is a somewhat limited answer, reducing the complexity of what it means to be human to a single feature, which is then used as the sole legitimate lens through which we view and judge humans and our artificial progeny.

Silicon Valley naturally tends towards the Hawking position in this debate. If a man's stock options are predicated on the idea that it is possible to turn data into knowledge and knowledge into artificial humans, it can be hard to convince him that his premises may be false. Moreover, if the only alternative view you are presented with is founded on the vague, self-fulfilling and apparently illusory notion of a soul, there is even less of an incentive to change your view.

But 'soul talk' need not simply be a cipher for substance dualism. People can talk about souls without believing there is a strange, immaterial thing drifting around their body, undetectable to all devices of medical science. After all, we commonly say that people have a moral compass, or a strong conscience or a sense of humour without imagining we will find those things on an MRI scan, let alone that they imply a kind of substance dualism.

None of these things – morality, consciousness, humour – is made up of a different kind of substance. Rather, they are the kinds of thing that emerge from ordinary material stuff when it reaches a certain level of complexity. Talk of them, just like talk of souls, is a near inevitability when you are dealing with creatures as complicated as humans, beasts that are incontestably material but whose material complexity gives rise to emergent qualities. Like talk of 'conscience' or 'morality', 'the soul' may simply be a necessary shorthand when dealing with creatures that use language, live in groups, engage in ritual, perceive the future, are cognisant of their changing nature and impending demise, agonise over moral truth, search for meaning, strive for a sense of purpose and are haunted by moments of transcendence. *Homo sapiens* currently occupies this particular niche on earth but there is in principle no reason why other organisms – carbon or silicon based – won't do so at some other point. Whether or not that is likely, those disciplines that have a long tradition of discussing such 'emergent' qualities as 'the soul' might just have something to contribute to the debates about how we should classify and relate to AI.

That contribution, at least that of the Abrahamic tradition in which modern science evolved, would nudge the discussion away from the idea that information, cognition and intelligence are the decisive dimensions within our humanity. In their place, there is an alternative view that, however important intelligence is, it is no more important than – and is certainly indivisible from – our creatureliness and our need for communion. To be human is to be clever, but it is also to be embodied, embedded, vulnerable, dependent and mortal.

We naturally prefer to focus on the first part of this. You don't have to be Stephen Hawking or Steven Pinker to feel a bit smug about our capacity for thought. Thinking elevates us above our mammalian peers and enables us to apprehend the world like a god. Reason may only be computation, at least in Hobbes's reckoning, but it is still what makes us the paragon of animals.

And yet so eager are we to celebrate our intelligence that, even as we genuflect before the Darwinian altar and intone that we are 'basically just animals', we are liable to forget that we *are* creatures, as much dust as quintessence. Reducing humanity to intelligence and failing properly to acknowledge our fragile, creaturely nature is what enables us to think that just because AI is super-intelligent, it will also therefore be super-human.

Humans are *irreducibly* embodied and embedded. Our existence in a unique, bounded body, with physical needs and personal ambitions, in a particular time and a particular place, is fundamental to our humanity, and the respect and rights we accord ourselves. Furthermore, to be embodied and embedded is to be vulnerable – to have needs and goals that are fundamental to our health and good, and which, if not met, threaten our very existence. To be vulnerable is to be dependent – to rely on other (equally vulnerable) creatures to protect, nurture, guide and support us. The solitary, self-sufficient creature, let alone human, is a contradiction in terms. We live only in dependent communion with others. The point at which AI may be considered human, or at least worthy of recognition and protection as quasi-human, is not the point at which it shows sufficient signs of intelligence or thought. It is the point at which it needs to feed itself, dispose of its own waste, find somewhere to rest, arrange for its own protection and repair, worry about its future, seek to perpetuate its own existence (and perhaps those of its progeny), struggle to avoid its demise and deliberate over the best way of achieving all these goals.

'We should not be irreverently usurping His power': taking the Test

The best known and arguably most important test intended to assess whether a machine is indistinguishable from a human is the Turing Test. In his seminal 1950 paper 'Computing Machinery and Intelligence', Alan Turing set out to answer the question 'Can machines think?' 'Think', Turing claimed, was an insufficiently precise term, however, and he argued that the question, to be answerable, should be translated into 'Are there imaginable digital computers which would do well in the imitation game?'[34] That game, which came to be known as the Turing Test, was one in which a human and a machine, isolated in separate rooms, were given questions to answer, ideally typewritten in order to discount tone of voice and handwriting, by a human 'interrogator'. If the interrogator was unable to distinguish between the answers given by machine and human, the former would have passed the test and could be considered to be 'thinking' (at least in the way Turing defined the term).

The majority of Turing's paper was not, in fact, occupied by setting out the test, which was done in a couple of pages, but by outlining his refutations to potential objections. The first of these was theological. 'I am not very impressed with theological arguments whatever they may be used to support', Turing remarked, namechecking Galileo in the process, but nonetheless he attempted to refute this one 'in theological terms'. Frustratingly, the theological argument he chose to refute was much like that of the faithful in the BBC report mentioned in the previous section: 'God has given an immortal soul to every man and woman, but not to any other animal or to machines.'

For all the circularity of this position, Turing offered a perceptive answer. The argument was, he pointed out, an arbitrary limitation of God's power. 'Should we not believe that He has freedom to confer a soul on an elephant if He sees fit?' And if an elephant, why not a machine? That prospect may be more difficult for humans to swallow, but that was simply for reasons of human pride. If the circumstances were right, there would be no reason why God should not 'confer' a soul on a machine, any more than on an elephant or a human. 'In attempting to construct such machines we should not be irreverently usurping His power of creating souls, any more than we are in the procreation of children: rather we are, in either case, instruments of His will providing mansions for the souls that He creates.'

Turing's demolition of the circular 'only humans have a soul' argument was more effective than he could have realised. Given what we now know about his preferred counterexample, elephants – not only their intelligence and sociality, but their sensitivity, memory, empathy and capacity to grieve – it is extremely tempting to use 'soulish' language when talking about them, in the same way we might do of some primates.[35] If elephants, why indeed not machines?

The strength of this argument is undermined, however, by his proposed means of assessment. Turing's talk of God 'conferring' souls, and of bodies (or machines) 'providing mansions' for such souls, reeks of the kind of substance dualism that Turing (rightly) rejected. That recognised, his test itself was predicated on its own kind of dualism, not of body and soul, but of body and mind.

The focus in his essay was on the mind or intellect *in isolation from* the body. The specific problem he set out, he claimed, had 'the advantage of drawing a fairly sharp line between the physical and the intellectual capacities of a man'. Even supposing some engineer or chemist could produce a material that was indistinguishable from the human skin, there would be 'little point in trying to make a "thinking machine" more human by dressing it up in such "artificial flesh" '. The test can be done without the body because the body is, effectively, an irrelevance.

And yet the body cannot be an irrelevance in any test that asks whether something passes as a human. Trying to do so assumes it is possible to draw a line between the intellectual and the physical capacities of a person or that it makes any sense at all to consider the intellectual outside the context of the physical. It is perhaps no surprise that the Turing Test works by comparing a machine with a human who has been isolated in a room, has impersonal questions fired at him by an interrogator and who is not even permitted to reveal his physical existence or personality or tone of voice by speaking or even writing. That is more or less what we do to humans when we want to *dehumanise* them – isolate, interrogate, depersonalise. By that reckoning, a machine might be able to pass as a human, but it's a pretty unpersuasive (and unappealing) understanding of a human.

The Turing Test has been enormously influential (last time I checked it had been cited nearly seventeen thousand times). It is clever, provocative and rightly celebrated, just the kind of creative thought experiment that you imagine the natural philosophers of late medieval Europe enjoying. But it is also, like so much AI talk, predicated on a problematically thin anthropology, which sees communication as simply a means of conveying information rather than as a way of entering into communion with (other) embodied, dependent, vulnerable creatures.

That is the kind of perspective that religious traditions might bring to this fizzing, speeding debate over the future metaphysical status of artificial intelligence. Of course, they could be wrong. It may be that the kind of metaphysical and spiritual concerns that are clearly characteristic in humans don't emerge when AI reaches a certain complexity, and that it will never be appropriate to use language of souls or rights for AI. Conversely, it may be that there *is* something different and external that is *added* to matter to make it 'soulish' (though I would bet against this). Alternatively, it may simply be that we are all getting rather overexcited about this stuff and that the AIpocalypse is not due now or at any foreseeable time.[36]

But it may also be that, as AI develops apace, these become precisely the conversations we must have, conversations that demand we grapple with the questions of what (or who) constitutes the human, and who (or what) gets to decide. If so, we will find that the entangled histories of science and religion have a long way still to run.

Notes

Introduction

1 Foremost among them David Lindberg, Ronald Numbers, John Brooke, Bernie Lightman, Peter Harrison, Edward Larson, Geoffrey Cantor, James Ungureanu, Elaine Howard Ecklund and Fern Elsdon-Baker.

2 For example, Catholicism is not considered to be the intellectual and political threat it was to the first historians of science and religion in the 1870s.

3 A point that has recently been made again with great force in James Hannam's *God's Philosophers: How the Medieval World Laid the Foundations of Modern Science* (London: Icon Books, 2009) and Seb Falk's *The Light Ages: A Medieval Journey of Discovery* (London: Penguin Books, 2020).

4 The closest thing to a big idea has been John Hedley Brooke's 'complexity thesis', which does pretty much what it says on the tin. As a replacement for the 'conflict thesis' that Brooke and other pioneers in the field were faced with, it does a good job. But Brooke himself has been clear that complexity 'is a historical reality not a thesis'. Bernie Lightman, *Rethinking History: Science and Religion* (Pittsburgh, PA: University of Pittsburgh Press, 2019), p.235.

Chapter 1

1 Voltaire, *L'Examen important de milord Bolingbroke*, in *Les Oeuvres Complètes De Voltaire, 1766–1767* (Oxford: Voltaire Foundation, 1987), pp.127–362.

2 Edward Gibbon, *The History of the Decline and Fall of the Roman Empire*, Vol. 3, ed. David Womersley (London: Allen Lane, 1994), chapter 47.

3 John William Draper, *History of the Intellectual Development of Europe*, Vol. I (London: George Bell and Sons, 1891), p.325.

4 Synesius, *Epistles*, 105, Sect. 10.

5 Seneca, *Natural Questions*, Sect. 6.4.2.

6 Pliny, *Natural History*, II. 1.

7 James 1:27.

8 Augustine, Letter 102, Sect. 19.

9 Augustine, Letter 102, Sect. 10.

10 Justin Martyr, *Dialogue with Trypho*, chapter 2.

11 Clement of Alexandria, *Stromata*, chapter 1.

12 Lindberg and Numbers, *God and Nature: Historical Essays on the Encounter between Christianity and Science* (Berkeley; London: University of California Press, 1986), p.24.

13 Lindberg and Numbers, *God and Nature*, p.24.

14 Tertullian, *Writings*, in *Ante-Nicene Fathers,* vol. III, p.246b, *On Prescription Against Heretics*, chapter 7.

15 Origen, *Contra Celsum*, 3.44.

16 Lindberg and Numbers, *God and Nature*, p.25.

17 Lindberg and Numbers, *God and Nature*, p.37.

18 Lindberg and Numbers, *God and Nature*, p.38.

19 Augustine, *On Christian Doctrine*, 29.45–6.

20 Augustine, *Confessions*, x.35.

21 Daniel Špelda, 'The Importance of the Church Fathers for Early Modern Astronomy', *Science & Christian Belief*, 26(1), 2014, p.28.

22 Augustine, *Confessions*, vi.3.

23 Book of Wisdom 11:20.

24 Quoted in Špelda, 'Importance', p.41.

25 Luke Lavan and Michael Mulryan (eds), *The Archaeology of Late Antique 'Paganism'* (Leidan; Boston: Brill, 2011), p.xxiv: 'As a result of recent work, it can be stated with confidence that temples were neither widely converted into churches nor widely demolished in Late Antiquity.'

26 Faith Wallis, 'Bede and Science', in *The Cambridge Companion to Bede* (Cambridge: Cambridge University Press, 2010), p.124.

27 Gibbon, *Decline and Fall*, chapter 40.

28 Cosmas Indicopleustes, *Christian Topography* 1.3–4. Quoted in Pablo de Felipe, 'Curiosity in the Early Christian Era – Philoponus's Defence of Ancient Astronomy against Christian Critics', *Science & Christian Belief*, 30, 2018, pp.38–56.

29 de Felipe, 'Curiosity', p.46.

30 de Felipe, 'Curiosity', p.46.

Chapter 2

1 Steven Weinberg, 'A Deadly Certitude', *Times Literary Supplement*, 17 January 2017; Jamil Ragep, 'Response to Weinberg', *TLS*, 24 January 2017; Steven Weinberg, 'Response to Jamil Ragep', *TLS*, 24 January 2017.

2 Alexander von Humboldt, *Cosmos: A Sketch of the Physical Description of the Universe* (London: Longman, 1848), 2:212.

3 Ernest Renan, 'Science and Islam', in Bryan Turner (ed.), *Readings in Orientalism* (London: Routledge, 2000).

4 Quoted in Jeff Hardin *et al.* (eds), *The Warfare Between Science and Religion: The Idea That Wouldn't Die* (Baltimore, MD: Johns Hopkins University Press, 2018), p.216.

5 Quoted in Hardin, *Warfare*, p.207.

6 David C. Lindberg, 'Lines of Influence on Thirteenth-Century Optics: Bacon, Witelo, and Pecham', *Speculum*, 46, 1971, pp.66–83.

7 John Watt, 'Syriac Translators and Greek Philosophy in Early Abbasid Iraq', *Journal of the Canadian Society for Syriac Studies*, 4, 2004, pp.15–26; id. George Saliba, 'Revisiting the Syriac Role in the Transmission of Greek Sciences into Arabic', *Journal of the Canadian Society for Syriac Studies*, 4, 2004. I am grateful to Jack Tennous for providing me with the statistics here.

8 Dimitri Gutas, *Greek Thought, Arabic Culture: The Graeco-Arabic Translation Movement in Baghdad and Early 'Abbāsid Society* (London: Routledge, 1998), p.98.

9 Michael H. Shank and David C. Lindberg, 'Introduction', in *Cambridge History of Science, Volume 2: Medieval Science* (Cambridge: Cambridge University Press, 2013), p.24.

10 For example, under the Mamluk Sultans in Egypt and Syria, between 1250 and 1517, there was ongoing study of astronomy and medicine. Timurid rulers encouraged scientific work in Persia and Central Asia after the Mongol conquest there, supporting the Maragha Observatory where the great astronomer al-Tusi worked, and the fifteenth century Ulugh Beg Observatory in modern-day Uzbekistan. Even in the Ottoman empire, which became a byword among early modern Europeans for decadence and intellectual incuriosity, there are signs of scientific activity, such as with the (rather short-lived) Taqi ad-Din Observatory in Constantinople. See Ekmeleddin İhsanoğlu (ed.), *History of the Ottoman State, Society & Civilisation* (Istanbul: IRCICA, 2001).

11 See Emilie Savage-Smith, 'Medicine in Medieval Islam', in Shank and Lindberg, *Cambridge History of Science*, p.149.

12 Toby E. Huff, *The Rise of Early Modern Science* (Cambridge: Cambridge University Press, 1993), p.101.

13 https://w2.vatican.va/content/benedict-xvi/en/speeches/2006/september/documents/hf_ben-xvi_spe_20060912_university-regensburg.html (Accessed 05 January 2023).

14 George Sabra, 'The Appropriation and Subsequent Naturalization of Greek Science in Medieval Islam: A Preliminary Statement', *History of Science*, 25(3), 1987, p.239.

Chapter 3

1 Philo, *Embassy to Gaius*, XXXI, 208.

2 G. F. Hegel, *Lectures on the Philosophy of Religion* (London: Routledge, & Kegan Paul, 1895), II.188.

3 Justin Marston, 'Jewish Understandings of Genesis 1 to 3', *Science and Christian Belief*, 12, p.131.

4 Isaiah 47:12–14.

5 Noah Efron, 'Early Judaism', in John Hedley Brooke and Ronald L. Numbers, *Science and Religion Around the World* (Oxford: Oxford University Press, 2011), p.23.

6 BT Pesachim 94b.

7 BT Pesachim 94b.

8 BT Sotah 49b.

9 See Gad Freudenthal, 'Stoic Physics in the Writings of R. Saadia Gaon al-Fayyumi and Its Aftermath in Medieval Mysticism', *Arabic Sciences and Philosophy*, 6, 1996, pp.113–36.

10 Y. Tzvi Langermann, 'Science in the Jewish Communities', in Lindberg and Shank, *Cambridge History of Science: Volume 2, Medieval Science* (Cambridge: Cambridge University Press, 2013), p.171.

11 Langermann, 'Science in the Jewish Communities', p.178.

12 Tamar Rudavsky, *Jewish Philosophy in the Middle Ages: Science, Rationalism and Religion* (Oxford: Oxford University Press, 2018), p.67.

13 Quoted in Gad Freudenthal, 'Abraham Ibn Daud, Avendauth, Dominicus Gundissalinus and Practical Mathematics in Mid-Twelfth Century Toledo', *Aleph*, 16(1), 2016, pp.81–2.

14 Y. Tzvi Langermann, 'Science in the Jewish Communities', p.187.

15 Gad Freudenthal, 'Introduction: The History of Science in Medieval Jewish Cultures: Toward a Definition of the Agenda', in *Science in Medieval Jewish Cultures* (Cambridge: Cambridge University Press, 2002), p.2.

16 Gad Freudenthal, 'Science in the Medieval Jewish Culture of Southern France', *History of Science*, 33(1), 1995, p.38.

17 Freudenthal, 'Science in the Medieval Jewish Culture', p.38.

18 Freudenthal, 'Science in the Medieval Jewish Culture', p.38.

19 Gad Freudenthal, 'Maimonides' Philosophy of Science', in K. Seeskin (ed.), *The Cambridge Companion to Maimonides* (Cambridge: Cambridge University Press, 2005), p.149.

20 Freudenthal, 'Science in the Medieval Jewish Culture', p.38.

21 Rudavsky, *Jewish Philosophy*, p.199.

22 Freudenthal, 'Maimonides' Philosophy of Science', p.157.

23 Rudavsky, *Jewish Philosophy*, p.98.

24 Freudenthal, 'Maimonides' Philosophy of Science', p.150.

25 Freudenthal, 'Science in the Medieval Jewish Culture', p.9.

26 Gersonides, *The Wars of the Lord*, V.I.

Chapter 4

1 Sa'id al-Andalusi, *Science in the Medieval World* (Austin: University of Texas Press, 1991).

2 Carl Sagan, *Cosmos* (London: Macdonald Futura, 1981), p.335.

3 Francis Bacon, *The Advancement of Learning*, Book I, iv, 5.

4 Tom McLeish, 'Beyond Interdisciplinarity to the Unity of Knowledge: Why we need both Medieval and Modern Minds', Ordered Universe Conference, April 2018.

5 Shank and Lindberg, 'Introduction', in *Cambridge History of Science*, p.10.

6 William of Auvergne, *The Universe of Creatures* (Milwaukee: Marquette University Press, 1998), p.139.

7 The following paragraphs are in great debt to Tina Stiefel's work, in particular 'The Heresy of Science: A Twelfth-Century Conceptual Revolution', *Isis*, 68(3), 1977 and 'Science, Reason and Faith in the Twelfth Century: The Cosmologists' Attack on Tradition', *Journal of European Studies*, 6(21), 1976, from which the source quotations come.

8 For example, 'the divine page says, "He divided the waters which were under the firmament from the waters which were above the firmament." Since such a statement as this is contrary to reason let us show how it cannot be thus.'

9 Edward Grant, *The Foundations of Modern Science in the Middle Ages: Their Religious, Instutional and Intellectual Contexts* (Cambridge: Cambridge University Press, 1996), p.21.

10 Grant, *The Foundations of Modern Science*, p.26.

11 Edward Grant, 'Science and the Medieval University', in *Rebirth, Reform, and Resilience: Universities in Transition, 1300–1700*, ed. James M. Kittelson and Pamela Transue (Columbus: Ohio State University Press, 1984), pp.68–70.

12 Grant, *The Foundations of Modern Science*, p.74.

13 Grant, *The Foundations of Modern Science*, p.173.

14 Grant, *The Foundations of Modern Science*, p.77.

15 Edward Grant, *A Source Book in Medieval Science* (Cambridge, MA: Harvard University Press, 1974), p.50.

16 Ronald L. Numbers, *Galileo Goes to Jail and Other Myths about Science and Religion* (Cambridge, MA; London: Harvard University Press, 2009), p.25.

17 Lindberg and Numbers, *God and Nature*, p.56.

18 Edward Grant, 'Late Medieval Thought, Copernicus, and the Scientific Revolution', *Journal of the History of Ideas*, 23(2), April–June 1962, p.200.

Chapter 5

1 Sigmund Freud, *A General Introduction to Psychoanalysis* (Ware, UK: Wordsworth Classics of World Literature), p.241.

2 Nicolas Copernicus, *De revolutionibus* (1543), Introduction.

3 F. J. Ragep, 'Copernicus and His Islamic Predecessors: Some Historical Remarks', *History of Science*, 45(1), 2007, p.68.

4 P. D. Omodeo, 'The Bible versus Pythagoras: The End of an Epoch', in *Copernicus in the Cultural Debates of the Renaissance* (Leiden, The Netherlands: Brill, 2014), p.273.

5 Joshua 10:12–14.

6 2 Kings 20:8–11, Isaiah 38:8.

7 Lindberg and Numbers, *God and Nature*, p.171.

8 It was not published at the time and appeared only in 1651.

9 Omodeo, 'Bible versus Pythagoras', p.274.

10 Owen Gingerich, *The Book Nobody Read: Chasing the Revolutions of Nicolaus Copernicus* (New York: Walker, 2004).

11 Numbers, *Galileo Goes to Jail*, p.53.

12 Numbers, *Galileo Goes to Jail*, p.54.

13 Lindberg and Numbers, *God and Nature*, p.88.

14 de Felipe, 'Curiosity', p.53.

15 Dorothea Waley Singer, *Giordano Bruno, His Life and Thought* (New York: Henry Schuman, 1950), chapter 7.

16 John William Draper, *History of the Conflict Between Religion and Science* (New York: D. Appleton, 1874), pp.178–80.

17 Michael White, *The Pope and the Heretic: A True Story of Courage and Murder at the Hands of the Inquisition* (London: Little, Brown, 2002), blurb.

18 John Gribbin, *Science: A History 1543–2001* (London: Penguin, 2009), p.18.

Chapter 6

1 Anna Beer, *Milton: Poet, Pamphleteer and Patriot* (London: Bloomsbury, 2011), p.98.

2 *Paradise Lost*, ed. Alastair Fowler (London: Longman, 1998), p.296.

3 *Paradise Lost*, p.78.

4 Jean Dietz Moss, 'Galileo's Letter to Christina: Some Rhetorical Considerations', *Renaissance Quarterly*, 36(4), Winter, 1983, p.575.

5 Snezana Lawrence and Mark McCartney (eds), *Mathematicians and their Gods: Interactions between mathematics and religious beliefs* (Oxford: Oxford University Press, 2015), p.79.

6 Lawrence and McCartney, *Mathematicians*, p.112.

7 J. K. Helibron, *Galileo* (Oxford: Oxford University Press, 2010), p.69.

8 Helibron, *Galileo*, p.106.

9 Helibron, *Galileo*, p.121.

10 Helibron, *Galileo*, p.165.

11 Lindberg and Numbers, *God and Nature*, p.118.

12 Helibron, *Galileo*, p.201.

13 Lindberg and Numbers, *God and Nature*, p.120.

14 Maurice A. Finocchiaro (ed.), *The Galileo Affair: A Documentary History* (Berkeley: University of California Press, 1989), pp.87–8.

15 Finocchiaro, *The Galileo Affair*, p.94.

16 Finocchiaro, *The Galileo Affair*, p.96.

17 Finocchiaro, *The Galileo Affair*, p.104.

18 Helibron, *Galileo*, p.215.

19 Helibron, *Galileo*, p.153.

20 Finocchiaro, *The Galileo Affair*, p.150.

21 Helibron, *Galileo*, p.220.

22 Galileo, *The Assayer* (1623).

23 Helibron, *Galileo*, p.225.

24 Helibron, *Galileo*, p.298.

25 Helibron, *Galileo*, p.301.

26 Stefano Gattei (ed. and tr.), *On the Life of Galileo: Viviani's Historical Account and Other Early Biographies* (Princeton, NJ: Princeton University Press, 2019), p.xxiii.

Chapter 7

1 John Herschel, *A Preliminary Discourse on the Study of Natural Philosophy* (London: Longman *et al.*, 1831), p.72.

2 Letter to John Herschel, 11 November 1859.

3 Herschel, *Preliminary Discourse*, p.114.

4 Herschel, *Preliminary Discourse*, p.105.

5 Herschel, *Preliminary Discourse*, pp.114–15.

6 Francis Bacon, Letter to Lord Burghley, 1591.

7 Francis Bacon, *The Advancement of Learning* (1605), I.vi.2

8 Anyone familiar with the academic literature on the topic will instantly recognise that what follows is profoundly in debt to the work of Peter Harrison.

9 David Wootton, *The Invention of Science: A New History of the Scientific Revolution* (London: Allen Lane, 2015), p.83.

10 Bacon, *Advancement*, I.v.16.

11 Bacon, *Advancement*, I.i.3.

12 Francis Bacon, *Novum Organum* (1620), 1.9.

13 Peter Harrison, *The Fall of Man and the Foundations of Science* (Cambridge: Cambridge University Press, 2007), p.153.

14 Bacon, *Novum Organum*, 1.37.

15 Harrison, *The Fall of Man*, p.132.

16 See Peter Pesic, 'Proteus Rebound: Reconsidering the "Torture of Nature"', *Isis*, 99(2), June 2008, pp.304–17.

17 Bacon, *Novum Organum*, 2.52.

18 Richard Popkin, *The History of Scepticism: from Savonarola to Bayle* (Oxford: Oxford University Press, 2003), p.94.

19 Pascal's wager is the idea that it is more logical to bet on God's existence given what you stand to win if you're right, than to bet against it given what you stand to lose if you're wrong. Popularly believed to have been intended as a logical demonstration of God's existence, Pascal really intended it as a way of revealing why people (or at least some of his contemporaries) didn't believe. As he wrote in his *Pensées*, 'At least get it into your head, that if you are unable to believe, it is because of your passions, since reason impels you to believe and yet you cannot do so.' See, Graham Tomlin, 'We've got Pascal all wrong', *Unherd*, 27 March 2019.

20 Harrison, *The Fall of Man*, p.135.

21 William R. Shea, *Designing Experiments & Games of Chance: The Unconventional Science of Blaise Pascal* (Canton, MA: Science History Publications, 2003), pp.107–9.

22 Newton's comment was recorded by John Craig, Cambridge University Library, Keynes MS 130.7, f. 1r.

23 Lindberg and Numbers, *God and Nature*, p.152.

24 Harrison, *The Fall of Man*, p.55.

25 Stephen Gaukroger, *Francis Bacon and the Transformation of Early-Modern Philosophy* (Cambridge: Cambridge University Press, 2001), p.5.

26 Thomas Sprat, *The History of the Royal-Society of London: For the Improving of Natural Knowledge* (London, 1667), p.347.

Chapter 8

1 Robert Iliffe, *Priest of Nature: The Religious Worlds of Isaac Newton* (Oxford: Oxford University Press, 2017), p.19.

2 Iliffe, *Priest of Nature*, p.4.

3 Newton, *Principia*, III. General Scholium (1713).

4 Edward B. Davis, 'Robert Boyle's Religious Life, Attitudes, and Vocation', *Science and Christian Belief*, 19(2), 2007, p.118.

5 The boundary between the two – and even, to an extent, the existence of physico-theology as an entirely distinct discipline – is much debated. See Ann Blair and Kaspar von Greyerz (eds), *Physico-theology: Religion and Science in*

Europe, 1650–1750 (Baltimore, MD: Johns Hopkins University Press, 2020) for more on this.

6 Friedrich Christian Lesser, *Insecto-theologie* (Edinburgh: William Creech; and London: T. Cadell, Jun. and W. Davies, 1799).

7 Carl Linnaeus, *Reflections on the Study of Nature* (London: George Nicol, 1785).

8 Brooke and Numbers, *Science and Religion*, p.197.

9 Blair and von Greyerz, *Physico-theology*, p.230.

10 Brooke and Numbers, *Science and Religion*, p.196.

11 Henry More, *Antidote against Atheism* (London: J. Flesher, 1655), Preface.

12 Davis, 'Robert Boyle's Religious Life', p.136.

13 Stephen Gaukroger, *The Emergence of a Scientific Culture: Science and the Shaping of Modernity 1210–1685* (Oxford: Clarendon, 2006), p.224.

14 John Ray, *Physico-Theological Discourses* (London: William Innys, 1713), p.116.

15 Newton, *Opticks*, III.i.378 (1704).

16 Peter Harrison and Jon Roberts, *Science without God?: Rethinking the History of Scientific Naturalism* (Oxford: Oxford University Press, 2019), p.67.

17 Isaac Newton to Richard Bentley, 189.R.4.47, ff. 7–8, Trinity College Library, Cambridge; Original letter from Isaac Newton to Richard Bentley (Normalized) (ox.ac.uk).

18 Harrison and Roberts, *Science without God?*, p.67.

19 Christopher Haigh, *The Plain Man's Pathways to Heaven: Kinds of Christianity in Post-Reformation England, 1570–1640* (Oxford: Oxford University Press, 2007), p.169.

20 Proverbs 6:6–8.

21 John Gribbin, *Science: A History, 1543–2001* (London: Penguin, 2009), p.138.

22 Jonathan Swift, *Gulliver's Travels*, III.4.

23 More, *Antidote*, p.115.

24 David Hume, *Dialogues Concerning Natural Religion*, XII.

25 'Monsieur Pascal's Thoughts', quoted in Blair and von Greyerz, *Physico-theology*, p.142.

Chapter 9

1 David Hartley, *Observations on Man, his Frame, his Duty, and his Expectations* (London: J. Johnson, 1801), p.62.

2 Hartley, *Observations*, p.500.

3 Ann Thomson (ed.), *La Mettrie, Machine Man and Other Writings* (Cambridge: Cambridge University Press, 1996), p.14.

4 Hartley, *Observations*, p.489.

5 Hartley, *Observations*, p.511.

6 Donald J. D'Elia, 'Benjamin Rush, David Hartley, and the Revolutionary Uses

of Psychology', *Proceedings of the American Philosophical Society*, 114(2), 1970, p.109.

7 *La Mettrie, Machine Man*, p.13.

8 *La Mettrie, Machine Man*, p.25.

9 *La Mettrie, Machine Man*, p.25.

10 La Mettrie, *L'Art de Jouir* (*The Art of Pleasure*) (1751).

11 John Locke, *First Treatise on Government* (1689), sect. 58.

12 Keith Thomas, *Man and the Natural World: Changing Attitudes in England 1500–1800* (London: Allen Lane/Penguin Books, 1983), p.41.

13 Thomas, *Man*, p.122.

14 Psalms 36:6, 104:11, 147:9, 148:10.

15 Thomas, *Man*, p.125.

16 Gomez Pereira, *Antoniana Margarita* (1554; Leiden, The Netherlands: Koninklijke Brill NV, 2019).

17 Javier Bandres and Rafael Llavona, 'Minds and Machines in Renaissance Spain: Gomez Pereira's Theory of Animal Behaviour', *Journal of the History of the Behavioral Sciences*, 28, 1992, p.168.

18 Pereira, *Antoniana Margarita*.

19 René Descartes, *Passions of the Soul* (1649), #31.

20 René Descartes, *Treatise on Man* (1662).

21 Gribbin, *Science*, pp.218–19.

22 Thomas Browne, *Religio Medici* (1643), I.xxxvi.

23 Thomas Willis, *The Anatomy of the Brain and the Description and Use of the Nerves* (1681), p.106.

24 *La Mettrie, Machine Man*, p.35.

25 Aram Vartanian, 'Trembley's Polyp, La Mettrie, and Eighteenth-Century French Materialism', *Journal of the History of Ideas*, 11(3), June 1950, p.259.

26 Vartanian, 'Trembley's Polyp', p.259.

27 Gribbin, *Science*, p.220.

28 La Mettrie, 'The System of Epicurus', in Ann Thomson (ed.), *La Mettrie*, p.98.

29 *La Mettrie, Machine Man*, p.24.

30 Jonathan Israel, *Enlightenment Contested: Philosophy, Modernity, and the Emancipation of Man, 1670–1752* (Oxford: Oxford University Press, 2006), p.719.

31 Thomas, *Man*, p.35.

32 John Locke, *Thoughts Concerning Education* (1693), sect. 116.

33 Milton, *Paradise Lost*, Book VIII.374.

34 Thomas, *Man*, p.127.

35 Pepys, *Diary*, Saturday 24 August 1661.

36 Thomas, *Man*, p.139.

37 Thomas Hobbes, *Leviathan* (1614), Introduction.

38 Thomas Hobbes, 'Six lessons to the Savilian professors of mathematics', in

Hobbes, *The collected English works of Thomas Hobbes*, ed. William Molesworth (London: Routledge/Thoemmes, 1997) Vol. 7, p.350.

39 Margaret Cavendish, *Philosophical Letters*, Sect. 4, Letter 2.

40 Cavendish, *Philosophical Letters*, 1:10.

41 Cavendish, *Philosophical Letters*, 2:1.

42 Cavendish, *Philosophical Letters*, 1:11.

43 Cavendish, *Philosophical Letters*, 4:10.

44 Cavendish, *Philosophical Letters*, 4:22.

45 Cavendish, *Philosophical Letters*, 3:19.

46 Voltaire, *Letters on the English* (1733), Letter 13.

47 Aram Vartanian, *La Mettrie's L'Homme Machine: A Study in the Origins of an Idea* (Princeton, NJ: Princeton University Press, 1960), p.95.

48 Peter Manseau, *The Jefferson Bible* (Princeton, NJ: Princeton University Press, 2020), p.31.

Chapter 10

1 Darwin's letter to Leonard Horner, 20 March 1861.

2 James Hutton, 'Theory of the Earth', *Transactions of the Royal Society of Edinburgh*, vol. I, Part II, p.304.

3 Lindberg and Numbers, *God and Nature*, p.312.

4 Charles Darwin, *The Autobiography of Charles Darwin 1809–1882* (London: Collins, 1958; repr. Penguin, 2002), p.102.

5 Lindberg and Numbers, *God and Nature*, pp.328, 337.

6 Lindberg and Numbers, *God and Nature*, p.329.

7 John Parker (ed.), *Essays & Reviews* (1860), p.341.

8 Thomas Browne, *Religio Medici* (1643), p.27.

9 John Playfair, *Works of John Playfair*, Vol. IV (Edinburgh, 1882), p.81.

10 George Combe, *The Constitution of Man and Its Relation to External Objects* (1828), p.1. This text is actually from the second paragraph of the 1847 edition of the book, from which the following quotations are taken. Earlier editions started differently.

11 Combe, *Constitution*, p.5.

12 Combe, *Constitution*, p.148.

13 A. Cameron Grant, 'Combe on Phrenology and Free Will: A Note on XIXth-Century Secularism', *Journal of the History of Ideas*, 26(1), 1965, p.142.

14 Combe, *Constitution*, p.228.

15 Combe, *Constitution*, p.193.

16 Combe, *Constitution*, p.450.

17 Combe, *Constitution*, p.174.

18 Combe, *Constitution*, pp.233, 235.

19 Combe, *Constitution*, p.418.

20 Combe, *Constitution*, p.26.

21 Jon H. Roberts, 'Psychology in America', in Gary Ferngren (ed.), *The History of Science and Religion in the Western Tradition: An Encyclopedia* (Abingdon: Routledge, 2000), p.504.

22 John van Whye, *Phrenology and the Origins of Victorian Science* (Farnham: Ashgate, 2004), p.143.

23 Whye, *Phrenology*, p.143.

24 F. A. Hayek, *The Counter-Revolution of Science* (Glencoe, IL: The Free Press, 1952), p.168.

25 Auguste Comte, *Course of Positive Philosophy*, Vol. 1. Freely translated and condensed by Harriet Martineau (London: George Bell & Sons, 1896), pp.1–2.

26 Comte, *Positive Philosophy*, I.8.

27 Comte, *Positive Philosophy*, II.113.

28 Comte, *Positive Philosophy*, I.459.

29 Comte, *The Catechism of Positive Religion* (London: John Chapman, 1858), p.253.

Chapter 11

1   Emma Wedgwood to Madame Sismondi, 15 November 1838.

2   Letter to William Fox, 9–12 August 1835.

3   Erasmus Darwin, *The Temple of Nature or, The Origin of Society* (London: J. Johnson, 1803).

4   Darwin, *Autobiography*, p.49.

5   Brooke and Numbers, *Science and Religion*, p.213.

6   William Paley, *Natural Theology: or, Evidences of the Existence and Attributes of the Deity* (London: J. Faulder, 1809), p.18.

7   Paley, *Natural Theology*, p.179.

8   Darwin, *Autobiography*, p.59.

9   Darwin, *Autobiography*, p.59.

10 Paley, *Natural Theology*, p.456.

11 Quoted in William E. Phipps, *Darwin's Religious Odyssey* (Harrisburg: Trinity International Press, 2002), p.49.

12 Adrian Desmond and James Moore, *Darwin* (London: Penguin, 1992), p.323.

13 Francis Darwin (ed.) *The Foundations of The Origin of Species. Two essays written in 1842 and 1844* (Cambridge: Cambridge University Press, 1909), p.51.

14 Darwin, *Foundations of The Origin*, p.51.

15 Darwin, *Foundations of The Origin*, p.52.

16 Darwin, *Foundations of The Origin*, p.52.

17 Thomas Robert Malthus, *An Essay on the Principle of Population* (Oxford: Oxford World Classics, 1999), chapter 1.

18 Darwin, *Foundations of The Origin*, p.51.

19 Darwin, *Autobiography*, p.88.

20 Darwin, *Autobiography*, p.89.

21 Darwin, *Foundations of The Origin*, p.52 [emphases added].

22 Letter to Emma Darwin, 18 April 1851.

23 Letter to Emma Darwin, 18 April 1851.

24 Letter to Emma Darwin, 19 April 1851.

25 Letter to Emma Darwin, 19 April 1851.

26 Letter to Emma Darwin, 20 April 1851.

27 Letter to Emma Darwin, 21 April 1851.

28 Letter to Emma Darwin, 21 April 1851.

29 Desmond and Moore, *Darwin*, p.383.

30 Letter to Emma Darwin, 23 April 1851.

31 Letter to William Fox, 29 April 1851.

32 Charles Darwin's reminiscence of Anne Elizabeth Darwin, 30 April 1851.

33 Letter to Thomas Huxley, 18 September 1860.

34 Letter to Charles Lyell, 18 June 1858.

35 Letter to Charles Lyell, 18 June 1858.

36 Letter to Alfred Russel Wallace, 22 December 1857.

37 Letter to J. D. Hooker, 22 November 1859.

38 Letter to J. D. Hooker, 3 January 1860.

39 Letter to J. Brodie Innes, 27 November 1878.

40 For the reception of *The Origin* see Janet Browne, *Charles Darwin: The Power of Place*, Volume 2 of a biography (London: Jonathan Cape, 2003), pp.82–125; also Janet Browne, *Darwin's Origin of Species: A Biography* (London: Atlantic Books, 2006).

41 J. R. Moore, *The Post-Darwinian Controversies* (Cambridge: Cambridge University Press, 1979), p.92.

42 Letter to J. D. Hooker (20? July 1860).

Chapter 12

1 At least intentionally published. In actual fact, his friend Henslow had collated and published some of the scientific material from his letters in Darwin's absence.

2 Florence C. Hsia, *Sojourners in a Strange Land: Jesuits and Their Scientific Missions in Late Imperial China* (Chicago, IL: University of Chicago Press, 2009).

3 Catherine Jami, *The Emperor's New Mathematics: Western Learning and*

Imperial Authority During the Kangxi Reign (1662–1722) (Oxford: Oxford University Press, 2012).

4   E. Carey, *Memoir of William Carey* (London: Jackson & Walford, 1836), p.18.

5   John Gascoigne, *Joseph Banks and the English Enlightenment: Useful Knowledge and Polite Culture* (Cambridge: Cambridge University Press, 1994), pp.43–4.

6   Joseph Banks, *Journal* (London: Macmillan & Co., 1896), p.108.

7   Sujit Sivasundaram, *Nature and the Godly Empire: Science and Evangelical Mission in the Pacific, 1795–1850* (Cambridge: Cambridge University Press, 2011), p.101.

8   Marwa Elshakry, 'The Gospel of Science and American Evangelism in Late Ottoman Beirut', *Past & Present*, 196, 2007, pp.178–9.

9   Sivasundaram, *Nature*, p.12.

10 Lightman, *Rethinking History*, p.76.

11 Sivasundaram, *Nature*, p.160.

12 Lightman, *Rethinking History*, p.75.

13 William Ellis, *Polynesian Researches During a Residence of Nearly Eight Years in the Society and Sandwich Islands*, Vol. 2 (London: Fisher, Son & Jackson, 1832–4), p.18.

14 Olaudah Equiano, *The Interesting Narrative and Other Writings*, ed. V. Caretta (London: Penguin, 2003), pp.1–34.

15 Thomas Dixon *et al.* (eds), *Science and Religion: New Historical Perspectives* (Cambridge: Cambridge University Press, 2010), p.184.

16 Edward Strachey, *Bijaganita* (London: W. Glendinning, 1813), p.10.

17 John Seeley, *The Expansion of England* (London: Macmillan & Co., 1883), p.256.

18 Marwa Elshakry, 'When Science Became Western: Historiographical Reflections', *Isis*, 101(1), 2010, p.103.

19 Elshakry, 'When Science Became Western', p.101.

20 Marwa Elshakry, *Reading Darwin in Arabic, 1860–1950* (Chicago, IL: University of Chicago Press, 2013), p.16.

21 Elshakry, *Reading Darwin*, p.208.

22 Elshakry, *Reading Darwin*, p.8.

23 Elshakry, *Reading Darwin*, p.217.

24 Elshakry, *Reading Darwin*, p.120.

25 Deepak Kumar, 'The "Culture" of Science and Colonial Culture, India 1820–1920', *The British Journal for the History of Science*, 29(2), June 1996, p.198.

26 Elshakry, 'When Science Became Western', p.104.

27 Dixon *et al.* (eds), *Science and Religion*, p.137.

28 David Hume, 'Of National Characters', *Essays, Moral and Political* (London: 1748).

29 Immanuel Kant, 'On the different races of man' (1775).

30 Prichard, *Researches into the Physical History of Man* (London: J. & A. Arch, 1813), p.233.

31 John Bird Sumner, *A Treatise on the Records of Creation and the Moral Attributes of the Creator* (London: J. Hatchard, 1816), I.377.

32 Denis Alexander and Ronald L. Numbers, *Biology and Ideology from Descartes to Dawkins* (Chicago, IL: University of Chicago Press, 2010), p.121.

33 Alexander, *Biology and Ideology*, p.133.

34 Alexander, *Biology and Ideology*, p.120.

35 Adrian Desmond and James Moore, *Darwin's Sacred Cause: Race, Slavery and the Quest for Human Origins* (London: Penguin, 2009).

36 Harrison, *Science without God?*, p.218.

37 Desmond and Moore, *Darwin*, p.332.

38 Desmond and Moore, *Darwin*, p.339.

39 Henry Chadwick, *The Secularization of the European Mind in the Nineteenth Century* (Cambridge: Cambridge University Press, 1990), p.166.

40 Darwin, *The Descent of Man, and Selection in Relation to Sex* (London: John Murray, 1871), I.230.

41 Letter from James Grant, 6 March 1878.

42 Anonymous letter, 13 June 1877.

43 Mitch Keller, 'The Scandal at the Zoo'. *New York Times*, 6 August 2006.

44 Dennis Sewell, *The Political Gene: How Darwin's Ideas Changed Politics* (London: Picador, 2009), p.2.

45 Phillips Verner Bradford and Harvey Blume, *Ota: The Pygmy in the Zoo* (New York: St. Martin's Press, 1992), p.185.

46 Bradford and Blume, *Ota*, p.182.

47 Jonathan Peter Spiro, *Defending the Master Race: Conservation, Eugenics, and the Legacy of Madison Grant* (Burlington: University of Vermont Press, 2008), pp.44–9.

Chapter 13

1 Letter to William Graham, 3 July 1881.

2 H. Bence Jones, *The Life and Letters of Faraday*, Vol. 2 (London: Longmans, Green and Co., 1870), pp.195–6.

3 John Tyndall, *Faraday as a Discoverer* (London: Longmans, Green and Co., 1879), p.185.

4 T. H. Levere, 'Faraday, Matter, and Natural Theology: Reflections on an Unpublished Manuscript', *The British Journal for the History of Science*, 4(2), 1968, p.102.

5 J[ames] R[orie] (ed.), *Selected Exhortations Delivered to Various Churches of Christ by the Late Michael Faraday, Wm. Buchanan, John M. Baxter, and Alex Moir* (Dundee: John Leng and Co. Ltd., 1910), p.5.

6 Tyndall, *Faraday*, p.178.

7 V. L. Hilts, 'A Guide to Francis Galton's English Men of Science', *Transactions of the American Philosophical Society*, 65(5), 1975, p.59.

8 Raymond Flood *et al.* (eds), *James Clerk Maxwell: Perspectives on His Life and Work* (Oxford: Oxford University Press, 2014), p.259.

9 James Clerk Maxwell, *A Student's Evening Hymn* (1853).

10 L. Campbell and W. Garnett, *Life of James Clerk Maxwell* (London: Macmillan & Co., 1882), pp.188–9.

11 Campbell and Garnett, *Life*, p.126.

12 Campbell and Garnett, *Life*, p.345.

13 P. M. Harman (ed.), *The Scientific Letters and Papers of James Clerk Maxwell*, Vol. 2 (Cambridge: Cambridge University Press, 1995), pp.786–93.

14 Campbell and Garnett, *Life*, p.323.

15 Harman (ed.), *Scientific Letters*, Vol. 1, p.426.

16 Anon. [J. C. Maxwell], [Review of Balfour Stewart's] 'The Conservation of Energy', *Nature*, 9, 1874, pp.198–200.

17 Harman (ed.), *Scientific Letters*, Vol. 2, pp.362–8.

18 Harman (ed.), *Scientific Letters*, Vol. 2, p.322.

19 Harman (ed.), *Scientific Letters*, Vol. 2, p.377.

20 Letter from J. D. Hooker, 23 September 1873.

21 Letter to J. D. Hooker, 27 September 1873.

22 Campbell and Garnett, *Life*, p.394.

23 Campbell and Garnett, *Life*, pp.405–6.

24 G. G. Stokes, 'On the bearings of the study of natural science, and of the contemplation of the Discoveries to which that study leads, on our religious ideas', *Journal of the Transactions of the Victoria Institute* (1881) 14, p.247.

25 J. Reddie, 'On current physical astronomy', *Journal of the Transactions of the Victoria Institute* (1870), 4, p.378.

26 G. G. Stokes, *Natural Theology: The Gifford Lectures Delivered Before the University of Edinburgh in 1891* (London and Edinburgh, UK: Adam and Charles Black, 1893), p.231.

27 Silvanus P. Thomson, *The Life of William Thomson, Baron Kelvin of Largs* (London: Macmillan and Co., 1910), Vol. II. p.1099.

28 'Address to the Christian Evidence Society, London, May 23, 1889', in Stephen Abbott Northrop, *A Cloud of Witnesses* (Portland, Oregon: American Heritage Ministries, 1987), pp.460–1.

29 William Thomson, 'On Geological Dynamics', *Transactions of the Geological Society of Glasgow* (1869), p.222.

30 Letter to J. D. Hooker, 24 July 1869.

31 Thomson, *Life*, p.1098.

32 Thomas Huxley, *Science and Christian Tradition* (London: Macmillan & Co., 1894), p.7.

33 Ruth Barton, ' "An Influential Set of Chaps": The X-Club and Royal Society Politics 1864–85', *British Journal for the History of Science*, 23, 1990, p.57.

34 Alvar Ellegard, *Darwin and the General Reader* (Göteborg, 1958), p.65.

35 A. S. Eve and C. H. Creasey, *The Life and Works of John Tyndall* (London: Macmillan and Co., 1945), p.187.

36 Revd Michael O'Ferrall, 'The New Koran', *Irish Monthly*, 2, 1874, p.659.

37 *New York Times*, 23 November 1874.

38 Hardin, *Warfare*, p.14.

39 Hardin, *Warfare*, p.14.

40 Hardin, *Warfare*, p.14.

41 John William Draper, *History of the Conflict between Religion and Science* (New York: D. Appleton, 1875), p.vi.

42 Draper, *History*, pp.ix–x.

43 Hardin, *Warfare*, p.20.

44 Draper, *History*, p.108.

45 S. P. Ragep (trans.) Ernest Renan lecture: 'Islam and Science', pp.2–3. [Online]. https://www.mcgill.ca/islamicstudies/files/islamicstudies/renan_islamism_cversion.pdf (Accessed 24 July 2022).

46 'Why there was no "Darwin's bulldog" '. *The Linnean Society*, 1 July 2019. [Online].https://www.linnean.org/news/2019/07/01/1st-july-2019-why-there-was-no-darwins-bulldog (Accessed 24 July 2022).

47 Thomas Huxley, 'Science and Religion', *The Builder*, 17, 1859.

48 T. H. Huxley, Letter to Charles Kingsley, 23 September 1860.

49 T. H. Huxley, *Science and Hebrew Tradition* (London: Macmillan & Co., 1893), p.101.

50 T. H. Huxley, *Life and Letters* (London: Macmillan & Co., 1900), Vol. 1, p.220.

51 T. H. Huxley, *Science and Christian Tradition* (New York: D. Appleton, 1894), p.74.

52 Brooke and Numbers, *Science and Religion*, p.265.

53 Francis Darwin, *The Life and Letters of Charles Darwin* (London: John Murray, 1887), Vol. 2, p.201.

54 'human reason is placed on a level with religion itself'

55 William Gladstone, *The Vatican Decrees in their bearing on Civil Allegiance* (London: Simpkin, Marshall & Co., 1874), p.6.

56 T. H. Huxley, *Darwiniana* (London: Macmillan & Co., 1893), p.147.

57 Huxley, *Darwiniana*, p.147.

58 John Tyndall, *Fragments of Science* (London: Longmans, Green, 1871), Vol. 2, p.207.
59 Bernard Lightman, 'Conan Doyle's Ideal Reasoner: The Case of the Reluctant Scientific Naturalist', *Journal of Literature and Science*, 7(2), 2014, p.24.
60 T. H. Huxley, *Methods and Results* (London: Macmillan and Co., 1894), p.65.
61 James C. Ungureanu, *Science, Religion, and the Protestant Tradition: Retracing the Origins of Conflict* (Pittsburgh, PA: University of Pittsburgh Press, 2019), p.18.
62 Ungureanu, *Science, Religion,* p.18.
63 Ungureanu, *Science, Religion,* p.xl.
64 Ungureanu, *Science, Religion,* p.71.

Chapter 14

1   Charles Hodge, *What is Darwinism?* (New York; Cambridge: Scribner, Armstrong, & Company, 1874), p.45.
2   See Edward J. Larson, *Summer for the Gods: The Scopes Trial and America's Continuing Debate Over Science and Religion* (New York: Basic Books, 1997), pp.20, 32.
3   David N. Livingstone, D. G. Hart and Mark A. Noll (eds), *Evangelicals and Science in Historical Perspective* (New York; Oxford: Oxford University Press, 1998), p.208.
4   George William Hunter, *A Civic Biology* (New York: American Book Co., 1914), p.263.
5   Larson, *Summer for the Gods*, p.93.
6   Hunter, *A Civic Biology*, p.263.
7   *Speeches of William Jennings Bryan* (New York: Funk and Wagnalls, 1909), Vol. 2, pp.266-7.
8   William Jennings Bryan, *In His Image* (New York: Fleming H. Revell Co., 1922), pp.103-4. Emphases original.
9   Bryan, *In His Image*, p.104.
10 Larson, *Summer for the Gods*, p.116.
11 Bryan, *In His Image*, p.118.
12 Larson, *Summer for the Gods*, p.45.
13 Bryan, *In His Image*, p.94.
14 Larson, *Summer for the Gods*, p.52.
15 Larson, *Summer for the Gods*, p.71.
16 Larson, *Summer for the Gods*, p.71.
17 Larson, *Summer for the Gods*, p.103.
18 Larson, *Summer for the Gods*, p.104.
19 Larson, *Summer for the Gods*, p.146.
20 Larson, *Summer for the Gods*, p.143.

21 Larson, *Summer for the Gods*, p.132.

22 'It isn't proper to bring experts in here to try to defeat the purpose of the people of this state by trying to show that this thing that they denounce and outlaw is a beautiful thing.' Larson, *Summer for the Gods*, p.135.

23 Larson, *Summer for the Gods*, p.156.

24 Larson, *Summer for the Gods*, p.176.

25 Larson, *Summer for the Gods*, p.157.

26 Larson, *Summer for the Gods*, p.171.

27 *The World's Most Famous Court Trial: A Compete Stenographic Report of the Famous Test of the Tennessee Anti-Evolution Act, Dayton, July 10 to 21, 1925* (Cincinnati, OH: National Book Company, 1925), p.113.

28 *World's Most Famous Court Trial*, p.175.

29 *World's Most Famous Court Trial*, p.180.

30 *World's Most Famous Court Trial*, p.179.

31 *World's Most Famous Court Trial*, p.197.

32 Larson, *Summer for the Gods*, p.180.

33 *World's Most Famous Court Trial*, p.284.

34 *World's Most Famous Court Trial*, p.286.

35 *World's Most Famous Court Trial*, p.288.

36 *World's Most Famous Court Trial*, p.304.

37 Larson, *Summer for the Gods*, p.200.

38 Larson, *Summer for the Gods*, p.207.

39 Larson, *Summer for the Gods*, p.145.

Chapter 15

1 Frederick Kuh, 'Ape Case Loosens up Tongue of Einstein', *Pittsburgh Sun*, 11 June 1925, quoted in Larson, *Summer for the Gods*, p.112.

2 Walter Isaacson, *Einstein: His Life and Universe* (London : Simon & Schuster, 2007), p.279.

3 Isaacson, *Einstein*, p.279.

4 R. B. Haldane, *The Reign of Relativity* (London: John Murray, 1921), p.x.

5 Lindberg and Numbers, *God and Nature*, p.428.

6 Arthur Eddington, *The Nature of the Physical World* (London: J. M. Dent & Sons, 1935), p.350.

7 Eddington, *The Nature of the Physical World*, p.350.

8 Arthur Eddington, *The Philosophy of Physical Science* (Cambridge: Cambridge University Press, 1939), p.9.

9 Graham Farmelo, *The Strangest Man: The Hidden Life of Paul Dirac* (London: Faber and Faber, 2010), p.121.

10 Paul Dirac, 'The Excellence of Einstein's Theory of Gravitation', in Maurice

Goldsmith, Alan Mackay and James Woudhuysen (eds), *Einstein: The First Hundred Years* (Oxford: Pergamon Press, 1980), p.44.

11 Paul Dirac, 'The Evolution of the Physicist's Picture of Nature', *Scientific American*, May 1963.

12 Quoted in Subrahmanyan Chandrasekha, 'The Perception of Beauty and the Pursuit of Science', *Bulletin of the American Academy of Arts and Sciences*, 43(3), 1989, pp.14–29.

13 James Jeans, *The Mysterious Universe* (Cambridge: Cambridge University Press, 1930), p.104.

14 'Sir James Jeans: God as a Mathematician', ran the *New Chronicle*'s review of his book.

15 Following quotes all taken from 'Religion and science' in *Physics and Beyond* (New York: Harper & Row, 1971).

16 Max Planck, 'The Nature of Matter', Speech at Florence (1944).

17 Werner Heisenberg, 'Scientific Truth and Religious Truth', *CrossCurrents*, 24(4), 1975, pp.463–73.

18 Max Jammer, *Einstein and Religion: Physics and Theology* (Princeton, NJ: Princeton University Press, 1999), p.19.

19 Jammer, *Einstein and Religion*, p.48.

20 Jammer, *Einstein and Religion*, p.49.

21 Jammer, *Einstein and Religion*, p.97.

22 A. Deprit, 'Monsignor Georges Lemaître', in A. Barger (ed.), *The Big Bang and Georges Lemaître* (New York: Reidel, 1984), p.370.

23 Rodney Holder, 'Lemaître and Hoyle: Contrasting Characters in Science and Religion', *Science and Christian Belief*, 24(2), 2012, p.119.

24 Holder, 'Lemaître and Hoyle', p.120.

25 Holder, 'Lemaître and Hoyle', p.125.

26 Pablo De Felipe, Pierre Bourdon and Eduardo Riaza, 'Georges Lemaître's 1936 Lecture on Science and Faith', *Science and Christian Belief*, 30(1), 2015, p.168.

Chapter 16

1 Werner Heisenberg, 'Science and Religion', in *Physics and Beyond: Encounters and Conversations* (New York: Harper & Row, 1971), p.85.

2 Darwin, *The Descent of Man, and Selection in Relation to Sex* (London: John Murray, 1871), I.65.

3 Edward B. Tylor, *Anahuac: or Mexico and the Mexicans, Ancient and Modern* (London: Longmans, 1861), p.288.

4 Edward B. Tylor, *Primitive Culture. Vol. 1* (London: John Murray. 1871), p.139.

5 Tylor, *Primitive Culture. Vol. 1*, p.121.

6 Tylor, *Primitive Culture. Vol. 2*, p.24.

7 Tylor, *Primitive Culture. Vol. 1*, p.426.

8 Tylor, *Primitive Culture. Vol. 1*, p.426.

9 Tylor, *Primitive Culture. Vol. 2*, p.189.

10 J. G. Frazer, *The Golden Bough: A Study in Magic and Religion* (1890), Vol 1.x.

11 Frazer, *The Golden Bough*, I.43.

12 J. G. Frazer, *Folk-Lore in the Old Testament* (London: Macmillan & Co., 1918), II.17.

13 Frazer, *Golden Bough*, 2nd ed. III.117.

14 Frazer, *Folk-Lore*, III.396.

15 J. G. Frazer, *Passages of the Bible Chosen for their Literary Beauty and Interest* (London: A. & C. Black, 1895), p.viii.

16 John B. Vickery, *The Literary Impact of The Golden Bough* (Princeton, NJ: Princeton University Press, 1973).

17 Philip Marcus, '"A Healed Whole Man": Frazer, Lawrence and Blood-Consciousness', in Robert Fraser (ed.) *Sir James Frazer and the Literary Imagination: Essays in Affinity and Influence* (London: Macmillan, 1990).

18 E. B. Tylor, *Researches into the Early History of Mankind and the Development of Civilization* (London: John Murray, 1865), pp.138–9.

19 Tylor, *Researches*, p.108.

20 Sigmund Freud, *New Introductory Lectures on Psychoanalysis* (London: Hogarth Press, 1933), p.212.

21 Letter to B'nai B'rith Lodge of Vienna, 6 May 1926, quoted in Michael Palmer, *Freud and Jung on Religion* (London: Routledge, 1997), p.5.

22 Letter to Oskar Pfister, 9 October 1918, quoted in Peter Gay, *A Godless Jew: Freud, Atheism and the Making of Psychoanalysis* (New Haven, CT; London: Yale University Press, 1987), p.37.

23 Letter to Carl Jung, 2 January 1910.

24 Sigmund Freud, *Leonardo da Vinci: A Memory of His Childhood* (London: Ark, 1984), p.73.

25 Sigmund Freud, *Totem and Taboo: Resemblances between the Psychic Lives of Savages and Neurotics* (New York: Moffat, Yard and Company, 1919), p.234.

26 Sigmund Freud, *The Future of an Illusion* (New York: Norton, 1989), pp.19, 33.

27 Freud, *Future of an Illusion*, p.19.

28 Freud, *Future of an Illusion*, p.23.

29 Sigmund Freud, *Civilisation and its Discontents* (New York: Norton, 1961), p.20.

30 Sigmund Freud, *Moses and Monotheism* (New York: Vintage Books, 1955) p.108.

31 E. E. Evans-Pritchard, *Theories of Primitive Religion* (Oxford: Clarendon Press, 1965), p.52.

32 Timothy Larsen, *The Slain God: Anthropologists and the Christian Faith* (Oxford: Oxford University Press, 2014), p.80.

33 E. E. Evans-Pritchard, *Witchcraft, Oracles and Magic among the Azande* (Oxford: Clarendon Press, 1976), pp.18, 159.

34 Evans-Pritchard, *Witchcraft*, p.150.

35 E. E. Evans-Pritchard, *Nuer Religion* (Oxford: Clarendon Press, 1956), p.vii.

36 Evans-Pritchard, *Nuer Religion*, p.322.

37 E. E. Evans-Pritchard, *Social Anthropology and Other Essays* (New York: Free Press of Glencoe, 1964), p.40.

38 Evans-Pritchard, *Social Anthropology*, p.36.

39 Larsen, *The Slain God*, p.97.

40 Ahmed Al-Shahi, 'Evans-Pritchard, Anthropology and Catholicism: Godfrey Lienhart's view', *Journal of Anthropological Society of Oxford*, 30(1), 1999, p.70.

Chapter 17

1 Freud, *Future of an Illusion*, p.55.

2 Larsen, *The Slain God*, pp.73–6.

3 Freud, *New Introductory Lectures On Psychoanalysis*, p.212; emphasis added.

4 Daniel Peris, *Storming the Heavens: The Soviet League of the Militant Godless* (Ithaca, NY; London: Cornell University Press, 1998), p.23.

5 Victoria Smolkin, *A Sacred Space is Never Empty: A History of Soviet Atheism* (Princeton, NJ: Princeton University Press, 2018), p.42.

6 Smolkin, *Sacred Space*, p.35.

7 David Powell, *Antireligious Propaganda in the Soviet Union: A Study in Mass Persuasion* (Cambridge, MA: MIT Press, 1975), p.22.

8 Peris, *Storming the Heavens*, p.31.

9 The posters can be seen in Roland Elliott Brown, *Godless Utopia: Soviet Anti-Religious Propaganda* (London: FUEL, 2019).

10 Peris, *Storming the Heavens*, p.94.

11 Victoria Smolkin-Rothrock, 'The Contested Skies: The Battle of Science and Religion in the Soviet Planetarium', in E. Maurer *et al.* (eds) *Soviet Space Culture: Cosmic Enthusiasm in Socialist Societies* (London: Palgrave Macmillan, 2011), pp.57–78.

12 Dekulakisation was the murderous Soviet campaign against wealthier peasants, who were stigmatised as class enemies, nearly two million of whom were deported and/or liquidated in the early 1930s.

13 Smolkin, *Sacred Space*, p.60.

14 *Pravda*, 27 March 1958, quoted in John Anderson, *Religion, State and Politics*

in the Soviet Union and Successor States (Cambridge: Cambridge University Press, 1994), p.15.

15 Smolkin, *Sacred Space*, p.87.

16 Smolkin, *Sacred Space*, p.88.

17 Smolkin, *Sacred Space*, p.89.

18 Smolkin, *Sacred Space*, p.90.

19 Smolkin, *Sacred Space*, p.84.

20 Smolkin, *Sacred Space*, p.84.

21 Smolkin, *Sacred Space*, p.89.

22 Christopher Marsh, *Religion and the State in Russia and China: Suppression, Survival, and Revival* (New York: Continuum, 2011), p.40.

23 Marsh, *Religion and the State*, p.41.

24 The quotation is from the author Vladimir Tendriakov and is used, adapted, as the title of Victoria Smolkin's excellent book . .

25 Smolkin, *Sacred Space*, p.90.

26 William Inboden, *Religion and American Foreign Policy, 1945–1960: The Soul of Containment* (Cambridge: Cambridge University Press, 2008), pp.258–9.

27 Harry S. Truman, 'Address at a Conference of the Federal Council of Churches', 6 March 1946.

28 Jonathan Herzog, 'America's Spiritual-Industrial Complex and the Policy of Revival in the Early Cold War', *Journal of Policy History*, 22(3), 2010, p.340.

29 Norman Mailer, *Of a Fire on the Moon* (New York: New American Library, 1971), p.244.

30 Kendrick Oliver, *To Touch the Face of God: The Sacred, the Profane and the American Space Program, 1957–1975* (Baltimore, MD: Johns Hopkins University Press, 2013), p.95.

31 Gordon Cooper (with Bruce Henderson), *Leap of Faith: An Astronaut's Journey into the Unknown* (New York: HarperCollins, 2000), pp.67–8.

32 Oliver, *To Touch the Face of God*, p.57.

33 Oliver, *To Touch the Face of God*, p.90.

34 Julie Zauzmer Weil, 'In space, John Glenn saw the face of God: "It just strengthens my faith" ', *Washington Post*, 8 December 2016.

35 'Ex-Astronaut James Irwin, 61, Founded Evangelical Organization', *Chicago Tribune*, 11 August 1991.

36 C. S. Lewis, 'The Seeing Eye', from *Christian Reflections*, ed. Walter Hooper (Grand Rapids, MI: Wm. B. Eerdmans Publishing Co., 1995), pp.167–9.

37 William Derham, *Astro-Theology; Or, a Demonstration of the Being and Attributes of God* (London: W. & J. Innys, 1721), xlii.

38 J. C. Whitcomb and Henry Morris, *The Genesis Flood: The Biblical Record and its Scientific Implications* (London: Evangelical Press, 1969), p.1.

39 Whitcomb and Morris, *The Genesis Flood*, p.118.

Chapter 18

1 Andrew Brown, *The Darwin Wars: The Scientific Battle for the Soul of Man* (London: Touchstone, 2000), p.2.

2 Brown, *The Darwin Wars*, p.220.

3 Michael Ghiselin, *The Economy of Nature and the Evolution of Sex* (Berkeley; London: University of California Press, 1974), p.247.

4 Richard Dawkins, *The Selfish Gene* (Oxford: Oxford University Press, 1989), p.3.

5 Dawkins, *Selfish Gene*, p.270. As Andrew Brown has observed, 'This lofty condescension – "popular, but erroneous" – is difficult for a popular writer to maintain. Who is he to tell us what the erroneous associations of the word "robot" are?' Andrew Brown, *The Darwin Wars: How Stupid Genes Became Selfish Gods* (London: Simon & Schuster, 1999), p.40.

6 Dawkins, *Selfish Gene*, p.2.

7 Dawkins, *Selfish Gene*, p.2.

8 Dawkins, *Selfish Gene*, p.196.

9 Dawkins, *Selfish Gene*, p.233.

10 Richard Dawkins, 'In defence of selfish genes', *Philosophy*, 56, 1980, pp.556–73. Andrew Brown: 'It's always hard to tell, when words, in his hands, mean more or less what he wants them to. He made his name proclaiming the self-ishness of genes, but when this view was challenged he explained that success-ful "selfishness" consisted largely in their capacity to co-operate with each other.' Brown, *Darwin Wars*, p.43.

11 Dawkins, *Selfish Gene*, p.266.

12 Except in so far as there as genes for everything. E. O. Wilson, *Sociobiology: A New Synthesis* (Cambridge, MA: Harvard University Press, 1975), p.555.

13 Against 'Sociobiology' by Elizabeth Allen, *The New York Review of Books* (nybooks.com).

14 Dawkins, *Selfish Gene*, p.10.

15 Richard Dawkins, *The Extended Phenotype* (Oxford: Oxford University Press, 1982), p.241.

16 Richard Dawkins, *Unweaving the Rainbow* (London: Allen Lane, 1998) p.308.

17 Dawkins, *Selfish Gene*, p.3.

18 Dawkins, *Selfish Gene*, p.201.

19 *McLean v. Arkansas Bd. of Ed.*, 529 F. Supp. 1255 (E. D. Ark. 1982).

20 Charles Darwin to J. F. W. Herschel in 1861: 'The point which you raise on intelligent Design has perplexed me beyond measure . . . I am in a complete jumble on the point. One cannot look at this Universe with all living produc-tions & man without believing that all has been intelligently designed; yet when I look to each individual organism, I can see no evidence of this.' Tyndale: 'The mechanical shock of the atoms being in his view the

all-sufficient cause of things, he combats the notion that the constitution of nature has been in any way determined by intelligent design.'

21 'Exploring Different Ways of Asking About Evolution'. Pew Research Center. [Online]. https://www.pewresearch.org/religion/2019/02/06/the-evolution-of-pew-research-centers-survey-questions-about-the-origins-and-develop-ment-of-life-on-earth/ (Accessed 27 July 2022).

22 'Views about human evolution – Religion in America: U.S. Religious Data, Demographics and Statistics'. Pew Research Center. [Online]. https://www.pewresearch.org/religion/religious-landscape-study/views-about-human-evolution/ (Accessed 27 July 2022).

23 Quoted in Ronald Numbers, 'Creationism Goes Global: From American to Islamic Fundamentalism', Lecture, 2 October 2009, '2009 and Earlier', Hampshire College, Amherst, MA.

24 Salman Hameed, 'Evolution and creationism in the Islamic world', in Dixon *et al.* (eds), *Science and Religion*, p.145.

25 'Muslim Views on Religion, Science and Popular Culture'. Pew Research Center. [Online]. https://www.pewresearch.org/religion/2013/04/30/the-worlds-muslims-religion-politics-society-science-and-popular-culture/ (Accessed 27 July 2022).

26 Dixon *et al.* (eds), *Science and Religion*, p.140.

27 Salman Hameed, 'Making sense of Islamic creationism in Europe', *Public Understanding of Science*, 24(4), 2015, pp.388–9.

28 Richard Dawkins, 'A Scientist's Case Against God', Speech at Edinburgh International Science Festival, 15 April 1992.

29 Richard Dawkins, 'A Reply to Poole', *Science & Christian Belief* 7, 1995, pp.45–50.

30 Dawkins, *Extended Phenotype*, p.181.

31 Richard Dawkins in conversation with John Cornwell, *Today* programme, 6 September 2007: 'I have likened it [religion] to a virus but that's a very special point.'

32 Pope John Paul II, 'Message to the Pontifical Academy of Sciences On Evolution', 22 October 1996.

33 Caroline Lawes, *Faith and Darwin: Harmony, Conflict, or Confusion?* (London: Theos, 2009).

34 Elaine Howard Ecklund, *Science vs. Religion: What Scientists Really Think* (Oxford: Oxford University Press, 2010), p.3.

35 Stephen Jay Gould, *Rocks of Ages: Science and Religion in the Fullness of Life* (London: Vintage Digital, 2011), p.3.

36 Gould, *Rocks of Ages*, p.6.

37 Subrena E. Smith, 'Is Evolutionary Psychology Possible?' *Biological Theory*, 15, 2020, pp.39–49.

38 Steven Pinker, *How the Mind Works* (London: Penguin Books, 1999), p.375.

39 Hilary Rose and Steven Rose (eds), *Alas, Poor Darwin: Arguments Against Evolutionary Psychology* (London: Jonathan Cape, 2000).
40 Dorothy Nelkin, 'Less Selfish than Sacred?', in Rose and Rose, *Alas, Poor Darwin*.
41 Simon Conway Morris, *Life's Solution: Inevitable Humans in a Lonely Universe* (Cambridge: Cambridge University Press, 2003), p.310.

Chapter 19

1 Quoted in Conway Morris, *Life's Solution*, p.324.
2 Dawkins, *Selfish Gene*, p.192.
3 Daniel Dennett, *Darwin's Dangerous Idea: Evolution and the Meanings of Life* (London: Penguin, 1996), p.349.
4 Hobbes, *Leviathan* (1614), III.32.
5 Fyodor Dostoyevsky, *The Idiot* (Oxford: Oxford Classics, 2008), p.237 'his brain seemed to flare up momentarily . . . the sensation of being alive and self-aware increased almost tenfold . . . his mind and heart were bathed in an extraordinary illumination . . . all his agitation . . . seemed to be instantly reconciled and resolved into a lofty serenity, filled with pure, harmonious gladness and hope.'
6 O. Devinsky, G. Lai, 'Spirituality and religion in epilepsy', *Epilepsy Behavior*, 12(4), 2008, pp.636–43.
7 Irene Cristofori *et al.*, 'Neural correlates of mystical experience', *Neuropsychologia*, 80, 2016, pp.212–20.
8 A. B. Newberg and M. R. Waldman, *How God Changes Your Brain* (New York: Ballantine Books, 2009).
9 Andrew B. Newberg *et al.*, 'A case series study of the neurophysiological effects of altered states of mind during intense Islamic prayer', *Journal of Physiology-Paris*, 109(4–6), 2015, pp.214–20.
10 Newberg *et al.*, 'A case series study', p.214.
11 Michael A. Ferguson *et al.*, 'Reward, salience, and attentional networks are activated by religious experience in devout Mormons', *Social Neuroscience*, 13(1), 2018, pp.104–16.
12 Ana Sandoiu (29 November 2016). 'Religious experience activates same brain circuits as "sex, drugs, and rock 'n' roll" '. MedicalNewsToday. [Online]. https://www.medicalnewstoday.com/articles/314433 (Accessed 4 November 2022).
13 R. R. Griffiths *et al.*, 'Psilocybin can occasion mystical-type experiences having substantial and sustained personal meaning and spiritual significance', *Psychopharmacology*, 187(3), 2006, pp.268–83.
14 Ramachandran *et al.*, 'The neural basis of religious experiences', *Society for Neuroscience Conference Abstracts* (1997), 1316.
15 J. L. Saver and J. Rabin, 'The neural substrates of religious experience', *Journal of Neuropsychiatry*, 9, 1997, pp.498–510.

16 Peter G. H. Clarke, 'Neuroscientific and psychological attacks on the efficacy of conscious will', *Science and Christian Belief*, 26(1), 2014, p.11.

17 Malcolm Jeeves, 'How Free is Free?: Reflections on the Neuropsychology of Thought and Action', *Science and Christian Belief*, 16(2), 2003, p.104.

18 Hobbes, *Leviathan* (1614), I.1.

19 Hobbes, *Elements of Philosophy* (1655), p.7.

20 George Boole, *An Investigation of the Laws of Thought on Which are Founded the Mathematical Theories of Logic and Probabilities* (London: Walton & Maberly, 1854).

21 Quoted in John Wyatt and Stephen N. Williams (eds), *The Robot Will See You Now: Artificial Intelligence and the Christian Faith* (London: SPCK, 2021), p.135.

22 Yuval N. Harari, *Homo Deus: A Brief History of Tomorrow* (London: Vintage, 2016), p.444.

23 Rory Cellan-Jones (2 December 2014). 'Stephen Hawking warns artificial intelligence could end mankind'. BBC News. [Online]. https://www.bbc.com/news/technology-30290540 (Accessed 24 July 2022).

24 James Barrat, *Our Final Invention: Artificial Intelligence and the End of the Human Era* (London: Thomas Dunne Books, 2013).

25 Martin Rees, *On the Future: Prospects for Humanity* (Princeton, NJ: Princeton University Press, 2018), p.152.

26 Harari, *Deus*, pp.460, 454.

27 Robert M. Geraci, 'Apocalyptic AI: Religion and the Promise of Artificial Intelligence', *Journal of the American Academy of Religion*, 76(1), p.13.

28 Jonathan Merritt (3 February 2017). 'Is Artificial Intelligence a Threat to Christianity?'. *The Atlantic*. [Online]. https://www.theatlantic.com/technology/archive/2017/02/artificial-intelligence-christianity/515463/ (Accessed 24 July 2022).

29 'God and robots: Will AI transform religion?' BBC News. [Online]. https://www.bbc.com/news/av/technology-58983047 (Accessed 24 July 2022).

30 Harari, *Deus*, p.328.

31 'Pope's November prayer intention: that progress in robotics and AI "be human" '. Vatican News. [Online]. https://www.vaticannews.va/en/pope/news/2020-11/pope-francis-november-prayer-intention-robotics-ai-human.html (Accessed 24 July 2022).

32 Linda Kinstler (16 July 2021). 'Can Religion Guide the Ethics of A.I.?' *New York Times*. [Online]. https://www.nytimes.com/interactive/2021/07/16/opinion/ai-ethics-religion.html (Accessed 24 July 2022).

33 Stephen Hawking. Speech from the launch of the Leverhulme Centre for the Future of Intelligence on 19 October 2016. [Online]. http://lcfi.ac.uk/resources/cfi-launch-stephen-hawking/ (Accessed 24 July 2022).

34 It is interesting to note that Turing straightaway in his paper dismisses the idea that the meaning of language lies in its usage. 'If the meaning of the words "machine" and "think" are to be found by examining how they are commonly used it is difficult to escape the conclusion that the meaning and the answer to the question, "Can machines think?" is to be sought in a statistical survey such as a Gallup poll. But this is absurd.' (A. M. Turing, 'Computing Machinery and Intelligence', *Mind*, LIX(236), 1950, p.433). This was precisely the idea that was then being formulated by, and would soon be published to great acclaim, in Wittgenstein's *Philosophical Investigations*.

35 Freddy Gray. 'If we have souls, then so do chimps'. Interview with Jane Goodall, *The Spectator*, 10 April 2010. [Online]. https://www.spectator.co.uk/article/-if-we-have-souls-then-so-do-chimps (Accessed 24 July 2022).

36 See, for example, 'Experts Doubt Ethical AI Design Will Be Broadly Adopted as the Norm Within the Next Decade'. Pew Research Center, 16 June 2021. [Online]. https://www.pewresearch.org/internet/2021/06/16/experts-doubt-ethical-ai-design-will-be-broadly-adopted-as-the-norm-within-the-next-decade/ (Accessed 24 July 2022).

Further Reading

Introduction/General

Denis Alexander and Ronald L. Numbers (eds), *Biology and Ideology from Descartes to Dawkins* (Chicago, IL: University of Chicago Press, 2010)

John Hedley Brooke, *Science and Religion: Some Historical Perspectives* (Cambridge: Cambridge University Press, 1991)

John Hedley Brooke and Ronald L. Numbers (eds). *Science and Religion around the World* (Oxford: Oxford University Press, 2011)

Thomas Dixon *et al.* (eds), *Science and Religion: New Historical Perspectives* (Cambridge: Cambridge University Press, 2010)

Gary Ferngren (ed.), *The History of Science and Religion in the Western Tradition: An Encyclopedia* (New York: Routledge, 2000)

Jeff Hardin *et al.* (eds), *The Warfare Between Science and Religion: The Idea That Wouldn't Die* (Baltimore, MD: Johns Hopkins University Press, 2018)

Peter Harrison and Jon H. Roberts (eds), *Science without God?: Rethinking the History of Scientific Naturalism* (Oxford: Oxford University Press, 2019)

David Hutchings and James C. Ungureanu, *Of Popes and Unicorns: Science, Christianity, and How the Conflict Thesis Fooled the World* (Oxford: Oxford University Press, 2021)

Snezana Lawrence and Mark McCartney (eds), *Mathematicians and their Gods: Interactions between Mathematics and Religious Beliefs* (Oxford: Oxford University Press, 2015)

Bernie Lightman (ed.), *Rethinking History, Science and Religion* (Pittsburgh, PA: University of Pittsburgh Press, 2019)

David. C. Lindberg and Ronald Numbers, *God and Nature: Historical Essays on the Encounter between Christianity and Science* (Berkeley; London: University of California Press, 1986)

Ronald L. Numbers, *Galileo Goes to Jail: And Other Myths about Science and Religion* (Cambridge, MA; London: Harvard University Press, 2009)

Part 1: Science and religion before science or religion

Jim Al-Khalili, *Pathfinders: The Golden Age of Arabic Science* (London: Allen Lane, 2010)

Ahmad Dallal, *Islam, Science, and the Challenge of History* (New Haven, CT; London: Yale University Press, 2010)

Seb Falk, *The Light Ages: A Medieval Journey of Discovery* (London: Penguin Books, 2020)

Edward Grant, *The Foundations of Modern Science in the Middle Ages: Their Religious, Institutional, and Intellectual Contexts* (Cambridge: Cambridge University Press, 1996)

James Hannam, *God's Philosophers: How the Medieval World Laid the Foundations of Modern Science* (London: Icon Books, 2009)

Toby E. Huff, *The Rise of Early Modern Science* (Cambridge: Cambridge University Press, 1993)

Tamar Rudavsky, *Jewish Philosophy in the Middle Ages: Science, Rationalism, and Religion* (Oxford: Oxford University Press, 2018)

Jeffrey Burton Russell, *Inventing the Flat Earth: Columbus and Modern Historians* (New York; London: Praeger, 1991)

Part 2: Genesis

Ann Blair and Kaspar von Greyerz (eds), *Physico-theology: Religion and Science in Europe, 1650–1750* (Baltimore, MD: Johns Hopkins University Press, 2020)

Maurice A. Finocchiaro (ed.), *The Galileo Affair: A Documentary History* (Berkeley: University of California, 1989)

Stephen Gaukroger, *The Emergence of a Scientific Culture: Science and the Shaping of Modernity 1210–1685* (Oxford: Clarendon, 2006)

Peter Harrison, *The Bible, Protestantism, and the Rise of Natural Science* (Cambridge: Cambridge University Press, 1998)

Peter Harrison, *The Fall of Man and the Foundations of Science* (Cambridge: Cambridge University Press, 2007)

J. L. Heilbron, *Galileo* (Oxford: Oxford University Press, 2010)

Robert Iliffe, *Priest of Nature: The Religious Worlds of Isaac Newton* (Oxford: Oxford University Press, 2017)

Jonathan Israel, *Enlightenment Contested: Philosophy, Modernity, and the Emancipation of Man, 1670–1752* (Oxford: Oxford University Press, 2006)

Ernan McMullin, *The Church and Galileo* (Notre Dame, IN: University of Notre Dame Press, 2005)

Richard Popkin, *The History of Scepticism: From Savonarola to Bayle* (Oxford: Oxford University Press, 2003)

Keith Thomas, *Man and the Natural World: Changing Attitudes in England 1500–1800* (London: Allen Lane/Penguin Books, 1983)

Aram Vartanian, *La Mettrie's L'homme machine: A Study in the Origins of an Idea* (Princeton, NJ: Princeton University Press, 1960)

David Wootton, *The Invention of Science: A New History of the Scientific Revolution* (London: Allen Lane, 2015)

Part 3: Exodus

Henry Chadwick, *The Secularization of the European Mind in the Nineteenth Century* (Cambridge: Cambridge University Press, 1990)

Adrian Desmond and James Moore, *Darwin* (London: Penguin, 1992)

Adrian Desmond and James Moore, *Darwin's Sacred Cause: Race, Slavery and the Quest for Human Origins* (London: Penguin, 2009)

Benjamin A. Elman, *On Their Own Terms: Science in China, 1550–1900* (Cambridge, MA: Harvard University Press, 2005)

Marwa Elshakry, *Reading Darwin in Arabic, 1860–1950* (Chicago, IL: University of Chicago Press, 2013)

Florence C. Hsia, *Sojourners in a Strange Land: Jesuits and Their Scientific Missions in Late Imperial China* (Chicago, IL: University of Chicago Press, 2009)

Catherine Jami, *The Emperor's New Mathematics: Western Learning and Imperial Authority During the Kangxi Reign (1662–1722)* (Oxford: Oxford University Press, 2012)

Dennis Sewell, *The Political Gene: How Darwin's Ideas Changed Politics* (London: Picador, 2009)

Sujit Sivasundaram, *Nature and the Godly Empire: Science and Evangelical Mission in the Pacific, 1795–1850* (Cambridge: Cambridge University Press, 2011)

Nick Spencer, *Darwin and God* (London: SPCK, 2009)

James C. Ungureanu, *Science, Religion, and the Protestant Tradition: Retracing the Origins of Conflict* (Pittsburgh, PA: University of Pittsburgh Press, 2019)

Part 4: The ongoing, entangled histories of science and religion

Peter J. Bowler, *Reconciling Science and Religion: The Debate in Early Twentieth-Century Britain* (Chicago, IL; London: University of Chicago Press, 2001)

Andrew Brown, *The Darwin Wars: The Scientific Battle for the Soul of Man* (London: Touchstone, 2000)

Roland Elliot Brown, *Godless Utopia: Soviet Anti-Religious Propaganda* (London: FUEL, 2019)

Elaine Howard Ecklund, *Science vs. Religion: What Scientists Really Think* (Oxford: Oxford University Press, 2010)

Stephen Jay Gould, *Rocks of Ages: Science and Religion in the Fullness of Life* (London: Vintage Digital, 2011)

Walter Isaacson, *Einstein: His Life and Universe* (London: Simon & Schuster, 2007)

Max Jammer, *Einstein and Religion: Physics and Theology* (Princeton, NJ: Princeton University Press, 1999)

Timothy Larsen, *The Slain God: Anthropologists and the Christian Faith* (Oxford: Oxford University Press, 2014)

Edward J. Larson, *Summer for the Gods: The Scopes Trial and America's Continuing Debate over Science and Religion* (New York: Basic Books, 1997)

David N. Livingstone, D. G. Hart and Mark A. Noll (eds), *Evangelicals and Science in Historical Perspective* (New York; Oxford: Oxford University Press, 1998)

Simon Conway Morris, *Life's Solution: Inevitable Humans in a Lonely Universe* (Cambridge: Cambridge University Press, 2003)

Kendrick Oliver, *To Touch the Face of God: The Sacred, the Profane and the American Space Program, 1957–1975* (Baltimore, MD: Johns Hopkins University Press, 2013)

Michael Palmer, *Freud and Jung on Religion* (London: Routledge, 1997)

Daniel Peris, *Storming the Heavens: The Soviet League of the Militant Godless* (Ithaca, NY; London: Cornell University Press, 1998)

Martin Rees, *On the Future: Prospects for Humanity* (Princeton, NJ: Princeton University Press, 2018)

Hilary Rose and Steven Rose (eds), *Alas, Poor Darwin: Arguments Against Evolutionary Psychology* (London: Jonathan Cape, 2000)

Victoria Smolkin, *A Sacred Space is Never Empty: A History of Soviet Atheism* (Princeton, NJ: Princeton University Press, 2018)

John B. Vickery, *The Literary Impact of The Golden Bough* (Princeton, NJ: Princeton University Press, 1973)

John Wyatt and Stephen N. Williams (eds), *The Robot Will See You Now: Artificial Intelligence and the Christian Faith* (London: SPCK, 2021)

Acknowledgements

I'm hugely grateful to many people for their kindness and encouragement over the years I have been working on this book.

Magisteria's origins lie in a series I presented on Radio 4 in 2019. I want to thank Christine Morgan for seeing the possibility of such a series and Dan Tierney for working with me on it. It turned out to be not only an educative experience but a thoroughly enjoyable one.

The series allowed me to meet Doug Young who subsequently became my literary agent and friend. He helped me knock the proposal into shape and took it to Sam Carter at Oneworld, who has proved, alongside Rida Vaquas, to be the very best of editors: erudite, diligent, literary, encouraging. I would also like to thank Paul Nash, Laura Mcfarlane and Matilda Warner for their help in getting the manuscript to its final form and alerting a waiting world to its arrival.

Others have read parts of the manuscript at different stages and offered valuable comments, and I want to thank Madeleine Pennington, Chris Oldfield, Denis Alexander, Andrew Brown, Bernie Lightman, Tom McLeish and the ever reliable Toby Hole. And then there is the incomparable John Hedley Brooke, whose writing first excited my interest in the history of science and religion, who was a key interviewee for the Radio 4 series and who, in spite of his many other commitments, read and commented on every page of the original draft. I could hardly dedicate the book to anyone else.

My friends and colleagues at Theos have proved themselves to be a constant source of inspiration and encouragement, and I have also benefited more than I can say from the writing and conversation of Peter Harrison.

Much of the first part of this book was written in the parents' room at the Royal Marsden Hospital in Sutton. Reading it even now brings about some powerful emotions. Never have I been so grateful for the progress of science. Never have I been so impressed with Ellen, her brother Jonny, and my wife Kate. Thank you.

Index

References to images are in *italics*; references to notes are indicated by n.

Nicholas Spencer is Senior Fellow at Theos, a Fellow of the International Society for Science and Religion and a Visiting Research Fellow at Goldsmiths, University of London. He is the author of a number of books, including *Darwin and God, The Evolution of the West* and *Atheists*. He has presented a BBC Radio 4 series on *The Secret History of Science and Religion*, and has written for the *Guardian, Telegraph, Independent, New Statesman, Prospect* and more. He lives in London.